Engineered Ceramics and Composite Materials

This book provides a comprehensive insight into the diversity of ceramics and their applications in producing clean devices and their capability of being recycled into useful in-demand products. The emphasis is on the role of ceramic industries to build a circular economy by reusing and recycling the ceramic waste for environmental, financial, and social sustainability. It summarizes the energy and environmental benefits of ceramic materials and composites, covering their complete life cycle, from synthesis to recycling and finally disposal.

Features:

- Focuses on the energy and environmental benefits and applications of ceramic materials with a coverage from their synthesis to recycling and disposal
- Discusses their suitability for photovoltaics and photo-electrocatalysis, membranes, sensors, and batteries
- Covers embedded circularity and life cycle thinking
- Provides relevant research results and case studies on various aspects of ceramics and composites with an emphasis on energy and environment
- Reviews bioceramics, wood ceramics, and nanoceramics.

This book is aimed at researchers and graduate students in materials science, ceramics, and composites.

Advanced Materials Processing and Manufacturing

Series Editor: Kapil Gupta

The CRC Press Series in Advanced Materials Processing and Manufacturing covers the complete spectrum of materials and manufacturing technology, including fundamental principles, theoretical background, and advancements. Considering the accelerated importance of advances for producing quality products for a wide range of applications, the titles in this series reflect the state-of-the-art in understanding and engineering the materials processing and manufacturing operations. Technological advancements for enhancement of product quality, process productivity, and sustainability, are on special focus including processing for all materials and novel processes. This series aims to foster knowledge enrichment on conventional and modern machining processes. Micro-manufacturing technologies such as micro-machining, micro-forming, and micro-joining, and Hybrid manufacturing, additive manufacturing, near net shape manufacturing, and ultra-precision finishing techniques are also covered.

Process Modeling and Optimization in Modern Manufacturing
Edited by Ravi Pratap Singh, Narendra Kumar, Vishal Francis and Ankur Jaiswal

Additive Manufacturing for Biomedical Implants and Prosthetics
Edited by Pawan Kumar Rakesh and Pravin Kumar

Manufacturing in the Digital Age
Edited by Mohan Kumar Pradhan, Mahendra U. Gaikwad, and Ragavanantham Shanmugam

Fused Deposition Modelling
Technology, Applications, and Developments
Edited by Anshuman Kumar Sahu and Siba Sankar Mahapatra

Manufacturing Tribology
Edited by Jibin T Philip, Chacko Preno Koshy, and MD Mathew

Machine Learning for Advanced Manufacturing
Edited by Nishant Ranjan, Rashi Tyagi, Ranvijay Kumar, Ashutosh Tripathi, and Amit Verma

Engineered Ceramics and Composite Materials
From Fundamentals to Applications
Edited by Neethu Bhaskar, Sathish Kumar Palaniappan, Sanjay Mavinkere Rangappa, and Suchart Siengchin

For more information about this series, please visit: www.routledge.com/Advanced-Materials-Processing-and-Manufacturing/book-series/CRCAMPM

Engineered Ceramics and Composite Materials

From Fundamentals to Applications

Edited by
Neethu Bhaskar, Sathish Kumar Palaniappan,
Sanjay Mavinkere Rangappa, and Suchart Siengchin

CRC Press
Taylor & Francis Group
Boca Raton London New York

CRC Press is an imprint of the
Taylor & Francis Group, an **informa** business

First edition published 2026
by CRC Press
2385 NW Executive Center Drive, Suite 320, Boca Raton FL 33431

and by CRC Press
4 Park Square, Milton Park, Abingdon, Oxon, OX14 4RN

CRC Press is an imprint of Taylor & Francis Group, LLC

ISBN: 978-1-032-80458-3 (hbk)
ISBN: 978-1-041-22315-3 (pbk)
ISBN: 978-1-003-73167-2 (ebk)

DOI: 10.1201/9781003731672

Typeset in Times
by Apex CoVantage, LLC

Contents

Editors' Biography

Neethu Bhaskar is a scientist at the Kerala State Council for Science, Technology, and Environment, India. She holds a PhD in environmental engineering from the Indian Institute of Technology Kharagpur. With expertise in microbial fuel cells, ceramic membrane synthesis, and clean energy technologies, she has published extensively, contributing several research papers and book chapters. She was awarded the Chief Minister's Nava Kerala Post-Doctoral Fellowship, a prestigious recognition by the Government of Kerala.

Sathish Kumar Palaniappan is a research scientist (Specialist 2) at King Mongkut's University of Technology North Bangkok, Thailand. He earned his PhD in mining engineering from the Indian Institute of Technology Kharagpur and has extensive experience in polymer nanocomposites, biomaterials, tribology, and surface modifications. With over 80+ research publications, 32+ book chapters, and a granted Indian patent, he is an active contributor to the field of material sciences. He has been ranked among the world's Top 2% scientists by Stanford University.

Sanjay Mavinkere Rangappa is an associate professor, principal research scientist, and advisor at King Mongkut's University of Technology North Bangkok, Thailand. He has authored over 600+ high-quality journal articles, 60+ book chapters, and edited 50+ books. His research spans natural fiber composites, polymer composites, and advanced material technology. Recognized as one of Stanford University's Top 2% Most-Cited Scientists, he has received numerous awards, including the Outstanding Young Researcher Award.

Suchart Siengchin is Director of KMUTNB TechnoPark and Ex-President of King Mongkut's University of Technology North Bangkok, Thailand. A globally recognized expert in polymer processing and composite materials, he has authored over 850+ peer-reviewed articles, 80+ book chapters, and 60+ books as an editor. He has been ranked among the world's Top 2% scientists by Stanford University and has made significant contributions to material science research and education.

Contributors

Arthur Behenck Aramburu
Federal University of Rio Grande do Sul
Brazil

Aryan Boora
Deenbandhu Chhotu Ram University of Science and Technology
Haryana, India

Sampa Chakrabarti
University of Calcutta
West Bengal, India

Rafael de Avila Delucis
Federal University of Pelotas
Brazil

Surender Duhan
D.C.R. University of Science & Technology
Haryana, India

Partha Sarathi Ghosal
Indian Institute of Technology Kharagpur
West Bengal, India

Antrakrate Gupta
Indian Institute of Technology
Kanpur, India

K. Ihjas
KSCSTE-Centre for Water Resources
Development and Management
Kerala, India

Jijo V. J.
Kerala State Council for Science
Technology and Environment
Kerala, India

Megha Sara Jacob
International Advanced Research Centre for Powder Metallurgy and New Materials (ARCI) IITM Research Park
Tamil Nadu, India

Shikhar Krishn Jha
Indian Institute of Technology
Kanpur, India

N. Mathan Kumar
KPR Institute of Engineering and Technology
Tamil Nadu, India

Ishita Kumari
Netaji Subhas University of Technology
Delhi, India

Abhradeep Majumder
Indian Institute of Technology Kharagpur
West Bengal, India

Priya Malik
D.C.R. University of Science & Technology
Haryana, India

Pubali Mandal
Birla Institute of Technology and Science Pilani
Rajasthan, India

Akhila Mathew
Nirmala College
Kerala, India

Pallavi Mitra
Rabindra Bharati University
West Bengal, India

N. Mohan Raj
Coimbatore Institute of Engineering and Technology
Coimbatore, Tamilnadu, India

Sahana B. Moodakare
International Advanced Research Centre for Powder Metallurgy and New Materials (ARCI) IITM Research Park

Ramasubramanian Lakshmi Narayan
Indian Institute of Technology
Delhi, India

Neethu Bhaskar
Kerala State Council for Science, Technology and Environment
Kerala, India

Aarati S. Panchbhai
Sharad Pawar Dental College and Hospital
Maharashtra, India

S. Paulsingarayar
NPR College of Engineering & Technology
Tamil Nadu, India

Poonam Pipil
Rajdhani College
University of Delhi
Delhi, India

Siju C. R.
Kerala State Council for Science, Technology and Environment
Kerala, India

Jyotsna Ratan
Zakir Husain Delhi College
University of Delhi
Delhi, India

Bhavna Rohilla
D.C.R. University of Science & Technology
Haryana, India

Anirban Roy
Maulana Abul Kalam Azad University of Technology
West Bengal, India

Supriya Sehrawat
D.C.R. University of Science & Technology
Haryana, India

Nidhi Sharma
Indian Institute of Technology
Kanpur, India

Shalini Singh
Indian Institute of Technology
Patna, India

Raman Vedarajan
International Advanced Research Centre for Powder Metallurgy and New Materials
IITM Research Park
Tamil Nadu, India

Karthickeyan Viswanathan
Dr. N.G.P. Institute of Technology
Kalapatti, Tamil Nadu, India

Manoj Kumar Yadav
Indian Institute of Technology
Patna, India

Preface

Ceramic materials and composites play a crucial role in sustainable development, offering exceptional thermal stability, mechanical strength, and chemical resistance. *Engineered Ceramics and Composite Materials: From Fundamentals to Applications* explores advancements in ceramics for environmental engineering, energy storage, and biomedical applications.

This book covers a broad spectrum of topics, beginning with ceramics used in environmental engineering, pollution control, and waste treatment. It delves into the sustainability aspects of dental ceramics and explores nanoceramics in pollution mitigation and energy solutions. The role of bioceramics in sustainable biomedical and energy applications is also highlighted.

With a focus on resource efficiency, the book discusses recycled ceramics, advanced environmental barrier coatings, and ceramic-based waste management strategies. Energy storage solutions are explored through ceramic-polymer composite electrolyte materials for lithium-based electrochemical systems. The industrial applications of ceramic membranes in water and wastewater treatment are also examined.

Further topics include advances in photocatalytic ceramic materials for energy conversion, ceramics in sensors and batteries, and supplementary cementing materials repurposing industrial waste into high-performance composites. The synergistic advancements in biopolymer-ceramic composites and their environmental applications are also discussed.

This compilation integrates research from diverse fields, providing both fundamental knowledge and practical insights into engineered ceramics. We hope this book serves as a valuable resource for researchers, engineers, and students, inspiring future innovations in sustainable materials and technologies.

Dr. Neethu Bhaskar
Dr. Sathish Kumar Palaniappan
Dr. Sanjay Mavinkere Rangappa
Dr. Suchart Siengchin

1 Ceramics Used in Environmental Engineering

Poonam Pipil, Ishita Kumari, and Jyotsna Ratan

1.1 INTRODUCTION

Ceramics have been used for tailoring engineering materials due to their properties such as high-temperature resistance, hardness, and chemical stability. Consequently, they are employed to create filters for procedures that include hot gases. By detecting oxygen in the exhaust gas, the lambda sensor, for instance, may be used to control the ratio of air to fuel, therefore lowering engine emissions using a three-way catalytic converter. These ceramics are resistant to heat shock, corrosion, and wear caused by heat at temperatures up to 1500 °C. They are perfect to use in water purification or waste treatment facilities, as they may reduce fuel consumption by up to 50 percent when used in heat exchangers. These properties make ceramics a choice for many industrial applications, including components, tools, and biomedical implants. The use of ceramics in engineering has been expanding rapidly, with new applications emerging in areas like energy conversion, environmental protection, and transportation. The adoption of ceramic materials has contributed to the development of novel technologies that have significantly improved the performance and quality of products. With ongoing research and development, there is an expectation that the role of ceramics in engineering will continue to increase, leading to the discovery of new exciting applications [1].

Ceramics have gained applications in a range of fields, including aerospace, electronics, and biomedical engineering. Ceramics can be classified into various categories based on their composition and application. For instance, ceramics, such as alumina, zirconia, and silicon dioxide, are typically used for structural components, while non-oxide ceramics, such as silicon carbide and boron nitride, are used at high temperatures. In addition, ceramics offer advantages over other materials, such as metals and polymers due to their inert behavior and low thermal expansion. In this, we explore the role of engineering applications and highlight their many benefits for modern. Ceramics play a special role in energy-related technologies, including nuclear and chemical energy, as well as light harvesting [1].

The term “ceramics” is used to describe a diverse group of materials that have been used by humans for thousands of years due to their unique properties. These are typically inorganic, non-metallic, and often made through the processes of heating

DOI: 10.1201/9781003731672-1

and cooling in a kiln. They are known for their durability, high melting point, and ability to resist wear and tear. While ceramics are commonly associated with pottery and decorative items, they have many important engineering features as well. Ceramics are employed in a variety of sectors, from electrical insulators to dental implants, owing to their distinctive mix of chemical and physical properties. Technical ceramics provide several advantages for eco-friendly goods. They are ideal for use in solar energy conversion, photovoltaic, and other power plant engine systems since they can endure temperatures beyond 1000 °C. Ceramic parts are also very resistant to thermal stress, wear, and corrosive assault, making them suitable for application in "environmental technology" in the water treatment and processing sector [2].

1.2 TYPES OF CERAMICS

Ceramics are classified into four primary types based on their properties and applications: oxide ceramics, non-oxide ceramics, composite ceramics, and bioceramics. Oxide ceramics are the most common type and are composed of oxygen-metal atoms. They have high temperature and chemical resistance but low mechanical strength. Non-oxide ceramics, such as carbides and nitrides, have excellent mechanical properties but are vulnerable to chemical reactions. Composite ceramics are made by blending different types of ceramics to produce a material that has unique properties in comparison to individual ceramics. Finally, bioceramics, which are used in medical applications, have excellent biocompatibility and bioactivity to address the properties of biological tissue. Overall, the type of ceramics used in environmental engineering varies based on the specific application and required properties. Ceramics are a diverse group of materials with an extensive range of physical and chemical properties that make them useful in many applications. There are different ways to classify ceramics, including based on their composition and specific applications. Here are some types of ceramics.

1.2.1 Traditional Ceramics

Traditional ceramics are one of the oldest and most widely used materials in human history, dating back to ancient times when early civilizations learned to create pottery vessels. These are typically made from natural materials such as clay and feldspar (Figure 1.1) and are processed at high temperatures to produce a hard, durable material. Traditional ceramics are commonly used for brick, tile, and pottery products, as well as for more specialized materials such as electrical insulators and refractory materials. They are valued for their low cost, ease of use, and high resistance to heat and corrosion. While they may have the high-tech properties of some modern materials, traditional ceramics remain an important material in the field of reengineering [2].

1.2.2 Advanced Ceramics

Advanced ceramics, also known as technical ceramics, refer to a wide range of high-performance ceramics that have been specially developed to meet specific industrial requirements. Unlike ceramics, which are brittle and have limited use, advanced ceramics are designed for extreme temperatures, corrosive environments,

FIGURE 1.1 Traditional ceramics derived from natural raw materials such as clay and feldspar.

Source: Open source image.

high pressures, and mechanical stress. They offer exceptional thermal, electrical, and chemical properties, making them ideal for a variety of engineering applications such as aerospace, automotive, biomedical, energy production, and electronics. Some examples of advanced ceramics include alum, zirconia, silicon carbide, and titanium diboride. The unique properties of advanced ceramics have led to demand for their use in different applications (Figure 1.2), making them crucial materials for modern engineering [2].

1.2.3 Silicate Ceramics

Silicates are substances that often include silicon and oxygen. Aluminosilicates and magnesium silicates are the two primary forms of silicate ceramics. Figure 1.3 illustrates components fabricated from alumino- and magnesium silicates, showcasing their typical geometric forms utilized in industrial applications. Silicate ceramics are often divided into coarse or fine varieties and, depending on the amount of water absorbed, dense (< 2% for fine and < 6% for coarse) or porous ceramics (> 2% and > 6%, respectively.

1.2.4 Oxide Ceramics

Examples of oxide ceramics include magnesia, silica, zirconia, and alumina, as well as other materials based on metal oxides. These are organic, nonmetallic compounds that could include nitrogen, oxygen, or carbon.

Oxide ceramics have the following properties:

- Low wear resistance
- High melting points

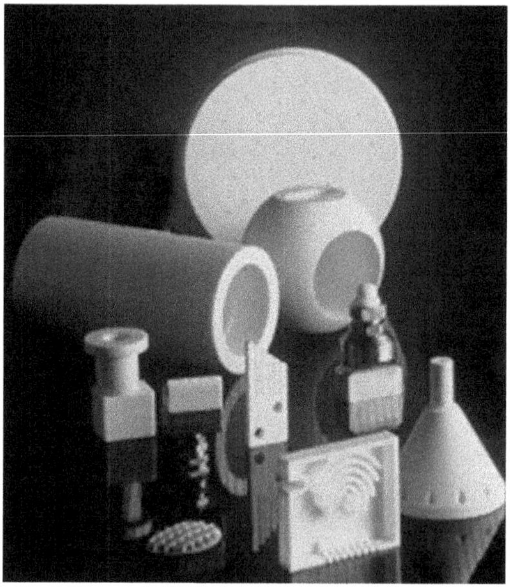

FIGURE 1.2 Representative components made from advanced ceramics (technical ceramics).

Source: Open source image.

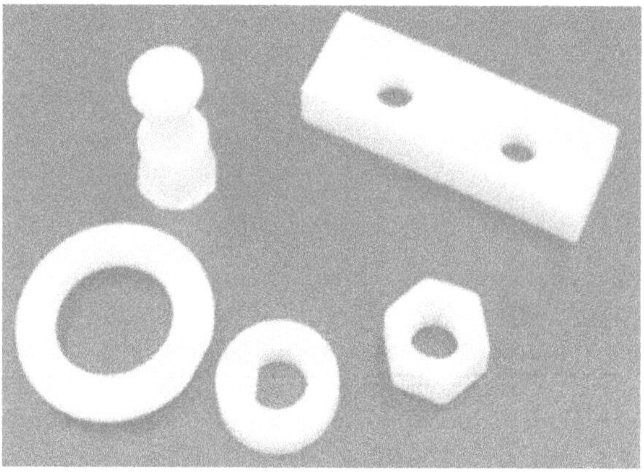

FIGURE 1.3 Examples of silicate ceramic components fabricated from alumino- and magnesium silicates.

These ceramics can have many unique properties. For instance, glazes and protective coatings seal porosity, promote connection to metals or other materials, and increase water or chemical resistance. There are several uses for oxide ceramics, including foundry and metal processing, microwave and radiofrequency uses, electrical and high-voltage power applications, and chemical and material processing. The most significant technical "oxide ceramic material" is aluminum oxide (Al_2O_3). This

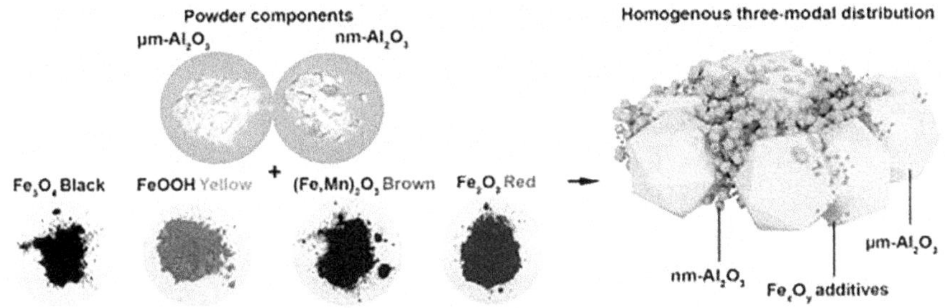

FIGURE 1.4 Schematic representation of oxide ceramic powder composition and microstructure formation.

synthetically created substance is made up of aluminum oxide to a degree of 80–99 percent (2). Figure 1.4 illustrates how the oxide ceramic powder is mixed and how its structure develops, showing an even spread of both large and small aluminum oxide (Al_2O_3) particles along with added iron oxide (Fe_xO_Y).

1.2.5 Non-Oxide Ceramics

Non-oxide ceramics include materials like carbides, nitrides, and borides. Even at high temperatures and under challenging thermal shock circumstances, severe wear and corrosion issues have been solved because of the use of non-oxide ceramics. These kinds of ceramics are used in a variety of fields, including the gas and oil sectors, medicine, wear plates, rotating parts, valves, seals, location pins for projection welding, metal-forming tools, abrasive powder blast nozzles, cutting tooltips, and so on [2].

1.3 PROPERTIES OF CERAMICS

Ceramics are increasingly being used in various fields of engineering owing to their exceptional physical properties. In environmental engineering, ceramics play a significant role in a variety of applications. One of the most significant contributions of ceramics in this field is their ability to effectively filter out impurities, water, and gases. They also show great resistance to high temperatures and harsh conditions, which makes them an ideal material for incineration and catalytic reactions. Ceramic materials are highly durable and possess excellent longevity, ensuring the sustainable use of resources. With these exceptional properties, ceramics have become an essential component in engineering [3].

1.3.1 Porous Ceramics Characteristics

Porous ceramics have the following characteristics in common:

- **Good Chemical Stability**: Porous goods may be made suitable for a variety of corrosive circumstances in which they are intended to be performed by selecting the right material species and manufacturing techniques.

- **Great Specific Strength and Rigidity**: Pores in porous ceramics will remain the same size and form even when subjected to liquid pressure, gas pressure, and other stress loadings.
- **Fine Thermal Stability**: High-temperature burning gas or molten steel can be filtered via porous items constructed of heat-resistant ceramics [4].

Porous ceramics have a bright future because of these outstanding qualities, which enable them to be employed in a variety of industries, like chemical engineering, environmental protection, energy production, metallurgy, and the electronic sector. Porous ceramics are useful in a variety of situations, depending on their composition and structure. Porous ceramics were initially employed as filter materials to remove germs from the microorganisms. Porous ceramics' fine pores can be controlled to a greater extent, which has led to an increase in the applications for the resulting products, such as separation, dispersion, and adsorption. These products are currently used in a variety of industrial sectors, such as the pharmaceutical, textile, petroleum, metal smelting, chemical engineering, and food machinery industries. Additionally, the application of these porous ceramics in artificial bones, dental root materials, sensitive components, and sound-absorbing materials has increased [4].

1.3.2 Mechanical Properties

Ceramics have superior mechanical properties, including hardness, high strength, and toughness, which make them ideal for use in various environmental engineering applications. They possess a high surface area, which makes them useful in applications that require high surface area-to-volume ratios, such as catalysis and filtration. They are resistant to stress and can withstand high temperatures, making them useful in applications that involve hot gases and high-temperature environments [3].

Components made of high-grade ceramic materials often contribute considerably to the realization of environmentally friendly technologies. Sometimes, they constitute a principal precondition when, for instance, the following properties are required:

(1) High mechanical stability at temperatures above 1000 °C;
(2) Thermal stability in oxidizing and reducing conditions as well as in a vacuum with temperatures sometimes well above 1500 °C;
(3) Resistance to thermal shock;
(4) Resistance to corrosive attack by melts, solutions, and gases, under reducing and oxidizing conditions, against supercritical fluids as well as in contact with solid bodies;
(5) Depending on the specific application, it has a high or low thermal and electrical insulating capacity (F);
(6) Ceramics are resistant to stress and can withstand high temperatures, making them useful in applications that involve hot gases and high-temperature environments [1].

1.4 ENVIRONMENTAL SUSTAINABILITY

Technical ceramics could be utilized to safeguard the environment and aid in the restoration of damaged ecosystems. Additionally, they may be utilized to develop novel elements intended to control energy generation, enhance energy storage, air pollution, environmental engineering, and green buildings, which can contribute to the apprehension of environment-friendly technologies. Ceramic rubble can be used as a sustainable construction material, providing an alternative to traditional building materials. They support energy; therefore, they are used in various clean energy applications, such as lambda sensor components and gas analysis factors. These materials can withstand the harsh conditions of combustion engines for long periods of time, making them suitable for use in clean energy generation [7].

1.4.1 Pollution Prevention, Control, and Remediation

Ceramics play a significant role in many environmental applications, such as the detection, monitoring, and quantification of pollutants and their management. They are used in environmental catalysis for pollution prevention, control, and remediation [8].

Ceramics have a wide range of applications in engineering; environmental engineering is no exception. Ceramics are used in the construction of filters, catalytic converters, and other systems that help mitigate pollution and reduce waste. The unique properties of ceramics, such as their high strength and chemical degradation, make them ideal materials for use in harsh environmental conditions. Additionally, ceramics offer a lower cost and environmental impact compared to other materials such as metals or plastics. Overall, ceramics are an important component in environmentally sustainable engineering practices [7].

1.5 APPLICATIONS OF TRADITIONAL CERAMICS: POROUS CERAMICS

Ceramics have special qualities that make them important in environmental engineering. Technical ceramics are widely utilized to create components that can tolerate harsh environments, resist corrosion, and withstand heat. These materials serve as sensors, photo catalysts, and carriers of catalysts in the prevention, control, and remediation of pollution. Refractoriness, filter ceramics, and dense-sintered monolithic oxides are among the range of ceramic materials that support environmentally friendly innovations that are long-lasting. Porous ceramics are traditional ceramics that are described by their high porosity and pore structure. These ceramics are made from a variety of materials, including alumina, zirconia, and silicon carbide. Porous ceramics are a subset of this type of material that offers a wide range of advantageous properties that are highly pertinent to science and industry for use in numerous and diverse applications. Some of their intriguing characteristics include strong mechanical strength, high permeability, abrasion resistance, chemical and thermal stability, and desirable selectivity. The common thermal, chemical, mechanical, and physical properties of these types of materials are examined in this chapter, along with new functional properties that can be attained by modifying and altering their formations and structures. Porous ceramic membranes and their applications are also examined. The applications of these membranes are then expressed in six primary categories, including catalyst carriers, ion exchange, functional materials, porous electrodes and membranes, filtration and separation, and finally combustion and fire retardance, after providing an overview of their synthesis processes and properties [8].

Additionally, they are used as catalyst supports in the chemical industry, as bone in medical implants, and as insulation materials in high-temperature environments. Porous ceramics have the potential to play an increasingly important role in environmental engineering and other industries due to their unique properties and versatility.

1.5.1 Glass Ceramics

Glass ceramics, which have a fine, crystal-uniform dispersion throughout the bulk material, are polycrystalline materials created by the controlled crystallization of foundation glass. By submitting appropriate glasses to a carefully controlled heat treatment schedule, crystallization is achieved, leading to the nucleation and development of crystal phases. The process of crystallization could frequently continue until it is almost finished, although, in a tiny percentage of situations, the remaining glass phase frequently persists. Many characteristics of both glasses and ceramics are shared by glass-ceramic materials. Glass ceramics are made by a process known as "controlled crystallization," as opposed to spontaneous crystallization, which is often undesirable in the glass manufacturing process. They have at least one crystalline phase as well as an amorphous phase [9].

1.5.2 Medical Applications

Glass ceramics are employed in medical applications because of their distinct interaction – or lack thereof – with human body tissue. Bioactive, passive, or resorbable ceramics are the traditional classifications for bioceramics based on their biocompatibility. As the name implies, passive (bioinert) ceramics are distinguished by their little contact with the surrounding "biological tissue." These "first-generation" biomaterials were historically utilized to replace missing or damaged tissues. The body's response to a foreign item was one concern with employing inert biomaterials; it was discovered that a phenomenon named "fibrous encapsulation" would happen, where tissues would develop around the implant to isolate it from the rest of the body [9]. This occasionally resulted in many issues, including implant sequestration or necrosis. Zirconia (ZrO_2) and alumina (Al_2O_3) are two frequently used bioinert materials (10).

1.5.3 White Ware Ceramic Properties and Applications of Advanced Ceramics

White ware is a general term for a variety of ceramic goods that are white to off-white in color and typically have a sizable vitreous, or glassy, component. White wares, which include items as varied as fine china dinnerware, toilet and bathroom sinks, dental implants, and spark-plug insulators, rely on a relatively small set of characteristics to function: imperviousness to liquids, low electrical conductivity, chemical inertness, and the capacity to be formed into complex shapes. These characteristics are influenced by the raw material composition used for the products and their shaping, as well as the firing techniques used in their production. The properties of white ware ceramics make them ideal for these applications; they can withstand harsh chemicals and high temperatures. Advanced ceramics, such as alumina and zirconia, are also used in environmental engineering for their thermal and electrical insulation properties, making them suitable for applications in power generation and distribution systems. In general, the special qualities of advanced ceramics make them the perfect option for a variety of environmental engineering applications [11].

1.5.4 Raw Materials: Clay, Flint, and Feldspar

Due to the three mineral forms – silica, clay, and feldspar – consistently present in their composition, white wares are sometimes called triaxial bodies. The plastic component, which gives the unfired product shape capabilities and also acts as a glass forming agent during firing, is clay. As filler, flint (the industry term for all varieties of silica) gives the formed body strength both before and after burning. The melting temperatures of the combination are lowered by the fluxing agent, feldspar.

The most crucial component is clay, and kaolin, commonly referred to as china clay, is the most crucial clay utilized in excellent white ware items. The only kind of clay that can be used to create white, transparent, vitreous pottery is kaolin. It is white-burning refractory clay, which means that it lends whiteness to the completed ware, and it can be burned at high temperatures without deforming. The main component of kaolin is the mineral kaolinite, which has a thin, platy structure and the perfect chemical formula "$Al_2(Si_2O_5)(OH)_4$." China clays are mostly made of impurity-free, well-ordered kaolinite. Ball clays, which include both ordered and disordered kaolinite along with other clay minerals and impurities, are typically used to make lower-grade white ware. The burnt pottery is colored off-white to gray or brown because of these impurities, especially iron oxides [11].

1.5.5 Structural Ceramics: Bioceramics and Electronic Ceramics

A class of ceramic materials, known as structural ceramics, is made to function as structural elements in a variety of devices that are frequently subjected to mechanical loading. These ceramics are appropriate for applications where strength and durability are essential because of their unique mechanical and physical characteristics. Their resistance to mechanical stress sets them apart from conventional ceramics, and they find application across numerous industries. Applications include mechanical system components, device structural members, and other situations where extreme durability and strength are necessary. Structural ceramics, bioceramics, and electronic ceramics are three important categories of ceramics used in environmental engineering.

1.5.5.1 Bioceramics

Ceramic substances that are biocompatible are known as bioceramics. An essential class of biomaterials is called bioceramics. The biocompatibility of bioceramics ranges from ceramic oxides, which are inert in the body, to the opposite extreme of resorbable materials, which, after aiding in healing, are finally replaced by the body. Many different kinds of medical treatments use bioceramics. Although certain bioceramics are flexible, they are commonly employed as stiff materials in surgical implants. The ceramic materials employed are not the same as ceramic materials of the porcelain kind. Instead, bioceramics are either exceptionally durable metal oxides or closely linked to the body's components. Additionally, their non-toxic, inert characteristics make them an ideal solution for point-of-use filtration applications, providing safe drinking water to people worldwide. A porous bioceramic scaffold with an open-cell structure is seen in Figure 1.5. This scaffold was developed for use in biomedical applications such as bone regeneration and tissue engineering.

1.5.5.2 Structural Ceramics

High-tech engineering ceramics, known as structural ceramics, can withstand a variety of harsh conditions, including corrosive environments, high temperatures, and high levels of mechanical, thermal, and tribological stress. Over time, improvements in processing methods have broadened the range of structural ceramics to include borides, silicates, nitride, carbides, and their composites. Initially, only oxide ceramics like alumina ceramics, zirconia ceramics, and so on attracted attention. The mechanical performance of these ceramics is generally outstanding in harsh conditions and at high temperatures, but their applications have always been constrained

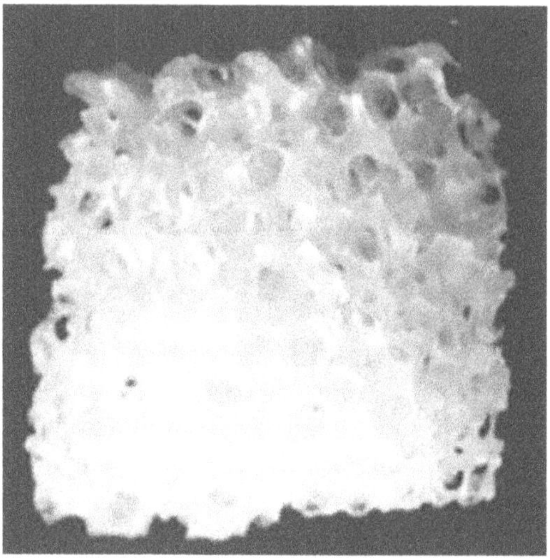

FIGURE 1.5 Porous bioceramic scaffold exhibiting an open-cell structure design.

Source: Open source image.

by their poor fracture toughness and brittle nature, as well as their high susceptibility to structural faults, fractures, and pores [13].

1.5.5.3 Electronic Ceramics

Materials are complicated in the field of electronic ceramics. They often make up a modest portion of a non-ceramic product and include many different businesses. One business starts with raw ingredients like oxides and then manufactures electronic ceramic. Next, another business fabricates these ceramics into components like capacitors or ceramic substrates. The third incorporates it onto an electrical subassembly like a PCB ("Printed Circuit Board"). The final firm integrates the subassembly into the consumer-facing final electronic product, such as a television or personal computer. The fourth business could not even be aware of the ceramic's composition, just knowing it as a sensor, capacitor, or disc drive head [14].

1.6 CERAMICS UTILITY

1.6.1 Water Treatment

Ceramic membranes have been shown to have better hydraulic performance than their polymeric counterparts, making them suitable for water and wastewater treatment. Ceramic water filters are a versatile method of water filtration that uses natural media to eliminate bacteria and sediment from drinking water. These filters can be used in various applications, such as under-sink installations, portable filters, and even for whole-house or light industrial processes. A ceramic filtration system uses tiny pores on a ceramic surface to filter bacteria and sediment out of drinking water. It is an affordable and versatile water filtration system that can be installed in households or used as a portable filtration device [15].

A ceramic filtration system uses a natural ceramic medium as the heart of the water filtration process. The ceramic filter cartridge, often referred to as a ceramic filter candle, processes the water and removes contaminants through a network of pores. Ceramic filtration systems exist for a variety of applications. They can be gravity-fed, under-sink, countertop, or, in some instances, used for a whole house or even light industrial processes. Ceramic alone is a powerful filter and handily removes many contaminants. The extent of the filtration varies and will depend on what additional filtration media the ceramic candle contains. The main categories of contaminants that basic ceramic filters remove are bacteria, sediment, and turbidity [11]. Membrane fouling is a significant challenge in the use of ceramic membranes for water and wastewater treatment. It occurs when organics, inorganic substances, and microorganisms clog the pores and pollute the membrane surface, greatly reducing the purification effectiveness. There are four main mechanisms for ceramic membrane fouling:

- Complete blocking
- Standard blocking
- Intermediate blocking
- Cake-filtration blocking

To control membrane fouling, various approaches have been explored, including ceramic membrane pretreatment, ceramic membrane surface modification, membrane cleaning, magnetization, ultrasonic, and nanobubbles. These methods aim to optimize the performance of ceramic membranes and ensure their efficient and stable operation in water treatment applications. By understanding and addressing membrane fouling, researchers and practitioners can develop more effective and reliable ceramic membrane systems for water and wastewater treatment. Figure 1.6 illustrates common fouling processes in ceramic membranes. These mechanisms include total, standard, intermediate, and cake-filtration blockage, which are produced by pollutants that are either reversible or irreversible. Examples of ways to manage fouling include using coagulation-based pretreatment, changing surfaces with dip coating, cleaning with chemical agents through backwashing, and using ozone for oxidation. In water and wastewater treatment systems, these methods enhance the lifespan of membranes and improve operational efficiency.

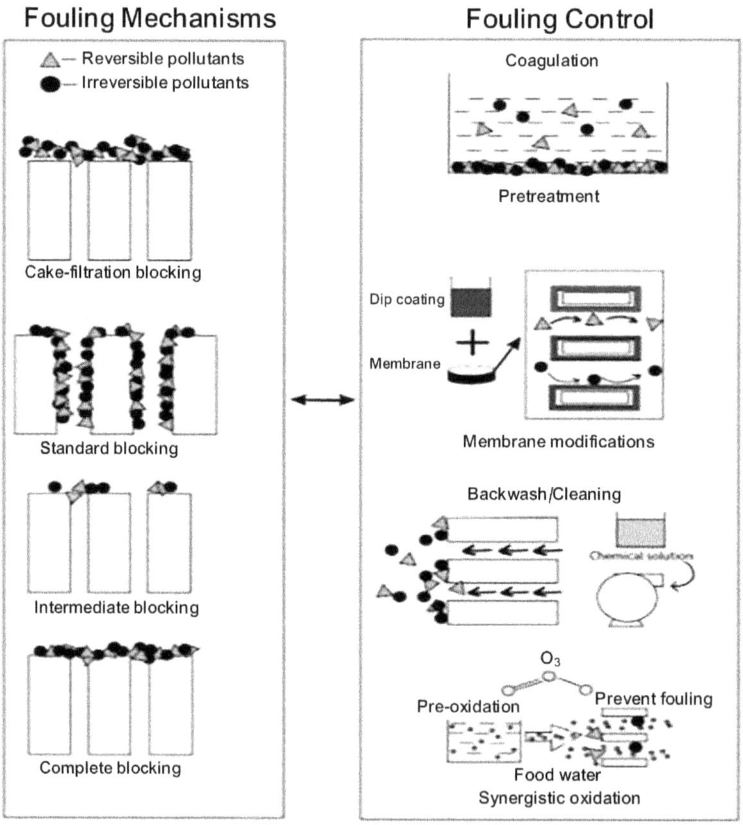

FIGURE 1.6 Illustration of common fouling mechanisms in ceramic membranes.

1.6.2 Air Treatment

Ceramic catalyst filters (CCFs) have been developed to simultaneously remove particulate matter (PM) and volatile organic compounds (VOCs) as primary air pollutants. These filters can be regenerated and used for long periods of time by simple water washing, making them a suitable option for air purification. Ceramic catalyst filters (CCFs) for air purification are designed to simultaneously remove particulate matter (PM) and volatile organic compounds (VOCs) as primary air pollutants. The key features of these CCFs include the following.

- **Long Lifetime and Renewability**: CCFs can be reused after simple water washing, retaining their initial PM and VOC removal performances.
- **High Efficiency**: The CCFs have a maximum dust loading capacity four times higher than conventional filters (approximately 20 g/L compared to 5 g/L).
- **Low cost**: The use of ceramic filters in automotive applications makes them a suitable candidate for next-generation filters supporting facile regeneration. The CCFs work by coating the inner and outer filter channels with an inorganic membrane and a Cu_2O/TiO_2 photo catalyst, respectively.

Ceramic filters offer several advantages, such as a large surface area over the volume of the filter porous wall, which enhances PM removal and the ability to decompose VOC gases with a coated photo catalyst operating under ambient conditions. The schematic representation of a CCF system for air purification is shown in Figure 1.7. This system has dual-layer filtering for the purpose of capturing PM and degrading VOC. Under UV LED irradiation, the Cu_2O/TiO_2 photocatalyst is able to enhance the oxidation of VOCs, therefore turning contaminants into carbon dioxide and water. Using deep-bed and cake filtering methods, the membrane layer can trap tiny particles (PM), allowing for effective air cleaning and easy maintenance with just water washing.

The illustration shows a rational design concept for CCF preparation. Two main mechanisms, namely wall-flow filtration of PM by porous ceramic filter with two types of filtration mechanism, namely deep-bed and cake-filtration, and photocatalytic perfect oxidation reaction for VOC removal by the developed Cu_2O / TiO_2 catalyst were applied to simultaneously remove the PM and VOC [18].

1.6.3 Soil Treatment

Advanced ceramic materials, such as alumina and zirconium, are used in various environmental applications, including soil stabilization and clean energy generation. Ceramic waste can be repurposed as an aggregate in structural concrete or as a soil stabilizer, providing novel solutions to advanced ceramic waste and facilitating more economical manufacturing and geotechnical engineering.

By appropriating ceramic waste and crushing it to use as an aggregate in structural concrete or as a soil stabilizer, the materials can provide novel solutions to advanced

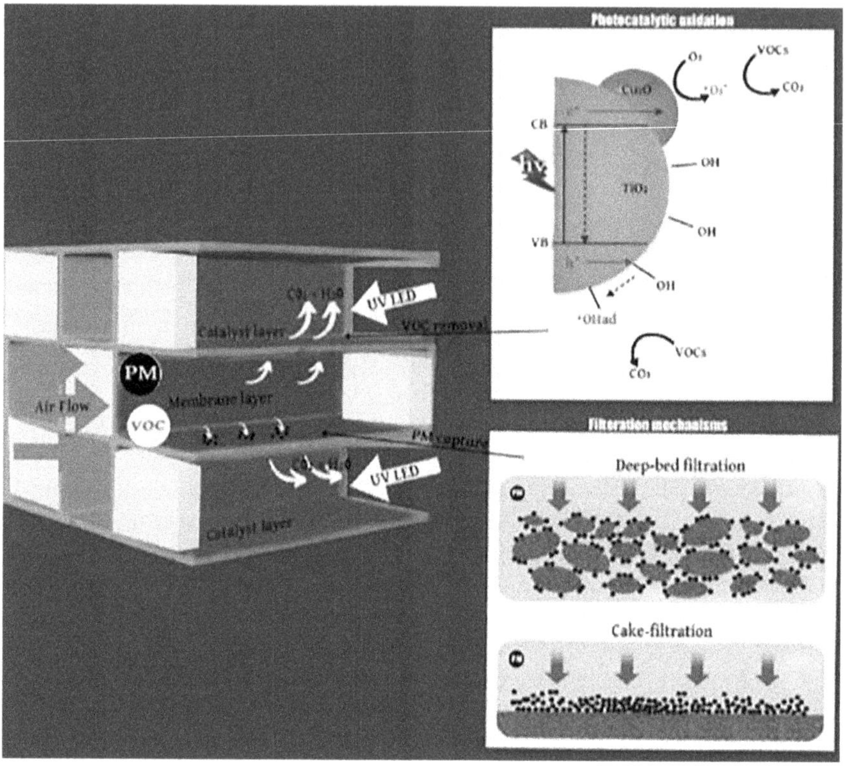

FIGURE 1.7 Schematic illustration of a CCF system for air purification, showing dual-layer filtration for PM capture and VOC degradation.

ceramic waste and facilitate more economical manufacturing and geotechnical engineering. Advanced ceramics like alumina and zirconium are used in various clean energy applications, such as lambda sensor components and gas analysis factors. These materials can withstand the harsh conditions of combustion engines for long periods of time, making them suitable for use in clean energy generation. Ceramic rubble can be used as a sustainable construction material, providing an alternative to traditional building materials [19].

1.7 CHALLENGES IN CERAMICS PROCESSING

Ceramics play a vital role in environmental engineering, but they encounter several challenges and limitations, and these challenges play a crucial role in developing strategies and innovations to enhance the performance and sustainability of ceramics in environmental engineering applications. Examining the difficulties and limitations that ceramics encounter in structural and functional applications shows nuances concerning statistical lifetimes, processing difficulties, and material limitations. This thorough review clarifies the various difficulties that scientists, engineers, and researchers encounter while working with ceramics. These difficulties include brittleness, limited ductility, and the difficulty of processing these materials. The scientific and engineering communities can work toward novel solutions to improve the functionality and suitability of ceramics in a variety of fields by being aware of these difficulties.

1.7.1 Recycling and Waste Management

Ceramic waste can be repurposed as an aggregate in structural concrete or as a soil stabilizer, providing novel solutions to advanced ceramic waste and facilitating more economical manufacturing and geotechnical engineering. However, the recycling process can be complex and may require additional resources and energy. High-porosity ceramics are also necessary for a variety of uses, ranging from military to transportation. You may adjust the porosity to fit every application. A circular mullite is a unique kind of ceramic material consisting of entwined needles with a distinct microstructure that can offer physical and performance properties far beyond what is possible with regular ceramics. These include tunable pores, high porosity and permeability, remarkable chemical stability, a high melting temperature, and exceptional strength. Porous ceramics, such as ceramic foams, are well-suited for contact processes that involve solids and fluids because they facilitate efficient fluid mixing, rapid mass transfer, and minimal pressure decreases all at once [1].

1.7.2 Environmental Impact

The production of ceramics can be resource-intensive and may contribute to environmental pollution. The long-term sustainability of advanced ceramic materials, such as alumina and zirconium, is still a matter of debate. Advanced ceramics are prized for their extraordinary durability, yet because of how well they withstand environmental deterioration, ceramic trash may sit in landfills for millennia. In an attempt to address this issue, sustainable engineers have taken ceramic waste and crushed it for use as a soil stabilizer or aggregate in structural concrete. These achievements might pave the way for new, more affordable manufacturing and geotechnical engineering eras, in addition to offering creative solutions to sophisticated ceramic waste [19].

1.7.3 Sustainable Manufacturing

Sustainable manufacturing of ceramics is an emerging trend that aims to reduce the environmental impact of ceramic production. Appropriating ceramic waste and crushing it to use as aggregate in structural concrete or as a soil stabilizer is an innovative solution that can facilitate more economical manufacturing and geotechnical engineering. Ceramics are made from several materials, such as glass and metals, that can be used to protect the environment and help regenerate damaged ecosystems. The use of technology created from these elements or materials for this purpose is recognized as environmental engineering. We can utilize high-temperature proton conductors in sensing devices for quantitative analysis of hydrogen molecules and humidity. Besides, these materials can be implemented through several other types of sensors, such as gas sensors and sensors for analyzing water, CO_2, and H_2.

Advanced ceramics are being exploited to create novel components designed to control air pollution, energy production, green buildings, improved energy storage, and environmental engineering. Ceramic catalyst filters (CCFs) have been developed to simultaneously remove particulate matter (PM) and volatile organic compounds (VOCs) as primary air pollutants [20].

1.8 CONCLUSION

The versatility of ceramics shows the potential of ceramics in environmental engineering, followed by their various applications in air treatment, water purification, waste treatment, and energy conversion and their potential that has contributed to the development of novel technologies that can address environmental challenges. The sustainability of the environment with ceramic utility makes potential technology for a green world. The properties of ceramics, such as high-temperature resistance, hardness, and chemical stability, make them suitable for these applications. The development of ceramic catalyst filters (CCFs) for air treatment has been designed to simultaneously remove particulate matter (PM) and volatile organic compounds (VOCs) as primary air pollutants. The ability to regenerate and reuse these filters through simple water washing makes them a sustainable option for air purification. Furthermore, the challenges and limitations encountered in the use of ceramics in environmental engineering include brittleness, limited ductility, and processing difficulties, emphasizing the need for innovative solutions to enhance the performance and sustainability of ceramics in various environmental engineering applications.

REFERENCES

[1] Thomas, Liji. 2023. Using Ceramic Materials for Clean Technology Alternatives. *AZoCleantech*, viewed 28 June 2023, https://www.azocleantech.com/article.aspx?ArticleID=966.

[2] Types and Applications of All Kinds of Ceramic Material, https://www.preciseceramic.com/blog/types-and-applications-of-all-kinds-of-ceramic-materials.html.

[3] Application of Ceramic Materials in Environmental Protection. 27 February 2021, https://www.earthreminder.com/ceramic-materials-in-environmental-protection/.

[4] Porous Ceramics, https://www.sciencedirect.com/topics/materials-science/porous-ceramics.

[5] https://www.ceramic-applications.com/environmental-engineering/; https://ceramics.org/about/what-are-engineered-ceramics-and-glass/structure-and-properties-of-ceramics/.

[6] Environmental Engineering. *Ceramic Applications*, https://www.ceramic-applications.com/environmental-engineering/?wpv-application=environmental-engineering&wpv_aux_current_post_id=276&wpv_aux_parent_post_id=276&wpv_view_count=12970 Opens a new window.

[7] Robinson, Isabelle. 2018. The Environmental Applications of Technical Ceramics. *AZoM*, viewed 28 June 2023, https://www.azom.com/article.aspx?ArticleID=15830.

[8] Labhsetwar, Nitin, P. Doggali, S. Rayalu, R. Yadav, T. Mistuhashi, and H. Haneda. 2012. Ceramics in Environmental Catalysis: Applications and Possibilities. *Chinese Journal of Catalysis*, 33(10), 1611–1621, https://www.cjcatal.com/EN/10.1016/S1872-2067(11)60440-3.

[9] Saremi, Omid, Mohammad Ghaani, Leila Keshavarz, and Niall English. 2022. *Application of Porous Ceramics*, https://www.researchgate.net/publication/356204175_Application_of_Porous_Ceramics.
[10] Wikipedia,https://en.wikipedia.org/wiki/Glass-ceramic#Nucleation_and_Crystal_Growth.
[11] Mason, Thomas O. 20 April 2010. White Ware. *Encyclopedia Britannica*, viewed 28 June 2023, https://www.britannica.com/art/whiteware.
[12] Bioceramic. *Wikipedia, The Free Encyclopedia, Wikimedia Foundation*, https://en.wikipedia.org/wiki/Bioceramic.
[13] Kaur, Dilraj Preet, and Meena Bhandari. 2022. *Advanced Ceramics for Versatile Interdisciplinary Applications*, https://www.sciencedirect.com/topics/materials-science/structural-ceramics.
[14] Young, W.S. 2001. *Encyclopedia of Materials: Science and Technology*, https://www.sciencedirect.com/topics/materials-science/electronic-ceramics.
[15] Asif, M.B., and Z. Zhang. 2021. Ceramic Membrane Technology for Water and Wastewater Treatment: A Critical Review of Performance, Full-Scale Applications, Membrane Fouling and Prospects. *Chemical Engineering Journal*, 418, 129481, https://doi.org/10.1016/j.cej.2021.129481.
[16] What Is Ceramic Filtration and How Does It Work? *Fresh Water Systems*, https://www.freshwatersystems.com/blogs/blog/what-is-ceramic-filtration-and-how-does-it-work.
[17] Cai, C., W. Sun, S. He, Y. Zhang, and X. Wang X. 2023. Ceramic Membrane Fouling Mechanisms and Control for Water Treatment. *Frontiers of Environmental Science & Engineering*, 17, 126, https://doi.org/10.1007/s11783-023-1726-9.
[18] Kwon, H.J., D.S. Yang, M.S. Koo, S.M. Ji, J. Jeong, S. Oh, S.K. Kuk, H. Heo, D.J. Ham, M. Kim, H. Choi, J. Lee, J. Shur, W. Lee, C. Bin, N. Timofeev, H. Wu, L. Wang, T. Lee, D.J. Jacob, and H.C. Lee. 2023. Long-Lifetime Water-Washable Ceramic Catalyst Filter for Air Purification. *Nature Communications*, 14(1), 1–10, https://doi.org/10.1038/s41467-023-36050-w.
[19] https://www.syalons.com/2021/03/25/using-advanced-ceramics-for-sustainable-innovations-an-applications-insight/.
[20] https://www.earthreminder.com/ceramic-materials-in-environmental-protection/.

2 Overview of Dental Ceramics vis-à-vis Environment-Friendly Approach

Aarati S. Panchbhai

2.1 INTRODUCTION

Composites are used in cosmetic dentistry extensively; the dental applications of ceramic materials have been augmented with advancing technology. Ceramics represents a class of material that is commonly used for the reconstruction of decayed, damaged, or missing teeth, as well as in the making of veneers, crowns, and braces. The dental applications of composites have been made possible owing to their various properties such as biocompatibility, aesthetics, insolubility in oral fluids, and hardness. The word Ceramic is the Greek term derived from "keramos", the meaning of which is "burnt stuff", that is material produced by burning [1–5]. When aesthetic is a concern, it's been a preferred material over conventional materials since a decade. The new generation of ceramic materials presents interesting options, both in terms of material selection and in terms of fabrication techniques.

The term *ceramic* refers to any product made essentially from a nonmetallic inorganic material usually processed by firing at a high temperature to achieve desirable properties. It can be an amalgamation of one or more metals with a nonmetallic element, usually oxygen. The American Ceramic Society has defined ceramics as inorganic, nonmetallic materials that are typically crystalline in nature and are compounds formed between metallic and nonmetallic elements such as aluminum and oxygen, calcium and oxygen, or nitrogen and silicon. The various elements that can be in blend would be lithium, aluminum, potassium, calcium, silicon, magnesium, phosphorus, zirconium, titanium, sodium, and so on [4–9].

A type of ceramic, known as porcelain, is specially used for dental purposes. The term *porcelain* is used for material made by firing a mixture of quartz, kaolin, and feldspar at high temperatures. Porcelain is essentially a white, translucent ceramic that is fired to a glazed state. Dental porcelain is primarily used as veneers, denture teeth, and minimal anterior tooth coverage. Dental porcelains may be classified based on their fusion temperature, microstructure, and processing technique [1, 2,

DOI: 10.1201/9781003731672-2

10–13]. Besides, nanocomposites have wide applications in dentistry and are paving the way toward better performance of ceramics as a material [4–10].

The ceramic system being used in oral cavities can be all ceramic or in combination with a metal. In all-ceramic systems, the proportion of the crystalline phase and glassy phase influences the optical and mechanical properties. Residual stress and microcracking determine the strength of ceramics [5–7]. Initially, the range of application of ceramics was narrow in the posterior areas of oral cavity due to their inability to bear up the functional forces of chewing. Further development has enabled their uses in posterior areas as well as in on lay restorations [2, 5, 6]. The all-metal restorations may be associated with issues such as allergies, gingival discoloration, and metal ion escape into the gingival tissue and fluid. When compared with metallic dental materials, the total dental ceramics restoration has lesser toughness [4, 5, 11–15]. The *metal-ceramic systems* possess the mechanical properties of metals as well as the esthetics of all-ceramics restoration. However, the recently introduced all-ceramic systems better fulfill the requirements of dental restorations. Computer-aided design and manufacturing (CAD-CAM) has enabled the acquisition of digital images of tooth preparations and computer-aided design of restorations [1, 2, 5, 6].

2.2 DIVERSE APPLICATIONS OF CERAMICS IN VARIOUS FIELDS

Ceramics have widespread applications. The composites are used in various other fields such as healthcare, nutrition and packaging, biomedical engineering, food science, transportation, automotive, and so on where they are being replaced by biocomposites. It is intended to replace synthetic composites with biocomposites. The toxicity and non-biodegradability of synthetic resins are the reasons of environmental hazards. Biocomposites are considered eco-friendly; thus they are considered to be less harmful to the society as well as to the environment. These may be considered as an alternative to replace glass-reinforced composites in future. The non-biodegradability of synthetic composites being used can be a matter of concern considering its consequences, such as climate change and a threat to a clean environment. The composites manufactured using thermoset resins systems cannot be recycled or reused. The dental ceramics and ceramics used in other industries altogether contribute to the vast landfills that are highly undesirable and impossible to manage. Moreover, the synthetic fibers can be a health hazard to the workers involved in their manufacture and pose an ecological threat [16–18].

2.2.1 Biocomposites: Composition and Renewable Resources

Renewable resources can be used for the manufacture of biocomposites; these may replace the traditionally used resins for restoration [16–22]. Various natural resources from plants and crops such as flax, kenaf, jute, henequen, sisal, soybean, hemp, and so on can be utilized for the manufacture of the natural composites. The composition of the biocomposite or green composite may contain resin phase or natural fibers (Table 2.1). Eco-friendly newer biocomposites derived from plants and crops would be a novel initiative considering the growing environmental hazard [16–18].

TABLE 2.1
Composition of Biocomposite/Green Composite

Components	• Biodegradable resin • Natural fibers
Natural/biofibers	• Non-wood fibers • Wood fibers

TABLE 2.2
Classification of Ceramics

Category	Types
Based on ceramic systems	• Silicate ceramics • Polycrystalline ceramics • Resin-based ceramics
Based on ceramic with or without metal	• Metal-ceramic systems • All-ceramic systems
Based on microstructure	• Category 1 – glass-based systems (mainly silica), • Category 2 – glass-based systems (mainly silica) with fillers, usually crystalline (leucite or lithium disilicate) • Category 3 – crystalline-based systems with glass fillers (mainly alumina) • Category 4 – polycrystalline solids (alumina and zirconia)
Based on processing technique	• Power-liquid, glass-based systems with or without crystalline fillers • Machinable or pressable blocks of glass-based systems • CAD/CAM or slurry, die-processed, mostly crystalline (alumina or zirconia) systems

2.3 CLASSIFICATION

Classification is constantly evolving with latest developments, presently, ceramics are categorized based on the fabrication technique and microstructural composition of the crystalline and glass phases [1–8, 23–26] (Table 2.2).

2.3.1 Based on Ceramic Systems

Under this, the various ceramic systems include silicate ceramics, polycrystalline ceramics, and resin-based ceramics.

2.3.2 Based on Ceramic With or Without Metal

2.3.2.1 Metal-Ceramics Systems

These systems combine metal and ceramic to use the advantages of both. This system is in vogue from 1960. In this, the ceramic layer is coved onto metal by the technique of firing. This will fulfill the requirement of aesthetics as well as strength of the material.

2.3.2.2 All-Ceramic Systems

The all-ceramic restoration has ceramic as its whole constituent without any metallic incorporation; however, the crystalline phase is in greater amount in all-ceramic restorations. The greater amount of crystalline phase, although augments the mechanical properties, may lead to opaque restoration owing to proportionate loss of translucency which may not be the desirable feature.

2.3.3 Based on Microstructure

On the basis of feasible glass-to-crystalline ratio, ceramics may be categorized as glass-based systems containing mainly silica or fillers, usually of crystalline nature or crystalline-based systems with glass fillers containing mainly alumina and polycrystalline solids, primarily containing zirconia or alumina.

2.3.4 Based on Processing Technique

2.3.4.1 Powder/Liquid, With or Without Crystalline Fillers

These are the porcelains that are made for veneering cores made from metal, alumina, or zirconia but can be used for porcelain veneers on either a refractory die or platinum foil technique.

2.3.4.2 Manufactured Blocks, With or Without Crystalline Fillers

These materials may be used for inlay and onlay restorations, anterior or posterior crowns and veneers. They can be used for full contouring and need bonding.

2.3.4.3 CAD/CAM or Slurry/Die-Generated or All-Crystalline Alumina- or Zirconia-Based Systems

These materials can be best used as core materials for making single crowns that are veneered with a powder/liquid glass-based material.

2.4 PROPERTIES OF CERAMICS

The properties of dental ceramics depend on their composition and microstructure. The nature, particle size, and amount of crystalline phases influence the mechanical and optical characteristics of the materials such as crack propagation or strength [1–6, 14]. While the translucency and opacity is the function of the amount of glassy phase in ceramic material.

As compared to polymers or metals, ceramics are considered to be brittle with less plastic behaviors. The presence of cracks, surface flaws, and brittle characteristics lead to the material or restoration with comparatively lower strength. Increased crystalline-filler content and its even distribution within the glass matrix may augment the flexural strength of the ceramic restoration. The performance of dental ceramics is compromised under conditions of tensile and shears stresses, although it can be resistant to compressive stresses [1, 11, 14, 22].

Silicate cements used earlier for the purpose of restorations had better aesthetic properties, but they were lacking the strength as compared to resin-based and polycrystalline ceramics. On the other hand, the mechanical properties of polycrystalline ceramics are superior; however, the aesthetic properties are undesirable. Being weaker, the silicate cements are now used for making veneers [1, 10].

Adhesion of ceramic restoration to the natural tooth and fatigue strength play an important role in the durability and longevity of dental ceramic restorations. Surface hardness of ceramics is optimum to the extent that they can abrade the natural teeth in the opposing dental arch [1, 11, 22]. The coefficient of thermal expansion of ceramics is approximately similar to the natural tooth. Glass ionomer cements and resin cements are most commonly used for luting of ceramic restorations. Studies have shown that chemical conditioning methods such as silanization increases the adhesion of the composite resin bond to the ceramic. The recent decades have been marked as an era of technological evolution for dental ceramics toward the development of zirconia-based all-ceramics. The remarkable improvement has been made in the clinical performance of ceramic restorations [1–6, 8, 27–39].

2.5 USES OF DENTAL CERAMICS [1–6, 25, 26]

- Artificial denture teeth
- Crowns
- Veneers
- Bridges
- Ceramic posts
- Abutments
- Implants
- Ceramic braces

Dental ceramics are materials that are used to replace missing or damaged dental structures. The use of ceramics in conservative dentistry is increasing; ceramics are used in restorative dentistry as inlay preparations, veneering porcelain, full- and partial-coverage crowns, denture teeth, and as particulate fillers for resin matrix composite filling materials. As a material, ceramics have definite advantages due to biocompatibility and aesthetics without identified adverse reactions as such. The blend of glassy phase and crystalline phase in proportion is critical to achieve required optical and mechanical properties for optimum performance of material in the oral cavity. The amount of the glassy phase decides the translucency while the amount of crystalline phase decides the strength. So altogether, the proportion of the two material phases needs to be balanced. Recent developments are targeted at removing the inadequacies of low fracture toughness and low tensile strength.

2.5.1 Aligning Ceramics to Environment-Friendly Practices

Currently, there is a greater preference toward the escalation of the materials which will maintain the greener environment. The production using raw materials or product which will get completely degraded and which can be recycled are desired these days. Apart from dental uses, composites are being used in sectors such as construction, transportation, automotive, biomedical engineering, medical, food science, nutrition and packaging, and so on [16–20]. Also, environmentally friendly nature of the green composite can be the emerging aspect in advancements.

Composites have been a part of cosmetic dentistry since long, wherein the materials are used with two purposes as how well they function in the mouth, that is, their clinical performance, and their aesthetic qualities. Glass-reinforced composites have occupied various fields since many years. Green composites or biocomposites are emerging as an alternative to glass-reinforced composites. Biocomposites are derived from renewable resources; fiber composites are gradually replacing conventional resin materials. Synthetic resins are non-biodegradable and toxic as they pose a greater threat to the environment as compared to biocomposites that may get easily composted [5, 16–20].

2.6 CHALLENGES IN USING GREEN COMPOSITES

The idea that natural fibers should be used as a resource material for green composite popped out because of its easy availability and abundance. In addition, it will be economic and biodegradable. The efforts to utilize the underused biomaterial resources have paved the way to exploration and significant research work in the field of green composites. However, there are certain challenges associated with the utilization of these naturally derived biocomposites (Table 2.3).

The challenges with the use of biocomposite could be many. Importantly, the materials or restorations in the oral cavity will come in contact with moisture, which may affect its functional efficiency and durability due to the propensity of biomaterials to absorb the moisture. In addition, the reliability of the biomaterial, its strength, and resistance would be a matter of concern. Besides, the inherent problems in the materials, such as the quality of natural fiber, consistency of composition, and poor bonding, may impact their exploitation. The weak bond between the two phases in the biocomposite such as natural fiber and polymer matrix may influence its performance and durability [18, 21].

TABLE 2.3
Challenges in the Use of Green Composites

- Dimensional instability
- Propensity to moisture absorption
- Susceptibility to environmental attack
- Inconsistency in processing
- Varied surface energy of natural fibers
- Incompatibility due to different surface energies
- Fiber–matrix interface adhesion
- Reliability of performance
- Inferior strength

2.6.1 Sustainability Aspects in the Production or Disposal of Dental Ceramics

Knowing the drawbacks of the synthetic materials, the efforts are now greatly focused on the production of ceramics from natural resources such as fiber, crop or plant. The research work is emphasized on plant-fiber derived sustainable 'green' composites with focus on their morphology, matrix-reinforcement combinations, and processing. At the moment, it would be hasty to comment about the composition of the green composites to be used in various sectors including dentistry. The composition of green composite and its dental applications would be the domain of further exploration [16, 21].

The hydrophilic nature of the fibers will enhance the moisture absorption in green composite. The proportion of matrix and fiber content in the composition of biocomposite would be crucial in deciding the susceptibility of restoration to the moisture. Keeping the fiber content limited in matrix-fiber ratio in biocomposite would help in controlling the dimensional instability due to moisture absorption. The poor mechanical properties of restoration, such as suboptimal strength, could be the considerable issue and hence make it a restoration with limited applications leaving the situations requiring higher strength in the restorations [18–21, 40, 41]. The features of raw materials used and the way these are processed will highly impact the properties of the final biocomposite. The arrangement of fibers, use of coupling agents, and the processing of natural fibers would affect the characteristics of the product. Furthermore, the inherent conditions needed for natural resources used, such as climate, maturity, harvesting, and further treatments, will affect the naturally derived composites. The various surface treatment and preprocessing techniques may help to improve the performance of the biomaterials [18–21, 40, 41].

The search for inexpensive material with high performance is still on. The development of bio-based materials over synthetic and completely degradable material over non-degradable material is being preferred. Thus, the quest for the eco-friendly biomaterial, which will get composted or disposed without difficulty so as not to damage the environment, is greatly desired.

2.6.2 Green Recommendation for Ceramics

The terms Green or Eco-friendly refer to energy efficiency, sustainability, or renewability. Green dentistry is an advanced way of enhancing wellness and an environment-friendly approach. It is the moral responsibility of every individual to preserve the climate, similarly giving best to the patient should not jeopardize the environment. It aligns human health with environmental health, and vice versa [42–44]. It reduces the amount of waste being produced and preserves the natural environment. CAD/CAM systems, enabling the single appointment dentistry, help to minimize the greenhouse gasses produced from patient and staff travel for the multiple appointments and the shipping of impressions and final restorations [16, 20, 42–47].

The current healthcare practices render advanced and multifaceted healthcare to the patient, although, with greater production of biomedical waste and landfills which has an unfavorable impact on the environment. Also, it leads to the contamination of water and aquifers. Incinerators are commonly used to dispose the waste, but they produce ash, methane, carbon dioxide, sulfur dioxide, etc. as its by-products that further contribute to global warming [44–55].

2.7 CONCLUSION

Ceramic materials have become a mainstream material in restorative dentistry; however, a closer understanding of the dynamics of the materials is needed for desired performance. The development of all-ceramic systems for dental restorations has been incredible with the use of novel processing techniques such as computer-aided design and computer-aided machining (CAD-CAM). Nowadays, there is a greater inclination toward the production of biocomposites owing to accumulation of waste and landfills the synthetic resins produce. Significant research is being conducted in developing plant- or fiber-derived resins. However, sustainability and clinical performance of the green composites are two key issues which need further exploration. It can be expected that coming years may render the biocomposites as the most desired materials to replace synthetic composites in healthcare fields as well as in other fields. Research work currently focuses on improving the suboptimal performance of biocomposites and to develop resources from which green composites can be produced. Environment-friendly approaches may promote sustainable ways by reducing the energy being consumed and the amount of waste being produced by the industry. The newer biocomposite derived from natural resources would be anticipated to be a novel material of the current century as a way out of growing environmental hazards.

REFERENCES

[1] Shenoy A, Shenoy N. Dental Ceramics: An Update. *J Cons Dent*, 2010; 13(4): 195–120

[2] Sakaguchi, R., Ferracane, J., and Powers, J. *Craig's Restorative Dental Materials, General Classes of Biomaterials*, Chapter 7, Ceramics, ed. Sakaguchi R, Feracane J, Powers J, 14th edition, Elsevier Publishers, 2019, 113–121

[3] Shi H, Pang R, Yang J Cai H, Jiang H, Han J. Overview of Several Typical Ceramic Materials for Restorative Dentistry. *BioMed Res Int*, 2022; 2022: 18 pages

[4] Babu J, Alla R, Alluri V, Datla S, Konakanchi A. Dental Ceramics: Part I – An Overview of Composition, Structure and Properties. *American J Mater Eng Technol*, 2015; 3: 13–18

[5] Denry I, Holloway M. Review Ceramics for Dental Applications: A Review. *Materials*, 2010; 3: 351–368

[6] Sukumaran VG, Bharadwaj N. Ceramics in Dental Applications. *Trends Biomater Artif Organs*, 2017; 20: 7–11

[7] Silva LE, de Lima R, de Paula Miranda S, Favero, U, Lohbauer, Cesar P. Dental Ceramics: A Review of New Materials and Processing Methods. *Brazilian Oral Res*, 2017; 31: 133–146

[8] Sakaguchi RL, Powers JM. *Craig's Restorative Dental Materials*, 12th edition, Elsevier, Mosby, A Division of Reed Elsevier India Pvt Ltd, New Delhi, India, 2007, 443–464

[9] Rashid H. The Effect of Surface Roughness on Ceramics Used in Dentistry: A Review of Literature. *Eur J Dent*, 2014; 8: 571–579

[10] Panchbhai A. *Nanocomposites: Past, Present and Future. Applications of Nanocomposite Materials in Dentistry*, ed. Asiri A, Mohammad IA, 1st edition, Elsevier, Woodhead Publishing, 2018, 364, eBook, ISBN: 9780128137598
[11] Rosenblum MA, Schulman A. A Review of All Ceramic Restorations. *J Am Dent Assoc*, 1997; 128: 297–307
[12] Rizkalla AS, Jones DW. Mechanical Properties of Commercial High Strength Ceramic Core Materials. *Dent Mater*, 2004; 20: 207–212
[13] Arvidson K, Wroblewski R. Migration of Metallic Ions from Screwposts into Dentin and Surrounding Tissues. *Scand J Dent Res*, 1978; 86: 200–205
[14] Venclíkova Z, Benada O, Bártova J, Joska L, Mrklas L. Metallic Pigmentation of Human Teeth and Gingiva: Morphological and Immunological Aspects. *Dent Mater J*, 2007; 26: 96–104
[15] Mehulic K, Prlic A, Komar D, Prskalo K. The Release of Metal Ions in the Gingival Fluid of Prosthodontic Patients. *Acta Stomatol Croat*, 2005; 39: 47–50
[16] Grigoras V. Thermal Behaviour and Crystallization of Green Biocomposites, Chapter 11. *In: Sustainable Polymer Composites and Nanocomposites*, Springer, 2019, Vol. 7, 1185–1231
[17] Kumar A, Sivakumar A, Tamilarasan R. Overview and Recent Advancements in Composite Materials. *Int J Res Appl Sci Eng Technol*, 2017; 5: 71–78
[18] Satyanarayana KG. Recent Developments in 'Green' Composites Based on Plant Fibers-Preparation, Structure Property Studies. *J Bioprocess Biotech*, 2015; 5: 206–212
[19] Chatzistavrou X, Fenno JC, Faulk D, et al. Fabrication and Characterization of Bioactive and Antibacterial Composites for Dental Applications. *Acta Biomate*, 2014; 10: 3723–3732
[20] Saba N, Tahir P, Jawaid M. A Review on Potentiality of Nano Filler/Natural Fiber Filled Polymer Hybrid Composites. *Polymers*, 2014; 6, 8: 2247–2273
[21] Shanmugasundaram OL. Green Composites: Manufacturing Techniques & Applications. *Indian Textile J*, October 2009; 37: 146–150
[22] Chatzistavrou X, Lefkelidou A, Papadopoulou L, Pavlidou E, Paraskevopoulos KM, Fenno JC, Flannagan S, González-Cabezas C, Kotsanos N, Papagerakis P. Bactericidal and Bioactive Dental Composites. *Front Physiol*, 2018; 9: 103–107
[23] Edward A, McLaren DS, Tran CP. *DDS Ceramics in Dentistry-Part I: Classes of Materials Inside Dentistry*. Available from: http://www.insidedentistry.net [Last accessed on October 2009]
[24] Mclean JW. Evolution of Dental Ceramics in the Twentieth Century. *J Prosthet Dent*, 2001; 85: 61–66
[25] Vianna L, Prado C, Bicalho A, da Silva Pereira R, das Neves F, Soares C. Effect of Cavity Preparation Design and Ceramic Type on the Stress Distribution, Strain and Fracture Resistance of CAD/CAM Onlays in Molars. *J Applied Oral Sci*, 2018; 26: 6 pages
[26] Bustamante-Hernández N, Montiel-Company J, Bellot Arcís C. Clinical Behavior of Ceramic, Hybrid and Composite Onlays: A Systematic Review and Meta-Analysis. *Int J Environ Res Public Health* 2020; 17: 7582, 171–181
[27] Seghi RR, Daher T, Caputo A. Relative Flexural Strength of Dental Restorative Ceramics. *Dent Mater*, 1990; 6: 181–184
[28] Seghi RR, Sorensen JA. Relative Flexural Strength of Six New Ceramic Materials. *Int J Prosthodont*, 1995; 8: 239–246
[29] Zelos L, Bevis RR, Keenan KM. Evaluation of Ceramic/Ceramic Interfaces. *Am J Orthod Dentofacial Orthop*, 1994; 106(1): 10–21
[30] O'Brien, W.J. *Dental Materials and Their Selection*, 3rd edition, Quintessence Publishing Co. Inc, 2002, 132–155

[31] Vallittu PK. Non-Metallic Biomaterials for Tooth Repair and Replacement. *In: Processing and Bonding of Dental Ceramics*, Woodhead Publishing Limited, Philadelphia, 2013, 125–160

[32] Morena R, Beaudreau GM, Lockwood PE, Evans AL, Fairhurst CW. Fatigue of Dental Ceramics in a Simulated Oral Environment. *J Dent Res*, 1986; 65(7): 993–997

[33] Gonzaga CC, Cesar PF, Miranda WG Jr, Yoshimura HN. Slow Crack Growth and Reliability of Dental Ceramics. *Dent Mater*, 2011; 27(4): 394–406

[34] Borges GA, Goes MF, Platt JA, Moore K, Menezes FH, Vedovato E. Extrusion Shear Strength Between an Aluminabased Ceramic and Three Different Cements. *J Prosthet Dent*, 2007; 98(3): 208–215

[35] Ravi RK, Alla RK, Shammas M, Devarhubli A, Dental Composties – A Versatile Restorative Material: An Overview. *Ind J Dent Sci*, 2013; 5(5): 111–115

[36] Dhillon J, Tayal SC, Tayal A, Amita, Kaur AD. Clinical Aspects of Adhesion of All Ceramics: An Update. *Ind J Dent Sci*, 2012; 4: 123–126

[37] Li R, Ren Y, Han J. Effects of Pulsed Nd:YAG Laser Irradiation on Shear Bond Strength of Composite Resin Bonded to Porcelain. *Surface Hua Xi Kou Qiang Yi XueZaZhi*, 2000; 18: 377–379

[38] Studart AR, Filser F, Kocher P, Gauckler LJ. In Vitro, Lifetime of Dental Ceramics Under Cyclic Loading in Water. *Biomater*, 2007; 28: 2695–2705

[39] Schmage P, Nergiz I, Herrmann W, Oscan M. Influence of Various Surface-Conditioning Methods on the Bond Strength of Metal Brackets to Ceramic Surfaces. *Am J Ortho Dentofacial Orthop*, 2003; 123(5): 540–546

[40] Davim JP. *Green Composites Materials*, Manufacturing and Engineering eBook (PDF), 2017, 149 pages, ISBN 978-3-11-043578-ch4

[41] Pickering K. *Properties and Performance of Natural-Fibre Composites*, Woodhead Publishing, Cambridge, UK, 2008. https://doi.org/10.1533/9781845694593

[42] Rekow ED, Silva NR, Coelho PG, Zhang Y, Guess P, Thompson VP. Performance of Dental Ceramics: Challenges for Improvements. *J Dent Res*, 2011; 90(8): 937–952

[43] Avinash B, Avinash BS, Shivalinga BM, Jyothikiran S, Padmini MN. Going Green with Eco-Friendly Dentistry. *J Contemp Dent Pract*, 2013; 14(4): 766–769

[44] Duane B, Kim C, Darshini R, Sara H, Inge S, Rachel S, Devika V. Environmental Sustainability: Measuring and Embedding Sustainable Practice into the Dental Practice. *Br Dent J*, 2019; 226: 891–896

[45] Shekar HSS, Ramachandra M. Green Composites: A Review. *Mater Today Proc*, 2018; 5: 2518–2526

[46] Thakur V, Singha AS, Thakur M. Green Composites from Natural Fibers: Mechanical and Chemical Aging Properties. *Int J Polym Anal Charact*, 2012; 17: 6 pages

[47] Shalwan A, Yousif BF. Investigation on Interfacial Adhesion of Date Palm/Epoxy Using Fragmentation Technique. *Mater Des*, 2014; 53: 928–937

[48] Denry IL, Holloway JA. Microstructural and Crystallographic Surface Changes After Grinding Zirconia-Based Dental Ceramics. *J Biomed Mater Res B-Appl Biomater*, 2006; 76B: 440–448

[49] ISO. *ISO-13356 International Standard – Implants for Surgery – Ceramic Materials Based on Yttria Stabilized Tetragonal Zirconia (Y-TZP)*, ISO, Geneva, Switzerland, 2008, 1–13

[50] Ban S. Reliability and Properties of Core Materials for All-Ceramic Dental Restorations. *Jpn Dent Sci Rev*, 2008; 44: 3–21

[51] Deville S, Chevalier J, Fantozzi G, Bartolome JF, Requena J, Moya J, et al. Development of Advanced Zirconia-Toughened Alumina Nanocomposites for Orthopaedic Applications. *In: Euro Ceramics VIII*, Pts 1–3, Trans Tech Publications, Zurich, Switzerland, 2004, 264–268, 2013–2016

[52] Nawa M, Nakamoto S, Sekino T, Niihara K. Tough and Strong Ce-TZP/Alumina Nanocomposites Doped with Titania. *Ceram Int*, 1998; 24: 497–506

[53] Ban S, Nawa, M. Application of Zirconia/Alumina Composite to All Ceramic Crown. *Dent Mater J*, 2005; 24: 70

[54] Shah K, Holloway JA, Denry IL. Effect of Coloring with Various Metal Oxides on the Microstructure, Color, and Flexural Strength of 3Y-TZP. *J Biomed Mater Res Appl Biomater*, 2008; 87B: 329–337

[55] Cetin S, Mahler L, Ramseie CA. Sustainable Dentistry – Part II: Sustainable Workplace and Waste Management. *Swiss Dent J*, 2022; 132(11): 781–787

3 Nanoceramics and Their Environmental Applications

Nidhi Sharma, Antrakrate Gupta, Ramasubramanian Lakshmi Narayan, and Shikhar Krishn Jha

3.1 INTRODUCTION

Lately, ceramics and ceramic-based materials have attracted much attention in the field of research and development owing to their ever-growing and emerging applications. Their characteristic features such as high hardness, superior wear resistance, lower thermal and electrical conductivity, high-temperature strength, and higher chemical stability have made them one of the significant engineering materials (Devarajan et al., 2023). Chemically, ceramics are inorganic, nonmetallic solids, composed of metallic or nonmetallic constituents, which can be shaped and densified by heating to higher temperatures with the process called sintering. In general, ceramics can be classified into two broad categories: traditional and advanced (or engineering ceramics). The engineering ceramics include oxides: zirconia (ZrO_2), alumina (Al_2O_3), and beryllia (BeO); non-oxides (borides, carbides, nitrides, and silicides); and composite ceramics (fiber- or particulate-reinforced combinations of oxides and non-oxides) (Ferraris et al., 2016). Technological advances, including novel synthesis and processing techniques, have produced a new class of nanoceramic materials with unprecedented control over structure, composition, defects, and other specific properties (Van Der Biest, 2013). Nanoceramics refer to the special class of ceramic materials that have been engineered at the nanoscale level and are known to possess high mechanical, chemical, and corrosion resistance with enhanced electrical, magnetic, and/or optical properties (Vunain et al., 2016). In addition to refractory properties and chemical inertness, nanoceramics with high surface area of the smaller particles are an attractive choice of material for a variety of applications, such as construction and transportation systems, drug delivery, bone repair, communication, energy supply, and storage to name a few (Palmero, 2015).

Nanoceramics are primarily made up of oxides (TiO_2, SiO_2, Al_2O_3, ZnO), carbonates ($CaCO_3$, $MgCO_3$), carbides (SiC, CaC_2), nitrides (HfN, TiN, TaN, ZrN), and phosphates of metals and metalloids (ZrP, TiP) (Basu & Balani, 2011). Recently, a range of mixed oxides and titanates, like indium–tin oxide $(In_2O_3 - SnO_2)$, barium titanate $(BaTiO_3)$, antimony-tin oxide (ATO), have also become imperative nanoceramics for

DOI: 10.1201/9781003731672-3

use in various applications such as transparent conductive coatings, electromagnetic shielding, electrochromic devices, touchscreens, LCDs and OLEDs, capacitors, sensors, energy harvesting, and so on (Basu, 2011). As nanoceramics are getting used in a large number of applications, environmental science and sustainability have also benefited. Some of the major environmental applications of nanoceramics include water purification, air pollution control, environmental remediation, energy efficiency, and waste management (Taran et al., 2021). Nanoceramic filters (Saleem & Zaidi, 2020) have shown great potential in water treatment and purification processes, which can remove contaminants such as heavy metals, organic pollutants, and microorganisms more efficiently than traditional filtration methods. The high surface area of nanoceramics enhances adsorption and catalytic processes, leading to improved water quality. Nanoceramic materials, such as titanium dioxide (TiO_2) nanoparticles, have photocatalytic properties that can be utilized for air purification. When exposed to ultraviolet light, nanoceramics can break down pollutants, including volatile organic compounds (VOCs) and nitrogen oxides (NOx), into harmless substances (Ren et al., 2017). This technology has also been used in self-cleaning coatings for buildings and air purification systems (Hamidi & Aslani, 2019). Nanoparticles of iron oxides (FeO, Fe_2O_3) are reported to facilitate the degradation of organic pollutants, such as chlorinated solvents and hydrocarbons, through processes like chemical oxidation or reduction (Khan & Lee, 2021). Nanoceramic coatings are found to improve the energy efficiency of buildings by enhancing their insulation properties. These coatings can be applied to windows, walls, and roofs to reduce heat transfer, resulting in lower energy consumption for heating and cooling (Ali et al., 2020). Nanoceramics have the potential to address challenges in waste management. For instance, incorporating nanoceramic materials in catalysts can enhance the efficiency of waste-to-energy conversion processes, such as the gasification or pyrolysis of organic waste materials, which can lead to higher energy recovery and reduced emissions (Needhidasan et al., 2014). Additionally, nanoceramics are also used in the development of advanced energy storage devices, such as high-performance batteries and supercapacitors, and contribute to renewable energy systems (Gogotsi, 2014).

3.2 CERAMICS IN NANOTECHNOLOGY: THE STATE OF ART

Ceramics play a significant role in nanotechnology and have contributed to advancements in various fields. An overview of the state of the art in ceramics within nanotechnology is summarized here.

(i) **Enhanced Mechanical Properties**: Nanoceramics exhibit exceptional mechanical properties due to their nanoscale structure. They demonstrate enhanced hardness, strength, and toughness compared to their bulk counterparts. The reduction in grain size and the presence of grain boundaries result in improved mechanical performance and fracture resistance (Jeong et al., 2022).

(ii) **Tailored Electrical and Optical Properties**: Ceramics at the nanoscale exhibit unique electrical and optical properties. For instance, nanoparticles of materials like titanium dioxide (TiO_2) and zinc oxide (ZnO) possess enhanced photocatalytic properties, making them useful in solar cells, sensors, and environmental applications. Nanoceramics also show enhanced optical transparency, enabling their utilization in transparent coatings, displays, and optoelectronic devices (S. F. Wang et al., 2013).

(iii) **Advanced Energy Storage**: Nanoceramics have made significant contributions to energy storage systems in terms of enhancing the specific capacities of the batteries and extending their cyclability. For example, lithium-ion batteries benefit from nanoceramic coatings on electrodes, enhancing their performance, safety, and lifespan. Ceramic materials like lithium titanate $\left(Li_4Ti_5O_{12}\right)$ nanoparticles are being explored for high-capacity and fast-charging batteries. Similarly, nanoceramics are used in the development of solid oxide fuel cells (SOFCs) and supercapacitors, enabling efficient energy conversion and storage (Wu et al., 2021).

(iv) **Biomedical Applications**: Nanoceramics have shown great promise in biomedical applications. For example, hydroxyapatite (HA) and bioactive glasses are used in bone tissue engineering and regenerative medicine. They can mimic the composition and structure of natural bone, promoting cell adhesion and bone growth. Nanoceramics such as porous silica-based drug delivery systems have also been developed for targeted and controlled release of therapeutic agents (Nandhini et al., 2021).

(v) **Wear and Corrosion Resistance**: Ceramic coatings at the nanoscale offer excellent wear and corrosion resistance. They are applied to surfaces to improve hardness, reduce friction, and protect against environmental degradation. Nanoceramic coatings find applications in cutting tools, automotive components, and biomedical implants (Attarzadeh et al., 2021).

The state of the art in ceramics within nanotechnology is continuously evolving, with ongoing research focusing on novel synthesis methods, improved properties, and innovative applications. These advancements hold great potential for various industries, including electronics, energy, healthcare, and environmental sectors.

3.2.1 Why Nanoscale?

Kamigatio (Cain & Morrell, 2001) suggested that nanoscale microstructures can lead to property changes if the structure size is smaller than a certain level, as shown in Table 3.1. Such potential changes in properties have spurred research in the field of nanoceramics, especially in the last decade, particularly in view of the property and performance limitations of conventionally processed materials.

Over the past three decades, particularly with respect to structural ceramics, efforts have been made to improve the strength and toughness, making ceramics more accessible and reliable for engineering applications. The understanding of strength σf in terms of Griffith 'flaws' related to grain size g, as embodied in the Hall-Petch relationship, $\sigma_f \propto g^{-1/2}$, suggests that finer grain sizes can lead to stronger materials (Nivas & Fulrath, 1970). This certainly happens in the metal industry, but in ceramics the strength is limited by the size of flaw at the surface formed during handling or machining, which extends the damage beyond several grain depths.

TABLE 3.1
Feature Size for Significant Changes in Properties (Cain & Morrell, 2001)

Technical Changes in Properties	Feature Size at Which Changes Can Be Expected (nm)
Catalytic activity	<5
Producing super-paramagnetism and other electromagnetic phenomena	<100
Producing strengthening and toughening	<100
Modifying matrix dislocation movement, hardness, and plasticity	<100
Making hard magnetic materials soft	<20
Producing refractive index changes	<50

However, fine grain size can be useful for other reasons, such as:

(i) Reduction of firing temperature, which saves energy and ensures removal of residual porosity;
(ii) Changes in optical properties, particularly, the improvement in transparency due to reduction in light scattering;
(iii) Reduction in microcracking due to crystallographic thermal expansion anisotropy;
(iv) Better polishing of surfaces;
(v) Improved wear resistance and surface finish, through changes to the surface deformation and material removal mechanisms;
(vi) Better strength and fracture toughness;
(vii) Possibility to develop phase assemblages and properties that are not available in coarser-grained versions.

3.3 TYPES OF NANOSTRUCTURED CERAMICS

The most commonly targeted ceramic nanostructures, as shown in Figure 3.1 include three basic types, designated as 'nano' (single nanophase); 'nano/nano' (two nanophases); and 'micro/nano' (the nanophase dispersed in the microphase) (Niihara, 1991). However, other subcategories can also be identified depending upon the different spatial relationships between the phases. In the case of the 'micro/nano' type, the dispersed nanophase component could be distributed in the bulk of the grains (i.e., left behind by grain growth), along the grain boundaries of the microphase (i.e., swept along with grain growth in sintering), at the grain boundary junctions only, or in all positions. Depending on this, the properties (structural, physical, technical) will vary for all the subtypes.

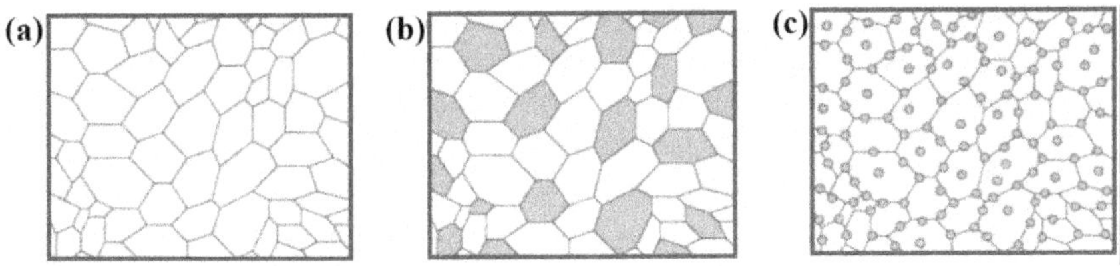

FIGURE 3.1 Basic types of nanoceramics: (a) all-nano, single phase; (b) all-nano, two phase; (c) micro/nano, two-phase type with the nanophase dispersed both within grains and at all places along the grain boundaries.

It is clear from these geometries that to utilize these nanostructures as bulk ceramic material, close attention must be paid to obtain:

(i) an understanding of interactions between different components during processing of mixed-species materials;
(ii) an understanding of the changes in internal stress distribution on addition of new species;
(iii) uniform fine particle size, to obtain uniform grain structure;
(iv) good dispersion of the components in multi-component materials;
(v) minimal agglomeration, to restrict large residual pores;
(vi) homogeneity, better density, and minimal residual defects.

3.3.1 Manufacturing Challenges

For utmost exploitation of nanoceramics in various applications, several key challenges have to be met and taken care of in an economic manner. Major consideration has to be given to the following factors:

(i) **Cost of Nano-Powders**: The cost of bulk technical ceramics directly relies on the cost of the raw materials and is significantly comparable to the preparation, processing, and consolidation cost. Any reduction in cost of nanoscale powders may come with its production in large volumes, but ultrafine powders are always likely to be more expensive than coarser ones.
(ii) **Powder Reactivity**: The high surface area of nanoceramics results in enhanced rate of reactions between the powder particles and their ambient environment. A few examples include oxidation of ultrafine non-oxides, hydration of silicates, and hydrolysis reactions between Si_3N_4 and water vapor. To avoid and inhibit these reactions, processing requires controlled environments, such as inert liquids or dry atmospheres, and thus adds to the processing costs.
(iii) **The problem of agglomeration/de-agglomeration**: The finer the powder, the higher its surface area, and the more likely it is to adhere to its adjacent

particles. Agglomeration leads to poor mixing, inhomogeneous packing, residual porosity, and, thus, poor density.

(iv) **Low green density**: Due to the tendency of agglomeration, the packing density while pressing the nanoceramic particles often remains poor, and to avoid that binders are added and tuned to attain the desired requirements. Poor packing density of systems leads to poor consolidation while sintering and contributes to distortions and reduced dimensional control.

(v) **Sintering inhibition**: When a non-sinterable additive is introduced to a sinter-active material (e.g., in a micro/nano material), the sinter-activity of the latter reduces. This phenomenon is called constrained sintering (Green et al., 2008). In such cases, particle rearrangement gets more difficult, and any change in shape requires a large amount of bulk diffusion under shear to allow the sintering of the matrix phase alone.

These reasons make the process control of nanoceramics more difficult than the conventional ceramics and thus require higher manufacturing cost to bring actual products to the market.

3.3.2 Materials

A broad classification of nanoceramic products is shown in Figure 3.2 (a), whereas the triaxial diagram in Figure 3.2 (b) illustrates the areas of commercial nanoceramic wares.

The chemical composition of raw materials used in the nanoceramic industry is shown in Table 3.2.

A brief description of constituent raw materials is as follows.

3.3.3 Quartz Sand

Quartz sand, also known as silica sand, can indeed serve as a raw material for the production of nanoceramic materials. Silica (SiO_2) is the primary component of

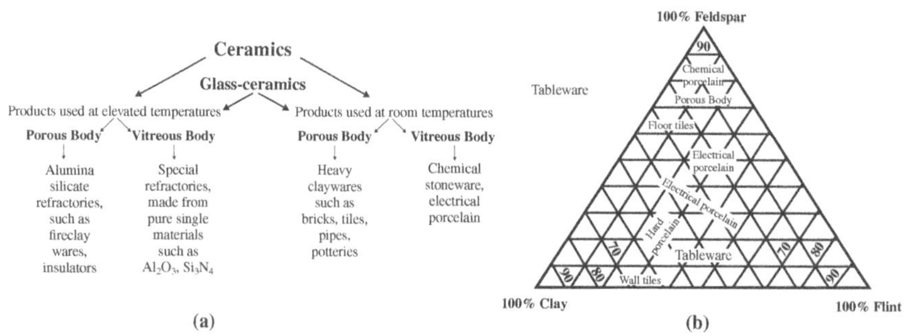

FIGURE 3.2 (a) Classification of ceramic products; (b) triaxial diagram showing areas of commercial wares.

TABLE 3.2
Chemical Composition of Nanoceramic Raw Materials (wt.%). (Mohamed Darweesh, 2018)

Materials Oxides	Clay	Feldspar	Quartz	Limestone
LOI	9.72	0.67	1.78	42.63
SiO_2	53.47	75.37	93.63	0.08
Al_2O_3	26.78	13.62	3.64	0.03
Fe_2O_3	3.99	0.41	0.08	0.04
CaO	0.60	0.53	0.18	56.84
MgO	1.38	–	–	0.10
MnO	0.03	0.03	0.02	–
SO_3	–	0.02	0.14	0.02
K_2O	1.18	5.84	0.14	0.05
Na_2O	1.15	3.44	0.17	0.12
Ti_2O	1.12	0.05	0.16	0.01
P_2O_5	0.51	–	–	–

quartz sand, and it is a widely abundant mineral found in the earth's crust. To obtain nanoceramic materials from quartz sand, several processing steps are typically involved, including the following.

(i) **Purification**: The quartz sand is purified to remove impurities, such as clay, minerals, and organic matter, which is usually achieved through processes like washing, sieving, and magnetic separation of foreign particles.
(ii) **Size Reduction**: The purified quartz sand is then subjected to size reduction techniques to obtain smaller particle sizes, achieved through methods like grinding, milling, or high-energy ball milling, to reduce the particle size down to the nanoscale range.
(iii) **Synthesis**: The nanoceramic materials can be synthesized using various techniques depending on the desired ceramic composition and properties. Common methods include sol-gel synthesis, chemical vapor deposition (CVD), and hydrothermal synthesis.
(iv) **Densification**: After the synthesis of nanoceramic powders, further processing steps like compaction and sintering are performed to enhance their density and mechanical properties.

The resulting nanoceramic materials derived from quartz sand can exhibit unique properties and find applications in a wide range of industries such as ceramic coatings, catalysts, sensors, and biomedical applications. Silica nanoparticles derived from quartz sand can be used as fillers to reinforce polymers, improve their mechanical properties, and enhance barrier properties (Akl et al., 2013).

3.3.4 Fluxes

In the nanoceramic industry, fluxes play a crucial role in the formulation and production of end-materials. Fluxes lower the melting point of ceramic compositions, making them easier to process and shaping the final properties of the product. They aid in the vitrification process by promoting the formation of a glassy phase when ceramics are fired at high temperatures. Major key aspects of fluxes in the nanoceramic industry include reduction in melting point, viscosity control, glass formation, and control of thermal expansion. Common examples of fluxing agents used in nanoceramic industry include feldspars (such as sodium, potassium, or calcium feldspars), borates (such as borax or boric acid), lithium compounds, and various other alkali and alkaline-earth metal compounds. It should be noted that each type of flux has its own specific effects on the ceramic composition and properties. Further, the selection and amount of flux used in nanoceramic formulations are determined by the factors such as the desired properties of the final product, firing temperature, composition, and the type of ceramic being produced. Different ceramic applications may require different fluxing agents and concentrations to achieve the desired results (Gokul Raja et al., 2021).

3.3.5 Feldspar

Feldspar is a vital ingredient of the nanoceramic industry, where it is used as a fluxing agent. It is composed of aluminum silicates combined with varying proportions of potassium, sodium, and calcium. It reduces the melting temperature of ceramic materials, promotes vitrification, and improves the sintering behavior. Feldspar, a common constituent in glazes, porcelain, earthenware, and sanitaryware, also contributes to the color, texture, and strength of nanoceramic products. Some key aspects of feldspar are given here.

(i) **Composition and Types**: Feldspar is a framework silicate mineral, having a three-dimensional structure of linked silicon–oxygen tetrahedra. The most common types of feldspar are potassium feldspar (orthoclase), sodium feldspar (albite), and calcium feldspar (anorthite). These different compositions lead to variations in their properties and applications.

(ii) **Nano-Glass Making**: Feldspar is an important component in the production of glasses and nano glasses. It acts as a flux, lowers the melting point of the glass batch, and improves its workability during the glass-forming process. Additionally, it also helps control the viscosity of the molten glass and influences its transparency, chemical durability, and thermal expansion properties.

(iii) **Construction Materials**: Feldspar is used as a nano-aggregate in the production of concrete, where it contributes to its strength and durability. Due to its hardness and resistance to weathering, it is also a suitable component for the development of various construction materials, such as ceramics tiles, countertops, and flooring.

It is important to note that the specific type of feldspar used and its concentration in nanoceramics can vary depending on the desired properties and requirements of the end product. The selection of feldspar is influenced by factors such as its mineralogical composition, purity, availability, and cost (Shui et al., 2023).

3.3.6 Talc

Talc, or magnesium silicate, is a member of the montmorillonite group having the formula $3MgO.4SiO_2.H_2O$. It acts as a cheap source of magnesia, which further acts as a flux. The presence of talc in nanoceramics results in the formation of cordierite when fired at high temperatures, which imparts low thermal expansion and provides improved thermal shock resistance. The only disadvantage of adding talc into nanoceramic systems is their tendency to easily distort due to limitation of the firing range. It has been reported that lime-bearing talc, in particular, gives low shrinkage ceramics, better strength, and improved glazing resistance (Ramamoorthy et al., 2021).

3.3.7 Nano-Clays

Nano-clays are typically derived from naturally occurring clay minerals, such as montmorillonite, kaolinite, or hectorite. These clay minerals are composed of stacked layers of silicate sheets held together by weak van der Waals forces. One of the significant characteristics of nano-clays is the ability to separate the individual layers. Through various processing techniques, the stacked layers of clay minerals can be separated into nanoscale platelets, resulting in improved dispersion and increased surface area for interactions with other materials. Nano-clays are widely used as fillers or reinforcements in nanoceramics and ceramic-based nanocomposites to enhance mechanical strength, stiffness, and dimensional stability. The incorporation of nano-clays has been found to improve properties such as tensile strength, modulus, and heat resistance of reinforced composites. Additionally, the layered structure of nano-clays creates tortuous diffusion paths, leading to improved gas barrier properties, fire resistance, and resistance to moisture transmission. Moreover, nano-clays can significantly affect the rheological behavior of composite systems. Due to their plate-like structure, they form networks or entangled structures within the matrix, influencing viscosity, shear thinning behavior, and processing characteristics. This property proves to be advantageous in applications like coatings, adhesives, and rheology modifiers (Kryuchkova et al., 2016).

3.3.8 Limestone

Limestone, primarily composed of calcium carbonate $(CaCO_3)$, itself is not typically used as a raw material for the production of nanoceramics. However, it is utilized as an indirect starting material for their production by serving as a source of calcium oxide (CaO) or lime as per the requirement of the desired application. Additionally, its derivatives, such as calcium oxide or hydroxide, also serve as precursors in the synthesis of nanoscale ceramic materials (Shashikumar et al., 2023). The general process for producing nanoceramics from limestone involves the following steps.

(i) **Calcination**: Limestone is heated at high temperatures (typically above 800°C) to drive off carbon dioxide (CO_2) and convert limestone into quicklime or calcium oxide (CaO).
(ii) **Slaking**: Quicklime obtained from calcination is then hydrated by adding water to produce calcium hydroxide [$Ca(OH)_2$].
(iii) **Synthesis**: Calcium hydroxide is used as a precursor for the synthesis of various ceramic materials. It can be further processed using techniques such as sol-gel synthesis, hydrothermal synthesis, or precipitation methods to produce nanoscale ceramic particles or coatings. The specific synthesis technique depends on the desired composition and properties of the nanoceramic.
(iv) **Densification**: Once the nanoceramic material is synthesized, additional processing steps such as compaction and sintering are employed to impart the density and mechanical properties.

3.3.9 Transparent Nanoceramics

Transparent nanoceramics are a class of ceramic materials that exhibit transparency in the visible light range while possessing the inherent properties of ceramics, such as high hardness, chemical resistance, and thermal stability. These materials find applications in various fields where both optical transparency and desirable ceramic properties are required (S. F. Wang et al., 2013). Some key points about transparent nanoceramics are given here.

(i) **Composition**: Transparent nanoceramics can be composed of different ceramic materials, including oxides, nitrides, and carbides such as alumina (Al_2O_3), yttrium aluminum garnet (YAG, $Y_3Al_5O_{12}$), magnesium aluminate spinel ($MgAl_2O_4$), and aluminum nitride (AlN).
(ii) **Grain Size Control**: One of the critical factors in achieving transparency in nanoceramics is controlling the grain size. By reducing the grain size to the nanometer scale, light scattering at grain boundaries is minimized, resulting in improved optical transparency.
(iii) **Optical Properties**: Transparent nanoceramics exhibit excellent optical properties, including high transparency, low light scattering, and a wide transmission range in the visible spectrum. They can be engineered to have specific optical characteristics, such as UV or IR transparency, by selecting the appropriate ceramic composition and controlling the processing parameters.
(iv) **Applications**: Transparent nanoceramics have a range of applications in fields such as optics, optoelectronics, photonics, and high-temperature environments. They are used for optical windows, lenses, laser gain media, waveguides, scintillators, transparent armor, and high-temperature

sensor components. Transparent nanoceramics are particularly valued in applications that require both optical transparency and durability, where traditional transparent materials like glasses or plastics may not be suitable.

It is important to note that the transparency of nanoceramics can depend on factors such as the composition, grain size, density, and the presence of impurities or defects. Further, the fabrication process and optimization of parameters play a crucial role in achieving the desired transparency and optical properties.

3.4 CHARACTERISTIC PROPERTIES OF NANOCERAMICS AS FUNCTIONAL MATERIALS

In general, nanoceramics are strong, flexible, and at the same time very lightweight. Due to their size and molecular structure, they have unique properties, which are often seen in terms of various magnetic, electrical, and physical phenomena (Shannigrahi & Sharma, 2018), such as the following.

(i) **Electro-Optical**: Change in the optical properties of nanoceramics when placed in an electric field.
(ii) **Magnetoresistive**: Change in the electrical resistance in accordance to an external magnetic field.
(iii) **Ferromagnetic**: Ability to sustain a magnetic field after being magnetized by the source.
(iv) **Superconductive**: Exhibition of zero electric resistance when cooled below a critical temperature.
(v) **Dielectric**: An electrical insulation that can be polarized by an electric field to store electric charge.
(vi) **Ferroelectric**: Dielectric nanoceramics which can remain polarized even after the electric field is removed.
(vii) **Piezoelectric**: Nanoceramics which accumulate an electrical charge when mechanical stress (pressure, strain etc.) is applied to it and vice versa.
(viii) **Pyroelectric**: Nanoceramics which can produce a temporary voltage on temperature change.

3.5 SYNTHESIS ROUTES OF NANOCERAMICS

Different techniques which allow the precise control over the size, shape, and composition of ceramic nanoparticles and lead to their tailored properties and improved performance are used for the synthesis and development of nanoceramics. Some major synthesis routes of nanoceramics are described here.

3.5.1 Microemulsion Method

The microemulsion method is a commonly used technique for the synthesis of nanoceramics. Within this system, the precursors for the desired ceramic material are dispersed and reacted to form nanoscale ceramic particles. A general overview of the microemulsion method for the synthesis of nanoceramics is given here.

(i) **Microemulsion Formation**: A microemulsion is formed by mixing water, oil, and surfactants. The choice of surfactants and oil phase depends on the specific ceramic precursor and desired properties. Typically, the surfactants are chosen to create a stable system with small droplet sizes.
(ii) **Incorporation of Precursors**: The ceramic precursors, such as metal salts or metal-organic compounds, are dissolved or dispersed in the water phase of the microemulsion. The choice of precursors depends on the desired composition of the nanoceramic material.
(iii) **Reaction and Nanoparticle Formation**: The microemulsion system is then subjected to specific conditions to induce the reaction between the precursors. This can include adjusting the pH and temperature or introducing additional chemicals or catalysts. The reaction leads to the nucleation and growth of nanoscale ceramic particles within the microemulsion droplets.
(iv) **Particle Stabilization**: Once the ceramic nanoparticles are formed, they are typically stabilized by coating them with surfactant molecules present in the microemulsion. This helps prevent particle aggregation and provides stability to the nanoceramic dispersion.
(v) **Separation and Drying**: After the nanoparticle formation and stabilization, the nanoceramic dispersion is typically separated from the microemulsion system, which can be achieved through processes such as centrifugation or filtration. The separated nanoceramic particles are then dried to obtain a powder form.
(vi) **Additional Processing**: The obtained nanoceramic powder undergoes additional processing steps, such as calcination or sintering, to further enhance its properties and achieve the desired microstructure.

The microemulsion method offers various advantages for synthesizing nanoceramics, including precise particle size control, narrow size distribution, and the potential for incorporating various dopants or additives. Also, it allows for the synthesis of complex compositions and tailored properties. However, optimization of parameters such as surfactant types and concentrations, reaction conditions, and post-processing steps are essential features to monitor while opting for this route of synthesis of nanoceramic particles (Ganguli et al., 2010). The schematic in Figure 3.3 represents processing in microemulsion method for the synthesis of nanoceramics.

3.5.2 Chemical Vapor Deposition (CVD) Method

Chemical vapor deposition (CVD) is a widely used technique for the synthesis of nanoceramics. It involves the deposition of ceramic materials onto a substrate through chemical reactions that occur in the gas phase. The CVD method allows precise control over the deposition process, resulting in the formation of nanoscale

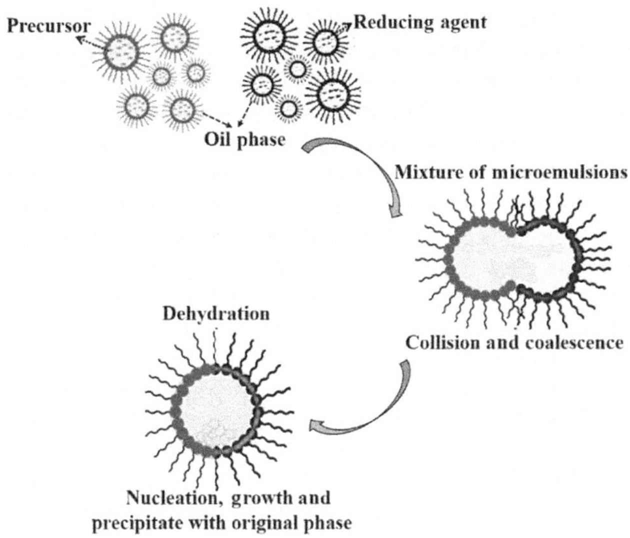

FIGURE 3.3 Schematic of microemulsion method for the synthesis of nanoceramics.

ceramic films or coatings with tailored properties. A general overview of the CVD method for the synthesis of nanoceramics is described in the following.

(i) **Choice of Precursors**: Selecting appropriate precursor chemicals is crucial in the CVD method. These precursors should be volatile at the reaction temperature and capable of decomposition or reactive enough to form the desired nanoceramic material. Common precursor types include metal organic compounds, metal halides, or metal alkoxides.
(ii) **Reactor Setup**: The CVD process requires a reactor chamber where the reaction takes place. The substrate onto which the ceramic material is deposited is placed inside the chamber, which is typically heated to temperatures ranging from several hundred to a few thousand degrees Celsius, depending on the specific ceramic material being synthesized.
(iii) **Carrier Gas and Precursor Delivery**: A carrier gas, such as hydrogen or nitrogen, is introduced into the reactor chamber. The precursor chemicals are then delivered into the chamber either by bubbling the carrier gas through a liquid precursor or by vaporizing solid or liquid precursors using vaporizers. The carrier gas transports the precursor molecules toward the heated substrate.
(iv) **Reaction and Deposition**: When the precursor molecules reach the heated substrate, they undergo chemical reactions or decomposition to form ceramic species. These species then adsorb onto the substrate surface, where they further react to deposit a thin film or coating of the nanoceramic material. The reactions occur through mechanisms such as thermal decomposition, reduction, or oxidation.

(v) **Growth and Control**: The deposition process is carefully controlled to achieve the desired thickness, composition, and microstructure of the nanoceramic film. Parameters such as precursor flow rates, reaction temperature, pressure, and reaction time are adjusted to control the growth rate, crystal structure, and other properties of the nanoceramic film.

(vi) **Post-Processing**: After deposition, the nanoceramic film may undergo additional processing steps, such as annealing or thermal treatment, to improve its crystallinity, density, and adhesion to the substrate. These post-processing steps can enhance the properties and performance of the nanoceramic film.

The CVD method offers advantages in terms of precise control over film thickness, composition, and microstructure. It allows for the deposition of nanoceramics onto various substrates, including metals, semiconductors, and other ceramics. This technique can be employed for the synthesis of a wide range of nanoceramic materials with tailored properties and is commonly used in industries such as electronics, coatings, and energy applications (Winterer & Hahn, 2022). The schematic in Figure 3.4 represents the general concept of CVD method for the synthesis of nanoceramics.

3.5.3 Sol-Gel Method

Another extensively used technique for the synthesis of nanoceramics is the sol-gel method. It involves the conversion of a precursor solution (sol) into a solid ceramic material (gel) through a series of chemical reactions and subsequent drying or heat treatment. The sol-gel method offers control over the composition, morphology, and structure (crystallinity) of the resulting nanoceramic, making it suitable for various applications. The following is a general overview of the sol-gel method for the synthesis of nanoceramics.

(i) **Preparation of Precursor Solution (Sol)**: The sol-gel process begins with the preparation of a precursor solution, which typically contains metal

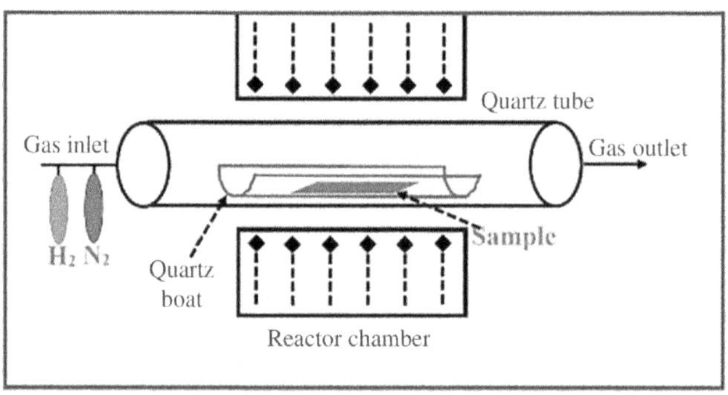

FIGURE 3.4 Schematic of chemical vapor deposition (CVD) technique for the synthesis of nanoceramics.

alkoxides, metal chlorides, or other metal salts dissolved in a suitable solvent. The choice of precursor depends on the desired ceramic composition.

(ii) **Hydrolysis**: The precursor solution undergoes hydrolysis, where water molecules react with the metal precursor to break the metal–oxygen bonds and form metal hydroxide species. This hydrolysis reaction is typically initiated by the addition of a catalyst or by adjusting the pH of the solution.

(iii) **Condensation**: After hydrolysis, the metal hydroxides undergo a condensation reaction, resulting in the formation of metal–oxygen–metal linkages, where the water molecules are eliminated from the system. The extent of condensation is controlled by adjusting the reaction conditions, such as temperature and time.

(iv) **Gelation**: As the condensation reaction proceeds, the precursor solution undergoes gelation, transforming into a three-dimensional network of interconnected particles or chains. In this process a solid gel with a porous structure is formed, which contains the precursor material in a partially reacted state.

(v) **Aging and Drying**: The gel is aged to enhance its stability and promote further condensation reactions. After aging, the gel is typically subjected to drying to remove the remaining solvent and water molecules, which is generally carried out through techniques such as air-drying, freeze-drying, or supercritical drying.

(vi) **Calcination or Sintering**: The dried gel is then subjected to a thermal treatment process known as calcination or sintering. This step involves heating the gel at high temperatures to convert it into a dense ceramic material. The thermal treatment facilitates further condensation, crystallization, and densification of the material, resulting in the formation of nanocrystalline ceramic particles.

(vii) **Post-Processing**: Depending on the desired properties and applications, the synthesized nanoceramic may undergo additional post-processing steps, such as grinding, milling, or surface treatments, to achieve the desired particle size, shape, and surface characteristics.

The sol-gel method offers several advantages for the synthesis of nanoceramics, including control over composition, purity, and uniformity of the resulting material. It further allows for the incorporation of dopants or additives, as well as the preparation of thin films, coatings, or bulk ceramics. This method is widely used in various fields, including optics, electronics, energy, and catalysis. Optimization of parameters, such as precursor concentration, reaction conditions, aging time, and thermal treatment, is crucial for obtaining the desired nanoceramic properties through this synthesis technique (Mishra, 2016). The schematic in Figure 3.5 represents the sol-gel method for the synthesis of nanoceramics.

3.5.4 Hydrothermal Method

The hydrothermal method involves the use of aqueous solutions under elevated temperatures and pressures to facilitate the formation of nanoscale ceramic particles.

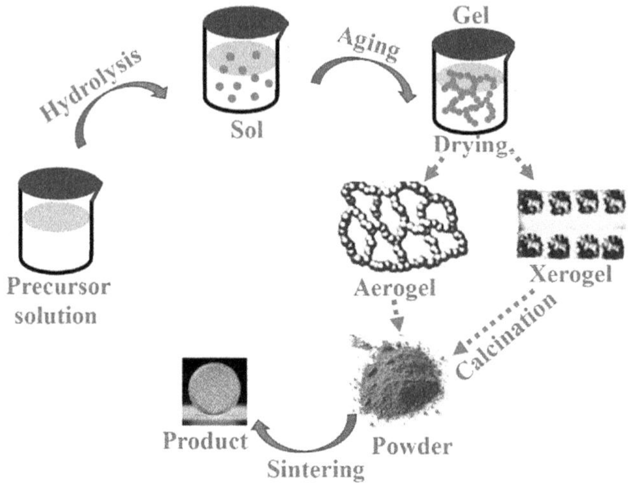

FIGURE 3.5 Schematic of the sol-gel method for the synthesis of nanoceramics. It starts with the precursor solution and undergoes hydrolysis to form sol. Aging of the sol will create the gel, which is then dried to form a powder.

This synthesis route offers precise control over the particle size, morphology, and crystallinity of the resulting nanoceramics. A brief overview of the hydrothermal method for the synthesis of nanoceramics is given here.

(i) **Precursor Solution Preparation**: A precursor solution is prepared by dissolving metal salts, metal hydroxides, or other ceramic precursors in water or an aqueous solvent.
(ii) **Reaction Vessel**: The precursor solution is transferred to a reaction vessel capable of withstanding high temperatures and pressures. The vessel is sealed to maintain the pressure and prevent evaporation of the solution during the process.
(iii) **Reaction Conditions**: The reaction vessel is heated to a specific temperature, typically ranging from 100 °C to 300 °C, and subsequently the pressure inside the vessel increases above the atmospheric pressure. The elevated temperature and pressure create a hydrothermal environment (or autoclave) within the vessel, which promotes the formation of nanoceramic particles.
(iv) **Reaction Time**: The precursor solution is allowed to react under hydrothermal conditions for a certain period, typically ranging from hours to several days. During this time, the ceramic precursor undergoes hydrolysis, dissolution, nucleation, and subsequent growth of nanoscale ceramic particles.

(v) **Cooling and Product Recovery**: After the desired reaction time, the hydrothermal reactor is cooled down gradually to room temperature. The resulting suspension or slurry contains the synthesized nanoceramic particles. The suspension can be further processed to isolate and purify the nanoceramic particles.

(vi) **Washing and Drying**: The nanoceramic particles are typically washed with a suitable solvent to remove any impurities or byproducts. They are then separated from the solution by filtration or centrifugation. The collected nanoceramic particles are dried to remove the remaining solvent and obtain a powder form.

(vii) **Post-Treatment**: Depending on the desired properties and applications, the synthesized nanoceramics may undergo additional post-treatment processes such as calcination or sintering to enhance their crystallinity, phase purity, and mechanical properties.

This technique offers advantages such as controlled particle size and morphology, as well as the ability to synthesize a wide range of ceramic compositions. It is particularly useful for producing nanoceramics with high crystallinity and uniform particle size distribution. The hydrothermal method is employed in various fields, including catalysts, ceramics for energy storage, environmental applications, and biomedical materials. Optimization of parameters such as temperature, pressure, reaction time, and precursor concentration is essential for obtaining the desired nanoceramic properties through this route (Ahmad, 2017). Figure 3.6 illustrates the steps followed in the hydrothermal method for the synthesis of nanoceramics.

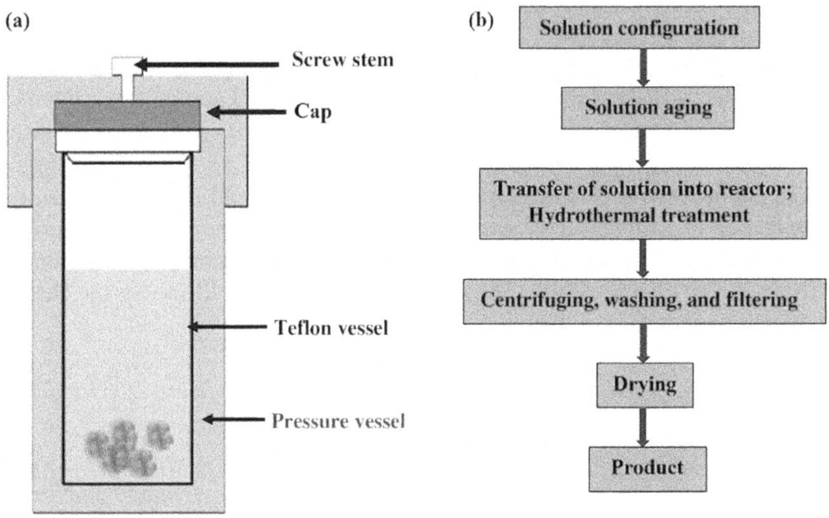

FIGURE 3.6 (a) Reaction vessel, (b) flowchart for the synthesis of nanoceramics by hydrothermal process.

3.5.5 Chemical Precipitation

For the formation of nanoceramics, chemical precipitation involves the precipitation of metal or metal oxide nanoparticles from a precursor solution by the addition of a precipitating agent. The process allows the control over particle size, composition, and morphology of the resulting nanoceramic. The synthesis of nanoceramics by this route is done through the following steps.

(i) **Precursor Solution Preparation**: A precursor solution is prepared by dissolving metal salts or metal complexes in a suitable solvent.
(ii) **Addition of Precipitating Agent**: A precipitating agent is added to the precursor solution to trigger the precipitation of the ceramic nanoparticles. The choice of precipitating agent depends on the specific chemistry of the system and matrix phase. Commonly used precipitating agents include ammonia, sodium hydroxide, or ammonium carbonate.
(iii) **Reaction and Particle Formation**: Upon the addition of the precipitating agent, a chemical reaction occurs between the precursor ions and the precipitating agent, which leads to the formation of insoluble ceramic nanoparticles. The reaction conditions, such as temperature, pH, and mixing rate, are closely monitored to achieve the desired nanoparticle properties.
(iv) **Particle Washing and Purification**: After the nanoparticle formation, the resulting suspension is typically subjected to a wash for the removal of any residual impurities or byproducts, which is generally done through centrifugation or filtration, which is then followed by rinsing with appropriate solvents.
(v) **Drying and Post-Treatment**: The collected nanoceramic particles are dried to remove the residual solvent, and a completely dry powder is obtained.

The chemical precipitation method offers several advantages for the synthesis of nanoceramics, including simplicity, cost-effectiveness, and scalability. It allows for the synthesis of a wide range of ceramic compositions and can produce nanoparticles with controlled size, shape, and surface properties. This method is commonly used in applications such as catalysts, sensors, coatings, and biomedical materials. Optimization of parameters such as precursor concentration, pH, temperature, and reaction time is crucial for obtaining the desired nanoceramic properties through this synthesis route (Li et al., 2018). The schematic in Figure 3.7 illustrates the chemical precipitation method of synthesis of nanoceramics.

3.6 APPLICATIONS OF NANOCERAMICS

Nanoceramics, due to their numerous exciting mechanical, electrical, and optical properties, have a large array of applications, such as energy supply and storage, communication, transportation systems, construction, nano-truss architecture, and medical industry, to name a few. A brief description of some of the key nanoceramic applications is given here.

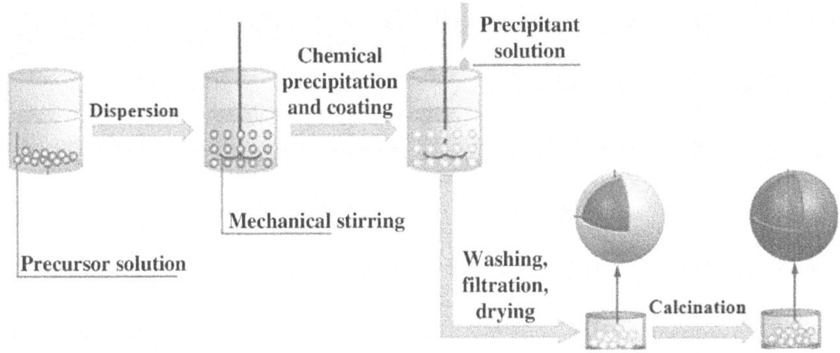

FIGURE 3.7 Schematic of the chemical precipitation method of synthesis of nanoceramics.

3.6.1 Environmental Applications

3.6.1.1 Environmental Nanoceramic Sensors (ENSs)

Environmental nanoceramic sensors (ENSs) are based on materials that are specifically designed for detecting and monitoring various environmental parameters. These sensors utilize the unique properties of nanoceramics to provide enhanced sensitivity, selectivity, and stability for environmental monitoring applications. Metal oxides such as titanium dioxide and zinc oxide or perovskites are generally used as the sensing elements in these sensors. Since nanoceramics offer a large surface-to-volume ratio, high surface area, and tailored surface properties, they are an attractive choice for efficient gas adsorption and interaction. ENSs are commonly used for gas-sensing applications, including detection and monitoring of pollutants, toxic gasses, volatile organic compounds (VOCs), and greenhouse gasses. The nanoceramic sensing elements interact with the target gasses through surface reactions or charge transfer processes, leading to changes in their electrical, optical, or catalytic properties. These sensors employ various sensing mechanisms, including resistance-based, capacitance-based, or optical-based techniques. Resistance-based sensors measure changes in electrical resistance when exposed to the target gas, while capacitance-based sensors detect changes in the dielectric properties. Optical-based sensors utilize changes in light absorption or emission properties of nanoceramics upon gas interaction. One of the key advantages of nanoceramic sensors is their ability to achieve selectivity and specificity for target gasses. The surface properties, crystal structure, and doping of nanoceramics can be tailored to enhance the sensor's response to specific gasses, making them suitable for detecting specific environmental pollutants. Moreover, nanoceramic sensors can be fabricated in miniature sizes, enabling their integration into portable or wearable devices for on-site or real-time environmental monitoring. The small size and low power consumption of these sensors make them suitable for deployment in various environmental settings. These sensors, moreover, offer improved stability and longevity compared to traditional sensors. The inherent stability of ceramic materials and the reduced grain sizes in nanoceramics help maintain sensor performance over extended periods, even in harsh environmental conditions. ENSs find applications in air quality monitoring, industrial emissions control, indoor air quality assessment, environmental impact assessment, and smart city initiatives and contribute to efforts aimed at pollution control, resource management, and sustainable development. Factors such as sensor architecture, operating conditions, calibration, and data analysis methods need to be carefully considered for reliable and accurate environmental sensing (Kapil et al., 2023).

3.6.1.2 Adsorption of Toxic Gasses

Nanoceramics have shown promising potential in the field of gas adsorption, including the adsorption of toxic gasses. Their unique properties, such as high surface area, tunable pore size, and chemical stability, make them attractive materials for capturing and removing harmful gasses from the environment. The adsorption process involves the adhesion of gas molecules to the surface of the nanoceramic material. The high surface area of nanoceramics provides a large number of active sites for the gas adsorption, allowing an efficient capturing of toxic gasses. Additionally, the tunable pore size of nanoceramics facilitates the selective adsorption of specific gasses and enables the removal of targeted pollutants. The most commonly used nanoceramics for the adsorption of gasses are metal-organic frameworks (MOFs) and activated carbon. MOFs are composed of metal ions or clusters coordinated with organic ligands, forming a highly porous structure, which provides a large internal surface area and enhances the adsorption capacity for the toxic gasses. The adsorption capacity of nanoceramics can be further enhanced through surface modifications or functionalization. Introduction of specific functional groups or metal nanoparticles onto the surface of nanoceramics improves both the selectivity and efficiency of gas adsorption (Oladipo & Gazi, 2016).

3.6.1.3 Water and Wastewater Treatment

Nanoceramics have emerged as promising materials for water and wastewater treatment by providing efficient and sustainable solutions to water pollution and ensuring access to clean and safe water resources. Some of the ways in which nanoceramics can be used in water and wastewater treatment include the following.

(i) **Filtration and Membrane Technology**: The high surface area and tunable pore size of nanoceramics allow for effective removal of contaminants, including suspended solids, bacteria, viruses, and organic pollutants. By controlling the pore size and surface chemistry of the nanoceramic membranes, selective filtration can be achieved, enabling the separation of specific contaminants.

(ii) **Catalysis**: Nanoceramics serve as catalysts in advanced oxidation processes (AOPs) for water treatment. By incorporating specific metal nanoparticles or oxides into the nanoceramic structure, they promote chemical reactions which degrade organic pollutants, disinfect water, and remove harmful compounds. The high reactivity and stability of nanoceramics make them valuable catalysts in various AOPs, such as photocatalysis and electrocatalysis.

(iii) **Water Desalination**: Nanoceramics play a role in desalination processes, such as reverse osmosis or nanofiltration. Their nanoscale pores and precisely controlled pore size distribution allow for efficient separation of salt ions from water, enabling the production of freshwater from seawater or brackish water sources.

(iv) **Antibacterial Applications**: Some nanoceramics such as ZnO, TiO_2, hydroxyapatite, $CuAlO_2$ possess inherent antibacterial properties, which can be beneficial for water and wastewater treatment. They inhibit the growth of bacteria and other microorganisms, reducing the risk of waterborne diseases and improving overall water quality.

However, it is important to consider the potential challenges associated with the use of nanoceramics in water and wastewater treatment, such as cost, scalability, and long-term stability (Zhang et al., 2014).

3.6.1.4 Nanoceramic Biomaterials

Nanoceramic biomaterials are a class of materials that combine the advantages of ceramics and nanotechnology for various biomedical applications. These biomaterials are designed to interact with biological systems, such as living cells and tissues, to promote specific therapeutic or diagnostic functions (Velu et al., 2020). Some key aspects and applications of nanoceramic biomaterials are listed here.

(i) **Composition and Structure**: Nanoceramic biomaterials are typically composed of inorganic compounds, such as calcium phosphates (e.g., hydroxyapatite), bio-glass, alumina, or zirconia. These materials possess unique properties, including high mechanical strength, biocompatibility, and chemical stability. The nanoscale structure of these materials facilitates enhanced surface reactivity and improved interactions with biological entities.

(ii) **Bone Tissue Engineering**: Nanoceramic biomaterials find extensive use in bone tissue engineering and regenerative medicine. Hydroxyapatite nanoparticles, for example, closely resemble the mineral composition of natural bone and are used to promote bone growth and repair. They can be incorporated into scaffolds or coatings to enhance the integration of artificial implants with the surrounding bone tissue.

(iii) **Drug Delivery Systems**: Nanoceramics such as mesoporous SiO_2, porous Al_2O_3, calcium phosphate, and zeolites can be utilized as carriers for controlled drug delivery. They can be engineered to have porous structures or functionalized surfaces to encapsulate and release therapeutic agents, such as drugs or growth factors, in a controlled manner. The high surface area of nanoceramics allows for efficient loading and release of drugs, enabling targeted and sustained delivery to specific sites in the body.

(iv) **Biosensing and Diagnostics**: Nanoceramic biomaterials can be engineered for biosensing and diagnostic applications. By modifying their surface with specific biomolecules or nanoparticles, they can selectively interact with target molecules or biomarkers in biological samples. This enables the development of sensitive and specific biosensors for detecting diseases, pathogens, or other analytes in body fluids or environmental samples.

(v) **Dental and Orthopedic Implants**: Nanoceramic biomaterials are used in dental and orthopedic implants due to their biocompatibility and mechanical properties. For example, nanoceramic coatings of hydroxyapatite can be applied to metal implants (stainless steel) to improve their wear resistance, reduce corrosion, and promote bone integration.

(vi) **Bioactive Coatings**: Nanoceramics can be applied as bioactive coatings on various medical devices or implants to enhance their biocompatibility and functionality. These coatings can improve cell adhesion, reduce inflammation, and facilitate tissue integration, thereby improving the performance and longevity of biomedical implants.

(vii) **Tissue Scaffolds**: Nanoceramic materials can be processed into porous scaffolds with tailored architectures and mechanical properties. These scaffolds provide a three-dimensional structure for cell growth and tissue regeneration. They can be used in tissue engineering to create artificial tissues or support the regeneration of damaged or diseased tissues.

3.6.1.5 Nanoceramic Coatings

Nanoceramic coatings refer to thin films or layers of ceramic materials that are applied onto the surface of various substrates, such as metals, polymers, or composites. These coatings offer a range of desirable properties, including high hardness, thermal and chemical stability, enhanced durability, and wear and corrosion resistance to name a few (Makhlouf & Guerrero, 2019). They are commonly used in industries such as automotive, aerospace, electronics, and energy for surface protection and performance enhancement. The major key aspects and applications of nanoceramic coatings are summarized here.

(i) **Deposition Techniques**: Nanoceramic coatings can be applied using different deposition techniques, including physical vapor deposition (PVD), chemical vapor deposition (CVD), sol-gel processes, and electrodeposition. These methods allow for precise control over the coating thickness, uniformity, and adherence to the substrate.

(ii) **Wear and Abrasion Resistance**: Nanoceramic coatings provide excellent wear and abrasion resistance, protecting the underlying substrate from surface damage and extending the lifespan of components. The hardness and high surface smoothness of nanoceramic coatings reduce friction and prevent the formation of wear debris, making them suitable for applications involving sliding or abrasive contact.

(iii) **Corrosion and Chemical Resistance**: Nanoceramic coatings of ZrO_2, TiO_2, and SiC exhibit superior corrosion resistance, protecting the substrate from chemical attack and environmental degradation. They act as a barrier against moisture, corrosive chemicals, and oxidation, making them particularly useful in applications where materials are exposed to harsh environments or corrosive substances.

(iv) **Thermal Barrier Coatings**: Nanoceramic coatings of Al_2O_3 and ZrO_2 function as thermal barrier coatings (TBCs), providing insulation against high temperatures. These coatings are used in gas turbines, jet engines, and other high-temperature applications to protect components from heat-related damage and improve their thermal efficiency.

(v) **Dielectric and Insulating Coatings**: Nanoceramic coatings of $BaTiO_3$, Al_2O_3, and lead zirconate titanate (PZT) with high dielectric strength and low electrical conductivity can be employed as insulating layers in electronic devices, capacitors, and sensors. These coatings provide electrical insulation, prevent short circuits, and offer protection against electrical arcing or breakdown.

(vi) **Optical and Decorative Coatings**: Certain nanoceramic coatings such as TiO_2, ZnO, SiO_2, and Al_2O_3 exhibit unique optical properties, such as high transparency, anti-reflective behavior, or specific color effects. These coatings find applications in optics, displays, architectural glass, and decorative finishes.

3.6.1.6 Nanoceramic Catalysts

Nanoceramic catalysts refer to catalytic materials composed of ceramic nanoparticles with dimensions in the nanoscale range. These catalysts leverage the unique properties of nanomaterials to enhance their catalytic performance, efficiency, and selectivity (Oladipo & Gazi, 2016). They find applications in various chemical reactions and industrial processes. The major aspects and applications of nanoceramic catalysts are given here.

(i) **Composition**: Nanoceramic catalysts can be composed of different ceramic materials, including metal oxides, mixed metal oxides, or metal-organic frameworks (MOFs). Common examples include cerium oxide (CeO_2), titanium dioxide (TiO_2), zinc oxide (ZnO), and various perovskite-type compounds. The choice of ceramic material depends on the specific catalytic reaction and the desired properties.

(ii) **Tunable Properties**: The properties of nanoceramic catalysts, such as surface composition, structure, and morphology, can be tailored by controlling the synthesis parameters. This tunability enables the optimization of catalytic performance for specific reactions, such as adjusting the acidity/basicity, redox properties, or surface-active sites.

(iii) **Catalytic Activity and Selectivity**: Nanoceramic catalysts exhibit high catalytic activity and selectivity, making them valuable in a wide range of chemical transformations. They can accelerate chemical reactions, increase reaction rates, and enable the production of desired products with high yields and purity. Additionally, nanoceramic catalysts provide improved control over reaction conditions, leading to enhanced selectivity and the ability to target specific reaction pathways.

(iv) **Heterogeneous Catalysis**: Nanoceramic catalysts are commonly used in heterogeneous catalysis, where the catalyst is in a different phase than the reactants. They can be supported on solid substrates or dispersed as colloidal suspensions. The nanoscale size and controlled dispersion of the catalyst particles allow for efficient mass transfer and accessibility of reactants, promoting catalytic activity.

(v) **Stability and Durability**: Nanoceramic catalysts often exhibit excellent stability and durability due to their high melting points, chemical resistance, and resistance to sintering or agglomeration. This stability ensures their longevity and sustained catalytic performance over prolonged reaction times.

3.6.1.7 Alternative and Renewable Energy

Nanoceramics play a significant role in alternative and renewable energy applications by enabling improved performance, efficiency, and sustainability. They offer unique properties and functionalities that can enhance various energy conversion and storage technologies (Yin et al., 2017) such as the following.

(i) **Solar Energy**: Nanoceramics have been employed in solar energy technologies, such as photovoltaics and solar thermal systems. For photovoltaics, nanoceramic materials like perovskites and quantum dots are being explored as light-absorbing layers or components due to their excellent light-harvesting properties and potential for cost-effective solar cell manufacturing. Nanoceramics can also be used in solar thermal systems as selective coatings or nanoparticles to improve the absorption and conversion of solar energy into heat.

(ii) **Fuel Cells**: Nanoceramic materials are crucial in fuel cell technologies, which generate electricity through electrochemical reactions. Proton exchange membrane fuel cells (PEMFCs) often utilize nanoceramic materials such as nafion, ceria (CeO_2), titania (TiO_2), yttria-stabilized zirconia (YSZ), and mixed oxide nanoceramics $(CeO_2 - ZrO_2)$ for catalyst supports, proton-conducting membranes, or electrode materials. Nanoceramics provide high surface area, enhanced ion conductivity, and improved electrocatalytic activity, contributing to better fuel cell performance and efficiency.

(iii) **Batteries and Energy Storage**: Nanoceramics play a role in advancing battery technologies for energy storage applications. They are used in electrode materials, electrolytes, and separators to enhance energy storage capacity, cycling stability, and charge-discharge rates. Additionally, nanoceramics are explored for supercapacitors and other energy storage devices, offering high capacitance and rapid charge-discharge characteristics.

(iv) **Hydrogen Production and Storage**: Nanoceramic materials are utilized in hydrogen production and storage processes. For example, MOFs are being used as catalysts for hydrogen evolution reactions, facilitating the efficient generation of hydrogen from water electrolysis or other renewable sources. Nanoceramics are also explored as hydrogen storage materials, such as metal hydrides, with enhanced hydrogen absorption/desorption kinetics and storage capacity.

(v) **Thermoelectric Energy Conversion**: Nanoceramics, such as $Ca_3Co_4O_9$ $SrTiO_3$, and ZnO, are employed in thermoelectric materials that convert waste heat into electricity. By optimizing the thermal conductivity and electrical properties at the nanoscale, nanoceramics enable efficient heat-to-electricity conversion. They enhance the figure of merit (ZT) of thermoelectric materials, enabling improved thermoelectric efficiency and performance.

3.7 VERSATILE PROSPECTS OF NANOCERAMICS

The usage of nanoceramics does have some environmental implications (Zhang et al., 2014) that need to be considered. While nanoceramics offer several benefits, including improved performance and functionality in various applications, their potential environmental impacts should be carefully evaluated. Some key considerations include the following.

(i) **Resource Consumption**: The production of nanoceramics typically involves the extraction and processing of raw materials. This process may require significant energy inputs and the consumption of natural resources, including minerals and water. The extraction and processing activities associated with nanoceramics production can have environmental impacts such as habitat destruction, soil erosion, and water pollution. It is important to implement sustainable practices and consider the life cycle impacts of nanoceramics to minimize resource consumption and associated environmental burdens.

(ii) **Release and Emission of Loose Nanoparticle**: During the production, use, and disposal of nanoceramic materials, there is a potential for the release of nanoparticles into the environment. These nanoparticles may have unique properties that differ from bulk materials, raising concerns about their potential toxicity and environmental safety. The release of nanoparticles into air, water, or soil could potentially impact ecosystems and human health. Proper handling, containment, and disposal practices should be implemented to minimize the release of nanoparticles and mitigate their potential impacts.

(iii) **Waste Disposal**: The disposal of nanoceramic waste, such as scraps or end-of-life products, should be carefully managed to prevent environmental contamination. The recycling and proper treatment of nanoceramic waste materials can help reduce the demand for raw materials and minimize the environmental burden associated with their production. It is important to develop appropriate waste management strategies to handle nanoceramic waste and prevent their accumulation in the environment.

(iv) **Ecotoxicity and Human Health Impacts**: Studies have shown that certain nanoceramic materials such as TiO_2 nanoparticles, particularly those in the anatase form, CeO_2 and Ag-based nanoceramics exhibit adverse effects on aquatic organisms, soil organisms, and human health on exposure. It is crucial to conduct comprehensive risk assessments and toxicological studies to understand the potential environmental and health risks associated with nanoceramics and ensure proper safety measures are in place.

(v) **Energy Consumption and Carbon Footprint**: The production of nanoceramics may require energy-intensive processes, contributing to greenhouse gas emissions and climate change. It is important to consider the energy efficiency of nanoceramic production and explore renewable energy sources to minimize carbon footprints. Additionally, efforts should be made to optimize the synthesis processes, reduce energy consumption, and implement sustainable manufacturing practices.

To mitigate the environmental implications of nanoceramics usage, a proactive approach should be adopted. This includes implementing sustainable design principles, conducting thorough life cycle assessments, promoting responsible manufacturing practices, and adhering to strict regulations and guidelines for the safe handling and disposal of nanoceramic materials.

3.7.1 Industrial Applications

Nanoceramics find numerous industrial applications across various sectors due to their unique properties and performance advantages. Some key industrial applications (Fujimoto, 2016; Huang et al., 2021; Kuti et al., 2022; X. Wang et al., 2021) of nanoceramics are mentioned here.

(i) **Automotive Industry**: Nanoceramic coatings are used in the automotive industry for applications such as corrosion protection, scratch resistance, and improved aesthetic appearance. These coatings provide enhanced durability, wear resistance, and chemical resistance, thereby extending the lifespan of automotive components and improving their performance.

(ii) **Aerospace and Defense**: Nanoceramics are employed in aerospace and defense applications due to their exceptional strength, hardness, and thermal stability. They are used in protective coatings for aircraft components, turbine blades, rocket nozzles, and heat shields. Nanoceramics also find applications in armor materials for personal protection and military vehicles.

(iii) **Electronics and Semiconductors**: Nanoceramics play a vital role in the electronics industry, particularly in semiconductor devices and electronic components. They are used as dielectric materials, insulators, and substrates. Nanoceramic films or coatings can provide electrical insulation, thermal management, and protection against moisture or environmental contaminants. Nanoceramics are also explored for advanced capacitor technologies and as high-κ gate dielectrics in microprocessors.

(iv) **Textiles and Coatings**: Nanoceramics are used in textile and coating industries to impart desired properties, such as stain resistance, UV protection, antimicrobial activity, and self-cleaning capabilities. Nanoceramic coatings on textiles provide water repellency, oil repellency, and enhanced durability. Also, they find applications in sportswear, outdoor fabrics, protective clothing, and architectural coatings.

These are just a few examples of the industrial applications of nanoceramics. The versatility and unique properties of nanoceramics continue to drive their adoption in numerous sectors, contributing to improved performance, durability, and efficiency of industrial processes and products.

3.8 FUTURE PROSPECTS OF NANOCERAMICS

The future prospects of nanoceramics are promising, with numerous potential advancements and applications. Some key areas where nanoceramics are expected to make significant contributions include advanced electronics and optoelectronics, energy storage and conversion devices, environmental and biomedical applications, and sustainable manufacturing. It is important to note that while nanoceramics hold significant potential, challenges remain in terms of large-scale production, cost-effectiveness, and environmental impact. Continued research and development efforts are essential to overcome these challenges and optimize nanoceramics' properties and manufacturing processes. Additionally, ensuring the safe handling and disposal of nanoceramics and understanding their potential environmental and health impacts will be crucial for their successful implementation in various industries.

3.9 CONCLUSION

Nanoceramics offer tremendous potential for various environmental applications. Their unique properties and characteristics make them valuable in addressing environmental challenges and promoting sustainable practices. However, it is important to consider the potential environmental implications associated with their usage and take proactive measures to mitigate any adverse effects. Nanoceramics have shown promise in pollution prevention, water and wastewater treatment, air purification, and waste management. They can efficiently adsorb toxic gases, remove contaminants from water sources, and facilitate the degradation of pollutants through photocatalytic processes. Nanoceramics also hold potential in alternative and renewable energy applications, such as energy storage, conversion, and photocatalysis, contributing to cleaner and more sustainable energy generation.

While reaping the benefits of nanoceramics, it is crucial to adopt responsible manufacturing practices, including resource efficiency, waste reduction, and proper disposal of nanoceramic waste. Environmental considerations should be integrated into the entire life cycle of nanoceramic materials, from production to end-of-life, to minimize their environmental footprint. Moreover, comprehensive risk assessments, toxicological studies, and regulations should be in place to ensure the safe handling and use of nanoceramics, protecting both the environment and human health. Research and development efforts are necessary to optimize nanoceramic synthesis, manufacturing techniques, and explore their long-term behavior in environmental systems. By harnessing the potential of nanoceramics in environmental applications and incorporating sustainable practices, we can address pressing environmental challenges and pave the way for a cleaner, greener, and more sustainable future.

ACKNOWLEDGMENT

The authors would like to acknowledge the support provided by the Science and Engineering Research Board, a statutory body of Department of Science & Technology (DST), Government of India for facilitating the research discussed in this chapter.

DECLARATION OF INTEREST

The authors claim no institutional, financial, or general conflicts of interest with any third party not involved in this research work.

REFERENCES

Ahmad, T. (2017). Hydrothermal synthesis, characterization and dielectric properties of zirconia nanoparticles. *Material Science & Engineering International Journal, 1*(3). https://doi.org/10.15406/mseij.2017.01.00017

Akl, M. A., Aly, H. F., Soliman, H. M. A., Abd ElRahman, A. M. E., & Abd-Elhamid, A. I. (2013). Preparation and characterization of silica nanoparticles by wet mechanical attrition of white and yellow sand. *Journal of Nanomedicine and Nanotechnology*, *4*(6). https://doi.org/10.4172/2157-7439.1000183

Ali, N., Sebzali, M., Bourisli, H., Safar, A., & Ebrahem, Z. A. (2020). Nanocoating: An energy efficient solution towards reducing buildings electrical consumption in the state of Kuwait. *2020 Advances in Science and Engineering Technology International Conferences, ASET 2020*. https://doi.org/10.1109/ASET48392.2020.9118309

Attarzadeh, N., Molaei, M., Babaei, K., & Fattahalhosseini, A. (2021). New promising ceramic coatings for corrosion and wear protection of steels: A review. *Surfaces and Interfaces*, *23*. https://doi.org/10.1016/j.surfin.2021.100997

Basu, B., & Balani, K. (2011). *Advanced Structural Ceramics*. John Wiley and Sons. https://doi.org/10.1002/9781118037300

Basu, K. B. (2011). *Advanced Structural Ceramics*. John Wiley and Sons.

Cain, M., & Morrell, R. (2001). Nanostructured ceramics: A review of their potential. *Applied Organometallic Chemistry*, *15*(5). https://doi.org/10.1002/aoc.153

Devarajan, D. K., Rangasamy, B., & Amirtharaj Mosas, K. K. (2023). State-of-the-art developments in advanced hard ceramic coatings using PVD techniques for high-temperature tribological applications. *Ceramics*, *6*(1). https://doi.org/10.3390/ceramics6010019

Ferraris, E., Vleugels, J., Guo, Y., Bourell, D., Kruth, J. P., & Lauwers, B. (2016). Shaping of engineering ceramics by electro, chemical and physical processes. *CIRP Annals – Manufacturing Technology*, *65*(2). https://doi.org/10.1016/j.cirp.2016.06.001

Fujimoto, Y. (2016). Structure, stabilities, and electronic properties of smart ceramic composites. In *Sol-Gel Based Nanoceramic Materials: Preparation, Properties and Applications*. Springer. https://doi.org/10.1007/978-3-319-49512-5_4

Ganguli, A. K., Ganguly, A., & Vaidya, S. (2010). Microemulsion-based synthesis of nanocrystalline materials. *Chemical Society Reviews*, *39*(2). https://doi.org/10.1039/b814613f

Gogotsi, Y. (2014). What nano can do for energy storage. *ACS Nano*, *8*(6). https://doi.org/10.1021/nn503164x

Gokul Raja, T. S., Balamurugan, S., & Reshma, A. (2021). Synthesis, different characterizations and pigment application of nanoceramic zirconate powder. *Journal of Nanoscience and Nanotechnology*, *21*(4). https://doi.org/10.1166/jnn.2021.19040

Green, D. J., Guillon, O., & Rödel, J. (2008). Constrained sintering: A delicate balance of scales. *Journal of the European Ceramic Society*, *28*(7). https://doi.org/10.1016/j.jeurceramsoc.2007.12.012

Hamidi, F., & Aslani, F. (2019). Tio2-based photocatalytic cementitious composites: Materials, properties, influential parameters, and assessment techniques. *Nanomaterials*, *9*(10). https://doi.org/10.3390/nano9101444

Huang, R., Zhang, S., Zhang, W., & Yang, X. (2021). Progress of zinc oxide-based nanocomposites in the textile industry. *IET Collaborative Intelligent Manufacturing*, *3*(3). https://doi.org/10.1049/cim2.12029

Jeong, B., Lahkar, S., An, Q., & Reddy, K. M. (2022). Mechanical properties and deformation behavior of superhard lightweight nanocrystalline ceramics. *Nanomaterials*, *12*(18). https://doi.org/10.3390/nano12183228

Kapil, N., Mayani, S. V., & Gopal Bhattacharyya, K. (2023). Environmental implications of nanoceramic applications. *Results in Chemistry*, *5*. https://doi.org/10.1016/j.rechem.2022.100724

Khan, S. B., & Lee, S. L. (2021). Nanomaterials significance; contaminants degradation for environmental applications. *Nano Express*, *2*(2). https://doi.org/10.1088/2632-959X/abf689

Kryuchkova, M., Danilushkina, A., Lvov, Y., & Fakhrullin, R. (2016). Evaluation of toxicity of nanoclays and graphene oxide: In vivo A paramecium caudatum study. *Environmental Science: Nano*, *3*(2). https://doi.org/10.1039/c5en00201j

Kuti, R., Szabó, Á. I., & Tóth, Á. D. (2022). Experimental investigation of tribological properties of two fully formulated engine oils with additional nanoscale spherical zirconia particles. *Lubricants*, *10*(10). https://doi.org/10.3390/lubricants10100246

Li, C., Jin, S., Guan, W., Tsang, C. W., Chu, W. K., Lau, W. K., & Liang, C. (2018). Chemical precipitation method for the synthesis of Nb2O5 modified bulk nickel catalysts with high specific surface area. *Journal of Visualized Experiments*, *2018*(132). https://doi.org/10.3791/56987

Makhlouf, A. S. H., & Guerrero, E. (2019). Toward smart and durable nanoceramic and nanobioceramic coatings and their medical and industrial applications. In *Advances in Smart Coatings and Thin Films for Future Industrial and Biomedical Engineering Applications*. Elsevier. https://doi.org/10.1016/B978-0-12-849870-5.00021-5

Mishra, A. K. (2016). *Sol-Gel Based Nanoceramic Materials: Preparation, Properties and Applications*. Springer. https://doi.org/10.1007/978-3-319-49512-5

Mohamed Darweesh, H. (2018). Nanoceramics: Materials, properties, methods and applications-part II. *Nanoscience*, *1*(1). https://doi.org/10.31058/j.nano.2018.11004

Nandhini, A., Sudhakar, T., & Premkumar, J. (2021). Ceramics and nanoceramics in biomedical applications. In *Handbook of Polymer and Ceramic Nanotechnology*. Springer. https://doi.org/10.1007/978-3-030-40513-7_71

Needhidasan, S., Samuel, M., & Chidambaram, R. (2014). Electronic waste – an emerging threat to the environment of urban India. *Journal of Environmental Health Science and Engineering*, *12*(1), 1–9. BioMed Central Ltd. https://doi.org/10.1186/2052-336X-12–36

Niihara, K. (1991). New design concept of structural ceramics. *Journal of the Ceramic Society of Japan*, *99*(1154), 974–982. https://doi.org/10.2109/jcersj.99.974

Nivas, Y., & Fulrath, R. M. (1970). Limitation of Griffith flaws in glass-matrix composites. *Journal of the American Ceramic Society*, *53*(4). https://doi.org/10.1111/j.1151-2916.1970.tb12068.x

Oladipo, A. A., & Gazi, M. (2016). Application of hydroxyapatite-based nanoceramics in wastewater treatment: Synthesis, characterization, and optimization. In *Sol-Gel Based Nanoceramic Materials: Preparation, Properties and Applications*. Springer. https://doi.org/10.1007/978-3-319-49512-5_8

Palmero, P. (2015). Structural ceramic nanocomposites: A review of properties and powders' synthesis methods. *Nanomaterials*, *5*(2). https://doi.org/10.3390/nano5020656

Ramamoorthy, R., Vanitha, S., Krishnadev, P., & Paramanantham, M. (2021). Synthesis and characterization of phyto mediated talc-based nanocomposite by wet chemical reduction method. *Materials Today: Proceedings*, *47*. https://doi.org/10.1016/j.matpr.2021.03.582

Ren, H., Koshy, P., Chen, W. F., Qi, S., & Sorrell, C. C. (2017). Photocatalytic materials and technologies for air purification. *Journal of Hazardous Materials*, *325*. https://doi.org/10.1016/j.jhazmat.2016.08.072

Saleem, H., & Zaidi, S. J. (2020). Developments in the application of nanomaterials for water treatment and their impact on the environment. *Nanomaterials*, *10*(9). https://doi.org/10.3390/nano10091764

Shannigrahi, S., & Sharma, M. (2018). Functional nanoceramics: A brief review on structure property evolutions of advanced functional ceramics processed using microwave and conventional techniques. *Nanomaterials*. https://doi.org/10.1201/b19168-6

Shashikumar, U., Chandra Jha, B., Chawla, S., Mustansar Hussain, C., Andaluri, G., Lin, Y. C., & Ponnusamy, V. K. (2023). Nanoceramics: Fabrication, properties and its applications towards the energy sector. *Fuel*, *336*. https://doi.org/10.1016/j.fuel.2022.126829

Shui, Y., Zhang, Y., Luo, T., Sun, M., Fang, T., Wu, Q., Sun, J., & Yu, H. (2023). The mechanical properties, tribological behaviors and color stability of a feldspar nanoceramics strengthening extrinsic stain for high-translucent zirconia. *Journal of the Mechanical Behavior of Biomedical Materials*, *140*. https://doi.org/10.1016/j.jmbbm.2023.105698

Taran, M., Safaei, M., Karimi, N., & Almasi, A. (2021). Benefits and application of nanotechnology in environmental science: An overview. *Biointerface Research in Applied Chemistry*, *11*(1). https://doi.org/10.33263/BRIAC111.78607870

Van Der Biest, O. (2013). Nanoceramics: Issues and opportunities. *International Journal of Applied Ceramic Technology*, *10*(4). https://doi.org/10.1111/ijac.12074

Velu, R., Calais, T., Jayakumar, A., & Raspall, F. (2020). A comprehensive review on bio-nanomaterials for medical implants and feasibility studies on fabrication of such implants by additive manufacturing technique. *Materials*, *13*(1). https://doi.org/10.3390/ma13010092

Vunain, E., Mishra, S. B., Mishra, A. K., & Mamba, B. B. (2016). Nanoceramics: Fundamentals and advanced perspectives. In *Sol-Gel Based Nanoceramic Materials: Preparation, Properties and Applications*. Springer. https://doi.org/10.1007/978-3-319-49512-5_1

Wang, S. F., Zhang, J., Luo, D. W., Gu, F., Tang, D. Y., Dong, Z. L., Tan, G. E. B., Que, W. X., Zhang, T. S., Li, S., & Kong, L. B. (2013). Progress in solid state chemistry transparent ceramics: Processing, materials and applications. *Progress in Solid State Chemistry*, *41*.

Wang, X., Gao, X., Zhang, Z., Cheng, L., Ma, H., & Yang, W. (2021). Advances in modifications and high-temperature applications of silicon carbide ceramic matrix composites in aerospace: A focused review. *Journal of the European Ceramic Society*, *41*(9). https://doi.org/10.1016/j.jeurceramsoc.2021.03.051

Winterer, M., & Hahn, H. (2022). Nanoceramics by chemical vapour synthesis. *International Journal of Materials Research*, *94*(10). https://doi.org/10.1515/ijmr-2003-0198

Wu, C., Wang, Y., Ma, G., & Zheng, X. (2021). Enhanced rate capability of Li4Ti5O12 anode material by a photo-assisted sol-gel route for lithium-ion batteries. *Electrochemistry Communications*, *131*. https://doi.org/10.1016/j.elecom.2021.107119

Yin, Y., Tudu, B., & Tiwari, A. (2017). Recent advances in oxide thermoelectric materials and modules. *Vacuum*, *146*. https://doi.org/10.1016/j.vacuum.2017.04.015

Zhang, N., Yang, F. Y., Zhang, T., Kan, H. M., Wang, X. Y., & Long, H. B. (2014). The role of nano ceramic materials with high specific surface in environment protection. *Advanced Materials Research*, 955–959. https://doi.org/10.4028/www.scientific.net/AMR.955-959.158

4 Ceramics From Recycled Materials and Recycled Ceramics

Nidhi Sharma, Antrakrate Gupta, Ramasubramanian Lakshmi Narayan, and Shikhar Krishn Jha

4.1 INTRODUCTION

Recycling of materials is a major matter of concern in the context of sustainability, as it helps to reduce the amount of waste that ends up in landfills, conserves natural resources, and reduces greenhouse gas emissions (Ahirwar & Tripathi, 2021). When someone talks about waste and recycling, the first thing that generally pops into our heads is 'plastic'. But, in reality, there are many other contenders, such as 'ceramic waste', that equally contribute to the ever-growing waste burden of the planet. Even though it takes millions of years for some plastics to decompose, different methods have been developed to recycle them several times (with the exception of single-use plastics). In contrast, ceramics are never considered for recycling despite the fact that they take a longer time, or perhaps forever, to decompose. Although ceramics do not affect the ecological balance adversely, they remain in the environment as unutilized dead weight and, if not recycled, get dumped from one place to another. Additionally, some ceramics in the waste stream, particularly, those from colorfully glazed products, often contain heavy metals that are detrimental for the environment (Aljerf, 2018). Similarly, waste generated from polishing media contains silicon carbide. This reacts with oxygen at high temperatures and leads to porosity and the emission of carbon dioxide, which is a greenhouse gas (Wattanasiriwech et al., 2009). Overall, the rapid growth of urbanization and industrialization, coupled with dwindling resources and waste pile-ups, has an undesirable impact on the environment, ecosystems, and human health. In that context, turning waste into functional ceramics is a potentially game-changing venture if it can lead to the development of application-oriented products that are recyclable.

Functionalization of wastes by the combination of heterogeneous substances has become an opportunity for the development of new, high-quality products with suitable magnetic, electrical, biological, adsorption, optical, and mechanical properties for technical and industrial purposes (Shrestha et al., 2013). By using recycled ceramic materials, we can significantly reduce our environmental impact and make a positive contribution to the planet. One of the main benefits of using recycled

DOI: 10.1201/9781003731672-4

ceramics is the conservation of natural resources. The production of new ceramics requires the mining of raw materials and other minerals, which has a detrimental impact on the environment, including the destruction of habitat and water pollution. By using recycled ceramics, we can significantly reduce our reliance on these natural resources and help in their preservation (Liu et al., 2020). In addition to conserving natural resources, using recycled ceramics also helps to reduce greenhouse gas emissions. The production of ceramics is energy-intensive, and the firing process releases significant amounts of carbon dioxide into the atmosphere. By using recycled ceramics, we can reduce the demand for new ceramics and, in turn, reduce the carbon footprint of the ceramics industry (Furszyfer Del Rio et al., 2022). Recycling ceramics also helps to reduce waste and decreases the strain on landfills. Landfills are a major source of methane, a potent greenhouse gas that contributes to climate change. By diverting ceramic waste from landfills and using it to create new products, we can help to reduce methane emissions and slow the rate of climate change (Huber-Humer et al., 2008). Furthermore, recycled ceramics used for aesthetic purposes often have a unique and authentic appearance that cannot be replicated with new materials, which adds to its character and sets it apart from other mass-produced ceramics.

Ceramics made from recycled materials and recycled ceramics are becoming increasingly popular as a sustainable and environmentally friendly alternative to traditional ceramics production. Some common examples of materials which can be incorporated in recycling of ceramics include glass, industrial by-products such as fly ash from power plants or slag from steel manufacturing, and post-consumer waste such as discarded ceramic products or porcelain fixtures (Valadares Folgueras et al., 2005).

4.2 REDUCE, RECYCLE, REUSE (3R): SUSTAINABILITY IN CERAMIC INDUSTRY

Since the dawn of time, humans have worked with ceramics and have used it for utilitarian and decorative purposes. Similar to glass, plastic, and wood, ceramics are present in our daily lives. But unlike these materials, these are not so easy to throw away and considered to be more valuable and more durable. The global industry of ceramics is structured around nine sectors (Ceramics Market Size, Share & Trends Analysis Report By Product [Traditional, Advanced], By Application [Abrasives, Tiles], By End-Use [Industrial, Medical], By Region, And Segment Forecasts, 2021), including construction materials, consumer goods, industrial processes, and advanced technologies, estimating a turnover of USD 239.5 billion and 2,00,00,00 direct or indirect jobs (Figure 4.1). According to a report published in 2014 (Ceramic Tiles [Floor Tiles, Wall Tiles and Other Tiles] Market – Global Industry Research Analysis, 2015), the production of the global ceramic tiles market stood at 13 billion square meters in terms of volume. Another report (Raval et al., 2013) states that approximately 2,50,000 tons of tiles wear out every year, and another ten crore tiles go into repairs. Further, one sanitaryware cluster in India carries out production using around 7.2 lakh tons of clay per year, out of which 21,600 tons of fired pieces turn into waste due to manufacturing defects.

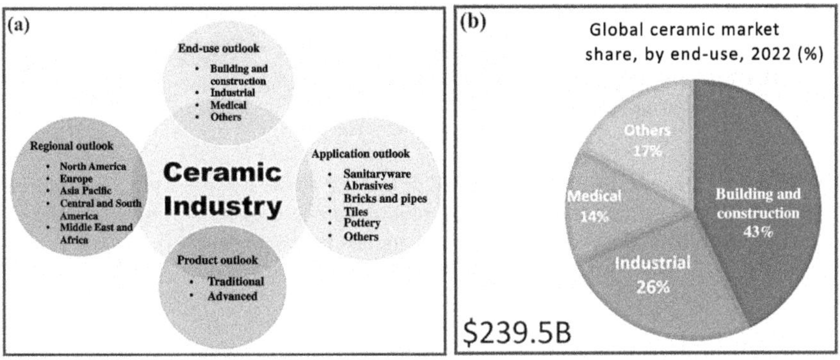

FIGURE 4.1 (a) A general perspective for the source and emergence of ceramic-derived waste materials; (b) its global revenue in the year 2022.

Source: Ceramics Market Size, Share & Trends Analysis Report by Product (Traditional, Advanced), by Application (Abrasives, Tiles), by End-Use (Industrial, Medical), by Region, and Segment Forecasts, 2021.

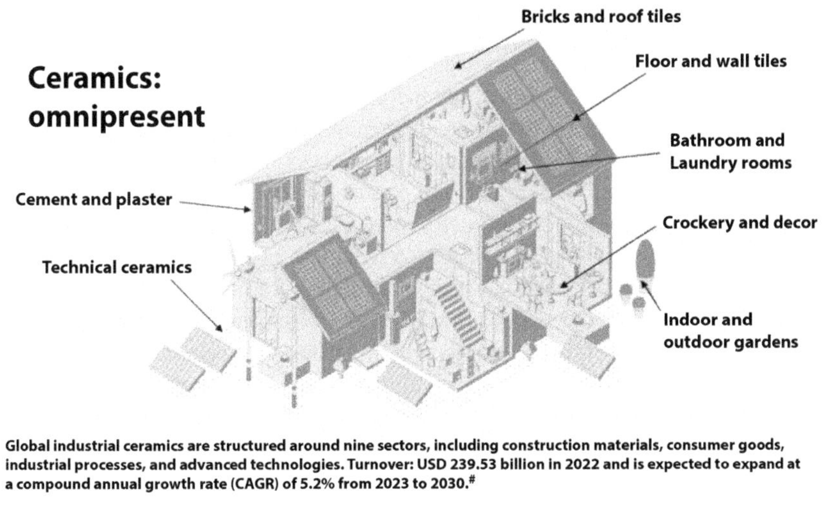

FIGURE 4.2 Utilization and usage of ceramics and ceramic-based products in various sectors.

Source: Image has been inspired from the stock image of Adobe [Adobe Stock, Energy self-sufficient house isometric banner, accessed on 20 March 2023 <https://stock.adobe.com/in/images>] and redesigned for this manuscript.

If ceramic is so used, it is because it is a technical material that caters to almost all new and emerging industrial, domestic, and medical needs along with required environmental standards (Colorado et al., 2020). Figure 4.2 illustrates the vast array of utilities in which ceramics are used in our day-to-day life.

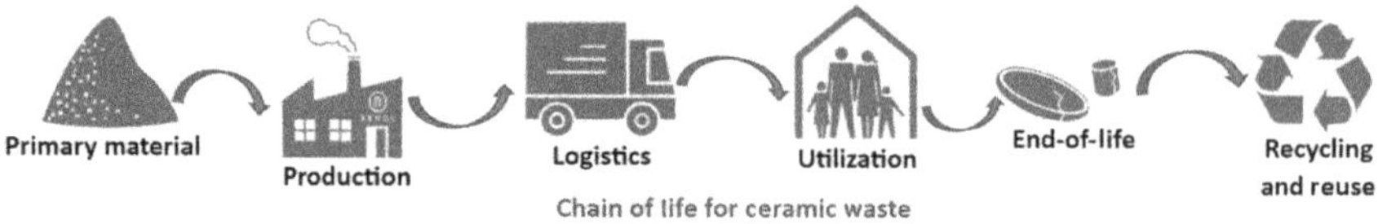

FIGURE 4.3 Chain of life for ceramics and its derived waste. It starts with primary materials mined from the ore deposits, followed by refining, processing, manufacturing, utilization, and finally end-of-life. Recycling can put the materials back into the cycle.

The use of ceramics is common in everyday life, but most people are not aware of how to dispose of them when they have reached their end-of-life (EOL). The EOL ceramic waste can be recycled and reformed into new multiple usable and valuable products, making ceramic waste a sustainable industry (Cooper & Gutowski, 2017). Besides being endowed with great binding and draining capacity, finely ground ceramic is a totally inert material which has the advantage of having no negative impact on the environment. Figure 4.3 shows a generalized chain of life for ceramics and their derived waste.

The ceramic industry sector is a large consumer of natural resources and energy, which leads to large carbon dioxide emissions (Rambaldi, 2021). The wastes produced by the ceramic industrial process are unfired scrap tiles, fired scrap tiles, mud produced by the washing lines, lapping and polishing mud, dried grinding residues, and exhausted lime by the fume abatement system. Almost all these wastes (pre-consumer wastes) are reused in the same process in a closed-loop cycle. Only the exhausted lime is landfill confined and acts as a hazard. The use of several kinds of waste in traditional ceramics has been investigated since the late 1990s. In the last few years, with the target of demonstrating a reduction in environmental impact and a significant reuse of waste materials, a new concept of traditional ceramics has been developed by replacing more natural raw materials with (i) new glassy raw materials for traditional ceramic from the vitrification of different wastes able to crystallize during firing (Bernardo et al., 2017) and (ii) different types of waste proportionately balanced to obtain a waste synergy during firing (Romero et al., 2001). As ceramics cannot be incinerated, and require dedicated and sophisticated handling, depending upon their individual and unique structural and physical properties, there is a need for identifying techniques that can promote their recycling and reuse. The production process for ceramics from recycled materials may vary depending on the specific materials used. However, it generally involves the following steps.

(i) **Collection and Sorting**: Recyclable materials are collected and sorted based on their composition and suitability for ceramic production.
(ii) **Processing**: The collected materials are processed, which may involve crushing, grinding, or purification to remove impurities.
(iii) **Mixing**: The recycled materials are mixed with new clay or other ceramic components to achieve the desired composition and properties.
(iv) **Shaping and Forming**: The ceramic mixture is shaped into the desired form using techniques such as molding, extrusion, or handcrafting.

(v) **Drying and Firing**: The shaped ceramic objects are dried to remove moisture and then fired at high temperatures in a kiln to achieve their final hardness and durability.

Recycled ceramics are accompanied with characteristics and performances that go far beyond traditional uses, as sustainability in multiple environments is the primary driving force behind these innovations.

4.3 RECYCLING CONTENT IN TRADITIONAL CERAMICS

Finding recyclable content in traditional ceramics can be a challenging task due to the nature of the materials involved. Traditional ceramics are typically made from clay, which is fired at high temperatures to create durable and nonporous objects such as pottery, tiles, and dishes. Unlike some other materials, such as paper or plastic, ceramics cannot be easily broken down and reprocessed into new ceramic materials. However, there are a few ways in which recycling of traditional ceramics can be done.

(i) **Reuse**: One common method of recycling ceramics is through reuse. Broken or unwanted ceramic items can be repurposed for other applications. For example, broken pottery pieces can be used in mosaic artwork or as decorative elements in gardens. Old tiles can be crushed and used as a base for new flooring or as aggregate in concrete.
(ii) **Crushing and Grinding**: Ceramics can also be crushed and ground into a fine powder, which can then be incorporated into new ceramic materials. This process requires specialized equipment to break down the ceramics into a usable form (Darweesh, 2019). The resulting powder can be mixed with fresh clay to create new ceramic objects, reducing the need for virgin clay resources.
(iii) **Ceramic Aggregate**: Crushed ceramics can be used as aggregate in construction materials such as concrete. The crushed ceramic pieces can replace a portion of the traditional aggregate, providing a sustainable alternative. This application is commonly used in road construction and building projects (Maier & Durham, 2012).
(iv) **Industrial Applications**: Some industries utilize ceramic waste as a raw material in their processes. For instance, ceramic waste can be used as a filler in manufacturing refractory bricks, insulation materials, or in the production of ceramic glazes (Silva et al., 2017).

It is important to note that while these methods allow for some degree of recycling, there are limitations and challenges associated with each approach. The high temperatures required in the production of traditional ceramics make it difficult to fully recycle them back into their original form. Additionally, contamination from other materials or glazes can also affect the recyclability of ceramics.

Further, ceramic waste can consist of various components, and the recyclability of each component depends on factors such as its composition and physical properties

(Lamba et al., 2022). Some potentially recyclable components of ceramic waste include the following.

(i) **Clay**: Ceramic waste often contains clay, which is the primary material used in traditional ceramics. Clay can be recycled by rehydrating it and reusing it in the production of new ceramic objects. The clay can be mixed with fresh clay to create a blend that maintains the necessary properties for forming and firing (Calatan et al., 2017).
(ii) **Glazes**: Glazes are applied to ceramics to provide color, texture, and surface finish. Some glazes may contain recyclable components, such as certain metal oxides or minerals. However, the recyclability of glazes depends on their composition and compatibility with the recycling process. Recycling glazes may involve separating them from the ceramic body and reusing or reformulating them for future applications (Strecker & Costa, 2014).
(iii) **Kiln Furniture**: Kiln furniture, such as shelves, posts, and setters, is used to support ceramic objects during the firing process. Kiln furniture is typically made from materials like cordierite or alumina, which can withstand high temperatures. These materials can be recycled by collecting and processing them separately from other ceramic waste. They can be crushed, ground, and mixed with fresh materials to create new kiln furniture or refractory products (Scheithauer et al., 2015).
(iv) **Broken Ceramic Pieces**: Broken or damaged ceramic objects can be recycled by crushing and grinding them into a fine powder. This powder, known as ceramic aggregate, can be used as a filler or additive in various applications. It can be incorporated into concrete, asphalt, or other construction materials to enhance their properties. The crushed ceramic pieces can also be used for artistic purposes, such as mosaic artwork (Magbool, 2022).
(v) **Porcelain Sanitary Ware**: Porcelain sanitary ware, such as toilets and sinks, contains ceramic components that can be recycled. Recycling processes for porcelain sanitary ware typically involve crushing and grinding the waste into a powder, which can then be used as a raw material in the production of new sanitary ware or as aggregate in construction materials (Silva et al., 2019).

It is worth noting that the recycling of ceramic waste can be complex due to factors such as the presence of contaminants, variations in composition, and the need for specialized types of equipment and processes. Recycling efforts may also vary depending on local infrastructure and available recycling technologies. As the field of sustainable materials and recycling continues to evolve, researchers and manufacturers are exploring new technologies and approaches to improve the recycling of ceramics. These advancements may lead to more efficient and effective methods of recycling traditional ceramics in the future.

4.4 APPLICATIONS OF RECYCLED CERAMICS

Recycled ceramics can be used in various applications, offering environmental benefits and reducing the demand for virgin materials. Some common applications of recycled ceramics include the following.

(i) **Construction Materials**: Crushed ceramic waste, also known as ceramic aggregate, can be used as a substitute for traditional aggregates in construction materials. It can be incorporated into concrete, asphalt, or other building materials to enhance their properties. Recycled ceramics can improve the strength, durability, and thermal properties of these materials while reducing the need for virgin aggregates (Ray et al., 2021).

(ii) **Road Infrastructure**: Ceramic waste can be used in road construction projects. It can be mixed with asphalt to create ceramic asphalt mixes, which provide enhanced durability and resistance to rutting and cracking. The use of recycled ceramics in road surfaces can reduce maintenance needs and extend the lifespan of the roads (Ali & Al-Mansoori, 2021).

(iii) **Landscaping and Decorative Purposes**: Broken or crushed ceramics can be used for landscaping and decorative applications. For example, they can be used as decorative elements in gardens, pathways, or as aggregates in decorative concrete surfaces. Recycled ceramics can add texture, color, and unique visual appeal to outdoor spaces (Kara & Karacasu, 2015).

(iv) **Art and Design**: Recycled ceramics can be a valuable resource for artists and designers. Broken pottery pieces can be used in mosaic artwork or sculptures, giving them a distinctive aesthetic. Artists can repurpose and transform discarded ceramics into new artistic creations, promoting sustainability and creativity (He, 2022).

(v) **Industrial Applications**: Certain industries utilize recycled ceramics in their manufacturing processes. For instance, ceramic waste can be used as a raw material in the production of refractory bricks, insulation materials, or ceramic glazes. Recycled ceramics can contribute to the circular economy within the ceramic industry by minimizing waste and conserving resources (Zabibulla Kamilovich, 2020).

(vi) **Fillers and Additives**: Ground ceramic waste can be used as fillers or additives in various products and materials. For example, it can be incorporated into plastics, rubbers, or composites to enhance their mechanical properties. Recycled ceramics provide reinforcement, thermal stability, or aesthetic qualities to these materials (Jackiewicz-Rek et al., 2015).

(vii) **Soil Amendments**: In some cases, ceramic waste can be used as a soil amendment. It can be added to soil to improve its structure, drainage, and nutrient-holding capacity. However, it is important to assess the composition and potential contaminants in the recycled ceramics to ensure they are safe for soil application (Lyckfeldt & Ferreira, 1998).

Apart from these examples, the specific applications of recycled ceramics may vary depending on their characteristics and the requirements of the target product or project, some of which are detailed in the following subsections.

4.4.1 Development of Structural and Nonstructural Concrete

The development of structural and nonstructural concrete incorporating recycled ceramics offers a sustainable solution for reducing waste and conserving natural resources. The utilization of recycled ceramics in concrete applications can be summarized as follows.

(i) **Structural Concrete:**
- ***Aggregate Replacement:*** Crushed ceramic waste can replace a portion of the traditional coarse aggregate in concrete mixtures. The ceramic aggregate can be used as a partial replacement for materials like gravel or crushed stone. It helps to enhance the strength, durability, and thermal properties of the concrete.
- ***Fine Aggregate Replacement:*** Ceramic waste can also be used as a partial replacement for fine aggregates, such as sand. The finely crushed ceramic particles can fill in the gaps between larger aggregates, improving the packing density and overall performance of the concrete.
- ***Binder Materials:*** In some cases, ceramic waste can be used as a partial replacement for cementitious materials. It can be processed and ground into a fine powder to be used as a supplementary cementitious material (SCM) or as a pozzolan. The ceramic powder reacts with calcium hydroxide in the presence of water, contributing to the strength and durability of the concrete (Li et al., 2020).

(ii) **Nonstructural Concrete:**
- ***Paving Blocks and Tiles:*** Recycled ceramics can be incorporated into the production of nonstructural concrete products such as paving blocks and tiles. The ceramic waste can be used as a partial replacement for both coarse and fine aggregates, depending on the specific requirements of the product. This application provides an opportunity to repurpose waste ceramics into functional and aesthetically pleasing materials.
- ***Decorative Concrete:*** In decorative concrete applications, crushed ceramics can be used as an aggregate to create unique visual effects. The various colors and textures of the recycled ceramics can be showcased in exposed aggregate concrete surfaces, providing an attractive and environmentally friendly design option.

It is important to note that incorporating recycled ceramics into concrete may require adjustments in mix design and production processes. Quality control measures should be in place to ensure the performance and consistency of the resulting concrete mixtures. Additionally, proper assessment and testing should be conducted to ensure the safety and suitability of the recycled ceramic materials for concrete applications.

4.4.2 Fabrication of Thermal Energy Storage Devices

The fabrication of thermal energy storage (TES) devices from recycled ceramics can be a sustainable approach that combines the benefits of energy storage with the reuse of ceramic waste. An overview of the process can be given as follows.

(i) **Ceramic Selection**: The first step is to identify suitable ceramic materials from the recycled waste stream. Ceramics with high thermal conductivity and stability at elevated temperatures are preferred for TES applications. Examples include alumina, cordierite, mullite, or zirconia-based ceramics. The recycled ceramics should be thoroughly cleaned and processed to remove contaminants or impurities.

(ii) **Ceramic Shaping**: The recycled ceramics can be shaped into suitable forms for TES devices. Depending on the specific application, different shapes may be utilized, such as bricks, tubes, plates, or honeycomb structures. Shaping can be achieved through processes like molding, extrusion, or casting. During this stage, additives or binders may be incorporated to improve the mechanical integrity of the formed ceramics.

(iii) **Thermal Energy Storage Medium**: The shaped ceramic components serve as the thermal energy storage medium within the TES device. The ceramic structures provide high heat capacity and can absorb and release thermal energy efficiently. The size, geometry, and arrangement of the ceramics will depend on the specific requirements of the TES system.

(iv) **Encapsulation and Insulation**: To prevent heat loss and ensure efficient thermal energy storage, the ceramic structures may need to be encapsulated or insulated. This can be done using materials such as metal casings, high-temperature coatings, or insulating materials. The encapsulation protects the ceramic components and minimizes heat transfer to the surrounding environment.

(v) **Integration and System Design**: The fabricated ceramic-based TES devices need to be integrated into larger systems to effectively store and retrieve thermal energy. This may involve connecting the TES devices to heat sources or heat sinks, incorporating control mechanisms and optimizing the overall system design for efficient energy storage and release.

4.4.2.1 Benefits of Using Recycled Ceramics in TES Devices

The benefits of using recycled ceramics in TES devices include the following.

- **Thermal Properties**: Ceramics have excellent thermal conductivity and can effectively store and release thermal energy.
- **High-Temperature Stability**: Ceramics can withstand high temperatures, making them suitable for TES applications in various industries.
- **Longevity**: Ceramics are durable materials that can withstand cyclic heating and cooling without significant degradation.

It should be noted that the fabrication process of TES devices from recycled ceramics may require specific expertise, testing, and quality control to ensure the performance, reliability, and safety of the final products.

4.4.3 Development of Glass Ceramics

Glass ceramics can be developed from recycled ceramics, offering a sustainable approach to repurposing ceramic waste. The process involves transforming the ceramic waste into a glassy phase, followed by controlled crystallization to create glass ceramics. An overview of the development process is given here.

(i) **Ceramic Waste Collection and Preparation**: Ceramic waste is collected and sorted based on its composition and properties. It undergoes a cleaning and processing stage to remove contaminants and impurities. The ceramics are then crushed into a fine powder or granules for further processing.
(ii) **Glass Formation**: The ceramic waste is mixed with other glass-forming materials, such as silica (SiO_2), alumina (Al_2O_3), and fluxing agents (e.g., sodium carbonate, boron oxide). This mixture is heated at high temperatures to melt and homogenize the ingredients. The molten material is then rapidly cooled to form an amorphous glass.
(iii) **Crystallization**: The glass is then subjected to controlled heat treatment, known as annealing, to induce controlled crystallization. This involves heating the glass to specific temperatures and holding it there for a designated period. During this process, specific crystalline phases nucleate and grow within the glass matrix, transforming it into a glass ceramic material.
(iv) **Characterization and Optimization**: The resulting glass ceramic material is characterized to assess its properties, such as crystalline phase composition, thermal stability, mechanical strength, and electrical properties. This data helps optimize the fabrication process and adjust the composition of the glass ceramic to meet desired specifications.

Glass ceramics from recycled ceramics can be used in various applications, such as kitchenware and tableware, combining the aesthetic appeal of glass with the durability and thermal properties of ceramics; building materials in architectural applications to offer high strength, resistance to staining, and withstanding temperature fluctuations; electronics and optics due to their excellent thermal and electrical properties. They can be used in substrates, insulators, and laser components.

4.4.3.1 Benefits of Developing Glass Ceramics From Recycled Ceramics

The benefits of developing glass ceramics from recycled ceramics include the following.

- **Enhanced Properties**: Glass ceramics offer improved mechanical strength, thermal stability, chemical resistance, and electrical properties compared to traditional ceramics.
- **Versatility**: Glass ceramics can be formed into various shapes and sizes, allowing for versatility in design and application.

It is noteworthy that the development of glass ceramics from recycled ceramics requires careful control of the processing parameters, including composition, heat treatment, and crystallization conditions, to achieve the desired properties and characteristics.

4.4.4 Production of Mineral Asphalt Mixtures

The development of mineral asphalt mixtures from recycled ceramics involves incorporating ceramic waste as a substitute for traditional mineral aggregates in the asphalt mixture. An overview of the process and its benefits are summarized here.

(i) **Ceramic Waste Collection and Processing**: Ceramic waste is collected, sorted, and processed to remove contaminants and impurities. The ceramics are then crushed and ground into fine particles, similar to the size range of traditional mineral aggregates used in asphalt mixtures.
(ii) **Mix Design Optimization**: The mix design of the asphalt mixture is optimized by determining the appropriate proportions of recycled ceramics, bitumen (asphalt binder), and other additives. The goal is to achieve a balance between the properties of the recycled ceramics and the overall performance requirements of the asphalt mixture.
(iii) **Incorporation of Recycled Ceramics**: The crushed ceramic waste is introduced into the asphalt mixture during the mixing process. The recycled ceramics act as a substitute for traditional mineral aggregates, providing the necessary particle structure and contributing to the overall volume and stability of the asphalt mixture.
(iv) **Performance Evaluation**: The performance of the mineral asphalt mixture containing recycled ceramics is evaluated through laboratory testing and field trials. Tests may include determining properties such as stability, durability, rutting resistance, fatigue resistance, and skid resistance. The performance should meet or exceed the specifications and requirements of the intended application.

4.4.4.1 Benefits of Using Recycled Ceramics in Mineral Asphalt Mixtures

- **Cost Savings**: Recycling ceramics can provide cost savings compared to purchasing virgin aggregates, contributing to more economically viable asphalt mixtures.
- **Performance Enhancement**: The properties of recycled ceramics, such as their hardness, thermal stability, and resistance to wear, can enhance the performance of the mineral asphalt mixture. This can lead to improved durability, increased resistance to rutting and cracking, and longer pavement lifespan.
- **Customization and Aesthetics**: Recycled ceramics may introduce unique colors, textures, or visual characteristics to the asphalt mixture, allowing for customized or aesthetically appealing pavement designs.

It should be noted that the incorporation of recycled ceramics in mineral asphalt mixtures may require adjustments in mix design, production processes, and quality control measures.

4.4.5 Improvement in Energy Efficiency and Flame Retardancy of Buildings

The use of recycled ceramics can contribute to the improvements in energy efficiency and flame retardancy in buildings. The aspects in which recycled ceramics help in these fields include the following.

(i) **Energy Efficiency:**
- **Insulation Properties**: Ceramic materials, including recycled ceramics, possess inherent thermal insulation properties. They have low thermal conductivity, which helps reduce heat transfer through building components. Incorporating recycled ceramics in insulation products, such as thermal insulation boards or coatings, can enhance the energy efficiency of buildings by minimizing heat loss or gain.
- **Reflective Coatings**: Ceramic-based coatings can be formulated with recycled ceramics to create reflective surfaces. These coatings have high solar reflectance, meaning they reflect a significant portion of the sun's heat, reducing the cooling load on buildings. By using these coatings on roofs, walls, or windows, the energy consumption for cooling can be reduced.

(ii) **Flame Retardancy:**
- **Fire-Resistant Materials**: Ceramics have excellent fire-resistant properties due to their high melting points and low flammability. Recycled ceramics can be incorporated into building materials, such as fire-resistant boards, fireproof coatings, or fire barriers, to enhance the flame retardancy of the construction elements. This can help prevent the spread of fire and increase the overall fire safety of buildings.

(iii) **Sustainable Construction:**
- **Environmental Impact**: By utilizing recycled ceramics in building materials, the demand for virgin materials is reduced, resulting in lower energy consumption and greenhouse gas emissions associated with their extraction and production. Using recycled ceramics supports sustainable construction practices and reduces the environmental footprint of buildings.
- **Waste Reduction**: Incorporating recycled ceramics diverts waste from landfills and promotes the circular economy by reusing materials that would otherwise be discarded. This contributes to more sustainable waste management practices and reduces the need for new raw materials.

It is worth noting that the specific benefits of recycled ceramics in terms of energy efficiency and flame retardancy depend on factors such as the composition of the recycled ceramics, their incorporation into building products, and their interaction with other components of the building system.

4.4.6 Development of Thermal and Acoustic Insulation Foam

The development of thermal and acoustic insulation foam from recycled ceramics can provide sustainable and effective solutions for enhancing the thermal and acoustic performance of buildings. An overview of the process is given here.

(i) **Ceramic Waste Collection and Processing**: Ceramic waste is collected, sorted, and processed to remove impurities and contaminants. The ceramics are then crushed into a fine powder or granules for further processing.
(ii) **Formulation and Mixing**: The recycled ceramic particles are combined with other components, such as foaming agents, binding agents, and additives. These materials are mixed together to form a homogenous mixture.
(iii) **Foaming Process**: The mixture is subjected to a foaming process, which typically involves introducing a blowing agent that creates gas bubbles within the mixture. The mixture expands and solidifies, forming a foam structure with interconnected voids.
(iv) **Curing and Solidification**: The foam structure is allowed to cure and solidify, forming a stable insulation material. This may involve subjecting the foam to controlled temperature and humidity conditions to facilitate the curing process.
(v) **Characterization and Testing**: The thermal and acoustic properties of the insulation foam are evaluated through laboratory testing. This includes measuring parameters such as thermal conductivity, sound absorption coefficient, and other relevant performance indicators. The foam should meet or exceed the required specifications and standards for thermal and acoustic insulation materials.

4.4.6.1 Benefits of Thermal and Acoustic Insulation Foam From Recycled Ceramics

The benefits of thermal and acoustic insulation foam from recycled ceramics include the following.

- **Thermal Insulation**: Ceramic-based foam provides excellent thermal insulation properties, reducing heat transfer through building components. This helps improve energy efficiency, reducing heating and cooling demands.
- **Acoustic Insulation**: The foam's porous structure and density help absorb sound waves, providing effective acoustic insulation. It can reduce noise transmission between spaces and improve indoor acoustic comfort.
- **Fire Resistance**: Ceramics inherently have high fire resistance properties such as non-combustibility, chemical stability and inertness, and low coefficient of thermal, which can enhance the fire safety of buildings when used as insulation foam.
- **Durability and Longevity**: Ceramic-based foam materials exhibit good durability and resistance to degradation, ensuring the long-term performance and lifespan of the insulation.

It is important to note that the specific formulation and manufacturing process for ceramic-based insulation foam may vary based on the desired properties and applications. Additionally, compliance with relevant building codes and standards is necessary to ensure the safety and effectiveness of the insulation foam in thermal and acoustic applications.

4.5 FUTURE PROSPECTS

The future prospects of ceramics from recycled materials and recycled ceramics are promising, driven by the growing need for sustainable and environment-friendly solutions. Some potential future developments and applications include the following.

(i) **Advanced Manufacturing Techniques**: Advancements in manufacturing technologies, such as additive manufacturing (3D printing), enable the production of complex ceramic components from recycled materials, which opens up new possibilities for custom-designed ceramic products with optimized properties and reduced waste.

(ii) **High-Performance Applications**: The use of recycled ceramics in high-performance applications like aerospace, defense, automotive, and electronics industries is gaining widespread attention, where ceramics offer exceptional properties like high strength, heat resistance, and electrical insulation. Recycling ceramics for these applications can contribute to reducing the demand for virgin materials and promoting sustainability.

(iii) **Energy Storage and Conversion**: Ceramics play a vital role in energy storage and conversion devices, such as batteries, fuel cells, and solar cells. Utilizing recycled ceramics in these applications can enhance their sustainability by reducing the reliance on non-renewable resources while maintaining or improving performance.

(iv) **Sustainable Construction Materials**: Recycled ceramics find increased use in construction materials, such as concrete, bricks, tiles, and insulation. These applications can contribute to energy-efficient buildings, reduced waste generation, and improved fire resistance.

(v) **Environmental Remediation**: Ceramic materials have the potential for environmental remediation applications, such as water purification, air filtration, and hazardous waste treatment. Recycling ceramics into tailored materials can help address environmental challenges and contribute to a cleaner and safer environment.

(vi) **Circular Economy Initiatives**: The adoption of circular economy principles encourages the reuse and recycling of materials to minimize waste generation. The future may see the development of comprehensive systems and infrastructure that enable efficient collection, sorting, and processing of ceramic waste, facilitating the widespread utilization of recycled ceramics.

(vii) **Research and Innovation**: Ongoing research and innovation efforts are likely to uncover new methods and applications for recycled ceramics. This includes the development of novel recycling techniques, characterization methods, and engineering solutions that optimize the properties and performance of recycled ceramics.

The future prospects of ceramics from recycled materials and recycled ceramics align with the global sustainability agenda. By leveraging the unique properties of ceramics and promoting recycling initiatives, we can move toward a more resource-efficient and environmentally conscious society.

4.6 CONCLUSION

Ceramics from recycled materials and recycled ceramics offer several benefits and opportunities in terms of sustainability, waste reduction, and resource conservation. Some key points regarding ceramics from recycled materials and recycled ceramics can be summarized as follows.

(i) **Waste Reduction**: Utilizing recycled materials and ceramics helps divert waste from landfills, reducing the environmental impact associated with waste disposal. By repurposing and recycling ceramics, we can minimize the amount of ceramic waste generated and promote a more circular economy.

(ii) **Resource Conservation**: Recycling ceramics reduces the demand for virgin raw materials, conserving natural resources such as clay, sand, and minerals. By incorporating recycled ceramics into new ceramic products, we can decrease the reliance on finite resources and contribute to sustainable resource management.

(iii) **Environmental Benefits**: The production of ceramics from recycled materials typically requires less energy and generates fewer greenhouse gas emissions compared to the production of ceramics from raw materials. Recycling ceramics helps lower carbon footprints and mitigate the environmental impact associated with traditional ceramic production processes.

(iv) **Product Development**: Recycled ceramics can be used to develop a wide range of ceramic products, including tiles, bricks, sanitaryware, kitchenware, and decorative items. These products maintain the inherent qualities of ceramics, such as durability, heat resistance, chemical stability, and aesthetic appeal while also incorporating sustainable elements.

(v) **Customization and Uniqueness**: Ceramics from recycled materials exhibit unique visual characteristics, as the recycled components may introduce variations in color, texture, or pattern. This allows for the creation of distinctive and customized ceramic products that appeal to consumers seeking sustainable and individually crafted items.

(vi) **Collaboration and Innovation**: The development of ceramics from recycled materials requires collaboration between ceramic manufacturers, recycling facilities, and research institutions. Continuous innovation in recycling technologies, material processing, and product design can unlock new possibilities and enhance the quality and performance of ceramics from recycled materials.

(vii) **Consumer Awareness and Demand**: The growing consumer awareness and concern for sustainability have created a demand for products made from recycled materials. Ceramics from recycled materials and recycled ceramics cater to this demand, providing eco-friendly options that align with consumer preferences for environmentally conscious choices.

(viii) **Regulatory Support**: Government regulations and policies aimed at promoting recycling, waste management, and sustainability can further drive the adoption and development of ceramics from recycled materials. Supportive policies, incentives, and standards can incentivize ceramic manufacturers to incorporate recycled materials into their production processes.

Overall, ceramics from recycled materials and recycled ceramics offer opportunities to reduce waste, conserve resources, and create sustainable and visually appealing ceramic products. The continuous development and integration of recycling technologies and sustainable practices in the ceramics industry can contribute to a more circular and environmentally friendly approach to ceramic production.

ACKNOWLEDGMENT

The authors would like to acknowledge the support provided by the Science and Engineering Research Board, a statutory body of Department of Science & Technology (DST), Government of India for facilitating the research discussed in this chapter. We would like to express our gratitude to Mr. Harshit Sharma, Krypt India Private Limited, Bengaluru, Karnataka, for helping out with the images illustrated in this chapter.

DECLARATION OF INTEREST

The authors claim no institutional, financial, or general conflicts of interest with any third party not involved in this research work.

REFERENCES

Ahirwar, R., & Tripathi, A. K. (2021). E-waste management: A review of recycling process, environmental and occupational health hazards, and potential solutions. *Environmental Nanotechnology, Monitoring and Management, 15*. https://doi.org/10.1016/j.enmm.2020.100409

Ali, A. M., & Al-Mansoori, T. (2021). Investigation of asphalt binder performance modified with ceramic waste powder. *IOP Conference Series: Materials Science and Engineering, 1090*(1). https://doi.org/10.1088/1757-899x/1090/1/012055

Aljerf, L. (2018). A gateway to metal resistance: Bacterial response to heavy metal toxicity in the biological environment. *Annals of Advances in Chemistry*. https://doi.org/10.29328/journal.aac.1001012

Bernardo, E., Scarinci, G., & Colombo, P. (2017). Vitrification of waste and reuse of waste-derived glass. In *Encyclopedia of Sustainability Science and Technology*. Springer. https://doi.org/10.1007/978-1-4939-2493-6_96-3

Calatan, G., Hegyi, A., Dico, C., & Mircea, C. (2017). Experimental research on the recyclability of the clay material used in the fabrication of adobe bricks type masonry units. *Procedia Engineering, 181*. https://doi.org/10.1016/j.proeng.2017.02.402

Ceramics Market Size, Share & Trends Analysis Report by Product (Traditional, Advanced), by Application (Abrasives, Tiles), by End-Use (Industrial, Medical), by Region, and Segment Forecasts. (2021). 115. https://www.grandviewresearch.com/industry-analysis/ceramics-market

Colorado, H. A., Velásquez, E. I. G., & Monteiro, S. N. (2020). Sustainability of additive manufacturing: the circular economy of materials and environmental perspectives. *Journal of Materials Research and Technology*, *9*(4), 8221–8234. https://doi.org/10.1016/J.JMRT.2020.04.062

Cooper, D. R., & Gutowski, T. G. (2017). The environmental impacts of reuse: A review. *Journal of Industrial Ecology*, *21*(1). https://doi.org/10.1111/jiec.12388

Darweesh, H. H. M. (2019). Recycling of glass waste in ceramics – part I: physical, mechanical and thermal properties. *SN Applied Sciences*, *1*(10). https://doi.org/10.1007/s42452-019-1304-8

Furszyfer Del Rio, D. D., Sovacool, B. K., Foley, A. M., Griffiths, S., Bazilian, M., Kim, J., & Rooney, D. (2022). Decarbonizing the ceramics industry: A systematic and critical review of policy options, developments and sociotechnical systems. *Renewable and Sustainable Energy Reviews*, *157*, 112081. https://doi.org/10.1016/J.RSER.2022.112081

He, Y. (2022). Research on innovative thinking of ceramic art design based on artificial intelligence. *Mobile Information Systems*, *2022*. https://doi.org/10.1155/2022/3381042

Huber-Humer, M., Gebert, J., & Hilger, H. (2008). Biotic systems to mitigate landfill methane emissions. *Waste Management and Research*, *26*(1). https://doi.org/10.1177/0734242X07087977

Jackiewicz-Rek, W., Załęgowski, K., Garbacz, A., & Bissonnette, B. (2015). Properties of cement mortars modified with ceramic waste fillers. *Procedia Engineering*, *108*. https://doi.org/10.1016/j.proeng.2015.06.199

Kara, Ç., & Karacasu, M. (2015). Use of ceramic wastes in road pavement design. *Proceedings of the World Congress on New Technologies (NewTech 2015)*, *2015*(226).

Lamba, P., Kaur, D. P., Raj, S., & Sorout, J. (2022). Recycling/reuse of plastic waste as construction material for sustainable development: A review. *Environmental Science and Pollution Research*, *29*(57). https://doi.org/10.1007/s11356-021-16980-y

Li, L., Liu, W., You, Q., Chen, M., & Zeng, Q. (2020). Waste ceramic powder as a pozzolanic supplementary filler of cement for developing sustainable building materials. *Journal of Cleaner Production*, *259*. https://doi.org/10.1016/j.jclepro.2020.120853

Liu, Y., Park, S., Yi, H., & Feiock, R. (2020). Evaluating the employment impact of recycling performance in Florida. *Waste Management*, *101*. https://doi.org/10.1016/j.wasman.2019.10.025

Lyckfeldt, O., & Ferreira, J. M. F. (1998). Processing of porous ceramics by "starch consolidation". *Journal of the European Ceramic Society*, *18*(2). https://doi.org/10.1016/s0955-2219(97)00101-5

Magbool, H. M. (2022). Utilisation of ceramic waste aggregate and its effect on eco-friendly concrete: A review. *Journal of Building Engineering*, *47*. https://doi.org/10.1016/j.jobe.2021.103815

Maier, P. L., & Durham, S. A. (2012). Beneficial use of recycled materials in concrete mixtures. *Construction and Building Materials*, *29*. https://doi.org/10.1016/j.conbuildmat.2011.10.024

Rambaldi, E. (2021). Pathway towards a high recycling content in traditional ceramics. *Ceramics*, *4*(3). https://doi.org/10.3390/ceramics4030036

Raval, A. D., Patel, Dr. I. N., & Pitroda, Prof. J. (2013). Re-use of ceramic industry waste for the elaboration of eco-efficient concrete. *International Journal of Advanced Engineering Research and Studies*, *2*(3).

Ray, S., Haque, M., Soumic, S. A., Mita, A. F., Rahman, M. M., & Tanmoy, B. B. (2021). Use of ceramic wastes as aggregates in concrete production: A review. *Journal of Building Engineering*, *43*. https://doi.org/10.1016/j.jobe.2021.102567

Romero, M., Rincón, J. M., Rawlings, R. D., & Boccaccini, A. R. (2001). Use of vitrified urban incinerator waste as raw material for production of sintered glass-ceramics. *Materials Research Bulletin*, *36*(1–2). https://doi.org/10.1016/S0025-5408(01)00501–3

Scheithauer, U., Schwarzer, E., Tscharntke, F., Schmidt, T., Jegust, S., Richter, H. J., Moritz, T., & Michaelis, A. (2015). New lightweight kiln furniture made by combination of ceramic green tapes and extrudates. *InterCeram: International Ceramic Review*, *64*(4). https://doi.org/10.1007/bf03401124

Shrestha, R. M., Varga, I., Bajtai, J., & Varga, M. (2013). Design of surface functionalization of waste material originated charcoals by an optimized chemical carbonization for the purpose of heavy metal removal from industrial waste waters. *Microchemical Journal*, *108*. https://doi.org/10.1016/j.microc.2012.11.002

Silva, R. V., de Brito, J., Lye, C. Q., & Dhir, R. K. (2017). The role of glass waste in the production of ceramic-based products and other applications: A review. *Journal of Cleaner Production*, *167*. https://doi.org/10.1016/j.jclepro.2017.08.185

Silva, T. H., Castro, A. C. M., Valente Neto, F. C., Soares, M. M. N. S., De Resende, D. S., & Bezerra, A. C. S. (2019). Recycling ceramic waste as a raw material in sanitary ware production. *Ceramica*, *65*(375). https://doi.org/10.1590/0366-69132019653752687

Strecker, K., & Costa, H. B. (2014). Formulation of ceramic glazes by recycling waste glass. *Materials Science Forum*, 775–776. https://doi.org/10.4028/www.scientific.net/MSF.775-776.635

Valadares Folgueras, M., De Novaes Oliveira, P., & Estevam Alarcon, O. (2005). Glass-ceramics obtained from processed slag and fly ash. *American Ceramic Society Bulletin*, *84*(11).

Wattanasiriwech, D., Saiton, A., & Wattanasiriwech, S. (2009). Paving blocks from ceramic tile production waste. *Journal of Cleaner Production*, *17*(18). https://doi.org/10.1016/j.jclepro.2009.08.008

Zabibulla Kamilovich, B. (2020). Waste from ceramic bricks, as a raw material for the production of restoration materials. *International Journal of Emerging Trends in Engineering Research*, *8*(8). https://doi.org/10.30534/ijeter/2020/56882020

5 Advanced Environmental Barrier Coating for Ceramic Matrix Composites

N. Mohan Raj, N. Mathan Kumar, S. Paulsingarayar, and Karthickeyan Viswanathan

5.1 INTRODUCTION

Environmental barrier-coated (EBC) ceramic matrix composites (CMCs) are broadly employed in the aerospace applications (gas turbine, combustion chamber, blades, vanes, shrouds, and afterburners) due to their capabilities (operated at high temperatures and highly resistant to corrosion) and eco-friendly nature. These accomplishments comprise lessening the weight of the turbine, lessening the amount of fuel it consumes, increasing the temperatures at which it operates, improving the thrust-to-weight ratio, and diminishing the emissions of dangerous gases. Even though these are real accomplishments, the reader usually doesn't get a good idea of how they all fit together. Also, "common sense" says that some of these technical feats, like higher operating temperatures and lower emissions, are actually at odds with each other [1]. This chapter's goal is to show the reader how these achievements are accomplished through the integration of creative engineering. It will describe the constant impetus for higher combustion temperatures, demonstrate the fundamental ideas behind CMCs, emphasise the critical importance of EBCs, and highlight the difficulty of developing EBC designs by air plasma spray (APS). Highlights of high-temperature testing procedures for EBCs/CMCs will be presented as a final topic of discussion. Anyone having a fundamental grounding in surface engineering and materials processing should have no trouble following the arguments presented in this work.

From a materials engineering perspective, the most game-changing technology for ceramic matrix composites (CMCs) and environmental barrier coatings (EBCs). The constant need for cleaner, greener forms of propulsion and energy generation is in making ceramic-based gas turbines a reality is the combination petus behind this innovation. These are essential components of a modern civilisation. EBCs and CMCs are now being added slowly to the parts of aerospace gas engines that are in hot zones (sections). The combustion tanks and shrouds are two of these parts. In future, it is likely that CMCs will also be used to make fans, blades, and afterburners

DOI: 10.1201/9781003731672-5

[2]. Today, Ni-based super-alloys are employed to make turbine parts, and a ZrO_2-7–8wt.% Y_2O_3 (YSZ) thermal barrier covering (TBC) protects them from high temperatures and environment. At elevated temperatures, gas turbine parts are alternatively changed as CMCs which are coated with EBCs that will protect from the environment. EBCs and CMCs are already changing the way gas engines are made and how well they work. They enable turbines to operate at higher combustion temperatures than present technology allows, boosting power [3]. These CMCs-based gas turbines will release less polluting gases than the present gas turbines, contrary to "conventional wisdom" that higher combustion temperatures indicate pollution. Finally, because CMCs have one-third the density of Ni-based superalloys, gas turbines with lower total mass and higher power levels will have thrust-to-weight ratios unsurpassed by present technology. Aerospace applications demand this [4].

5.2 DEVELOPMENT OF EBCS

As can be seen in Figure 5.1, EBCs can be broadly classified into distinct age bands based on the primary elemental composition being used in their development. In this section, we'll take a closer look at the various generations, comparing and contrasting their features and benefits.

Mullite was one of the first materials to garner attention due to its advantageous coefficient of thermal expansion (CTE) and chemical compatibility with CMC's ceramic matrix. This was due to the fact that mullite exhibited all three of these characteristics. Mullite, however, does not have a particularly lengthy lifespan in steam environments. Barium–strontium–aluminosilicate (BSAS) exhibits improved thermal cycle durability despite its low silica activity, good CTE match with SiC, and low modulus. Besides, the use of a Si bond coat is able to significantly increase the coating's ability to withstand repeated cycles. As a result, BSAS/BSAS EBCs + Si/mullite were utilised for fabricating the initial generation of SiCf/SiC CMC. EBCs with the capacity to withstand higher temperatures are required for the next generation of aeroengines [5–7]. Due to their stability in water vapour, high melting temperatures, and close CTE match with CMC, rare earth silicates were top contenders. Monosilicates of rare earth elements $\left(RE_2SiO_5\right)$ and disilicates of rare earth elements $\left(RE_2Si_2O_7\right)$ are the two possible classifications for rare earth silicates, where RE stands for rare earth elements. As a consequence of this, Si/mullite/RE_2SiO_5 and $RE_2Si_2O_7$ were decided to be the materials that would make up the second generation of EBCs.

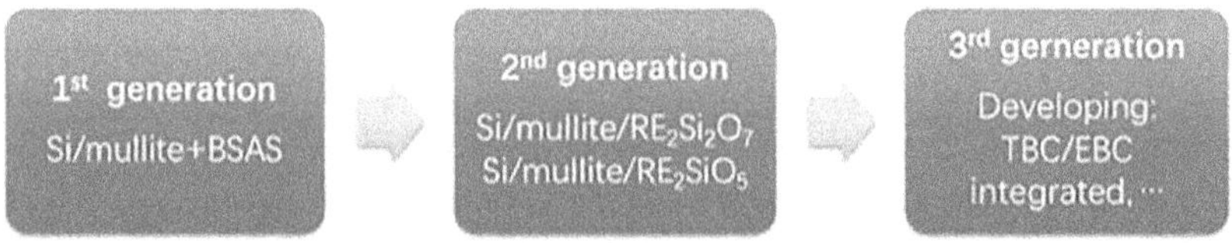

FIGURE 5.1 EBC development.

Aeronautical engines continue to see their working temperatures increase to produce ever-greater thrusts. This design concept offers thermal/environmental barrier coatings (T/EBCs) in combination with TBC coating as the next generation of environmental barrier coatings (EBCs), with the goal of facilitating the development and deployment of CMC components in future aero-engines. To further improve temperature resistance and safeguard EBCs and CMCs in a steam environment up to 1650 °C [8], a thermal barrier layer is incorporated into the structure.

5.3 INTEGRATION OF CMC/EBCS

Thermal barrier coatings (sometimes known as EBCs) are not the same as EBCs. TBCs are utilised primarily for the purpose of providing the superalloy component with protection. In comparison to CMC, the superalloy substrate possesses superior oxidation resistance. Insulating materials known as TBCs are being utilised in the production of superalloys to further expand the temperature range across which they may be used. TBCs often contain materials with coatings that have a poor thermal conductivity, which can lower the superalloy component's surface temperatures that have been coated with TBCs [9, 10]. In the past, the primary purpose of EBCs was to prevent corrosion of the CMC substrate caused by the presence of combustion gases in the environment [11, 12].

As a result of the continued growth and improvement of technology, a greater number of functions have been available, as depicted in Figure 5.2. The current state of the art in the design of EBCs involves the application of four distinct coating sublayers. The top coat is a specially formulated coating that can withstand high temperatures. After then, the thermal–mechanical energy is allowed to be released through the intermediate coat. Following that, the environment coat serves as a barrier against oxidation and corrosion. At the point where the coating meets the CMC, the bond coat works as a connection between the two parts. Because of these functions, the coatings and the base are able to work together more closely. At present, it's getting harder to combine the CMC substrate with the EBCs. The latest work is focused on improving the EBC technologies and making sure that the coatings work with the CMC base. Aside from this, there is a consistent push to create methods that can develop CMCs and EBCs jointly [13, 14].

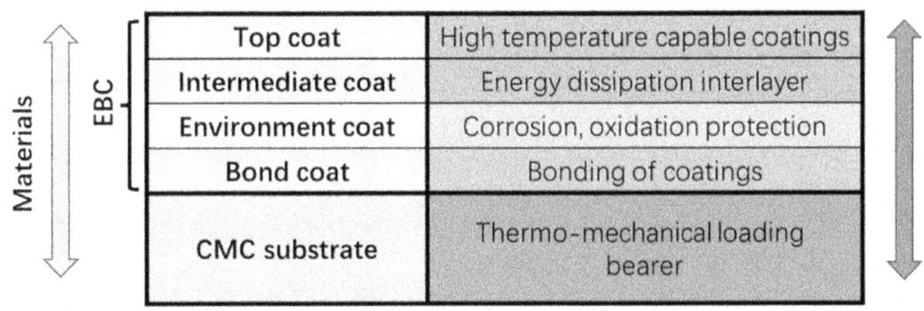

FIGURE 5.2 Materials and functions of a typical CMC/EBC system.

5.3.1 Failure Mechanism of EBCs

One must research the physical and chemical properties to design and optimise EBC coating materials and microstructure, damage modes, driving force, and other factors. As a result, most of the efforts that have been put into the study of EBC have been concentrated on the failure process. This pertains to the preparation of materials, the progression of damage, the mode of failure, and the prediction of lifespan. In spite of the many years of research that have been put into it, researchers have not made any significant headway in understanding how EBC coatings fail [15, 16]. Accordingly, the findings just indicate that the failure of EBC coatings is very certainly brought on by the following factors:

Thermal residual stress caused by a difference in the coefficient of thermal expansion (CTE) between the CMC substrate and the coating sublayers; coating and substrate defects, including vacancies, microcracks, and other flaws; layer of oxide; sintering and contraction of the coating; layer of oxide; wear and tear on the bonding layer; damages caused by mechanical forces, including delamination and spalling.

5.3.2 EBC Requirements

Both the environmental and chemical processes and the thermal and thermomechanical processes can be categorised as types of EBC degradation. The deterioration resulting from environmental factors and chemical agents encompasses three main types: water vapour-induced oxidation, water vapour-induced recession, and degradation that results from the accumulation of calcium-magnesium-aluminium-silicate (CMAS) deposits. Degradation caused by foreign object damage (FOD), particle erosion, and thermal and thermomechanical stresses are all included in thermal and thermomechanical degradation. Figure 5.3 presents a schematic representation of the

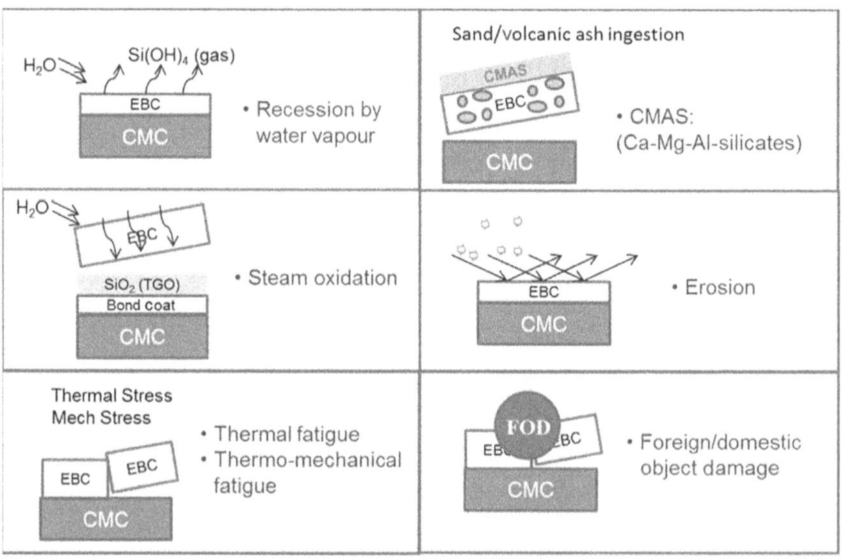

FIGURE 5.3 Diagrams depicting the most common causes of EBC failure.

various modes of EBC degradation and failure. When it comes to services, a number of different deterioration modes can interact with one another, which can lead to complicated EBC failure scenarios. The purpose of EBCs is to safeguard EBC/CMC systems against a variety of degrading mechanisms and, as a result, to extend the lifespan of EBC/CMC systems [17, 18].

To achieve the stated goal, it is imperative to fulfil the subsequent EBC prerequisites: (i) maintenance of environmental stability in H_2O and CMAS; (ii) minimal residual stresses; (iii) adherence to CMC; and (iv) chemical compatibility with CMC. To minimise residual stresses, it is imperative that the material exhibits a favourable thermal expansion that is similar to the CMC, sinter resistance, a low modulus, and phase stability. Often, EBCs fail because of stresses, and this is true irrespective of the type of degradation that occurs (thermal, chemical, environmental, or thermo-mechanical). The EBC stress, which is also referred to as EBC, exhibits three distinct components upon application to CMC substrates, as represented by the ensuing equation: the equation for rEBC is expressed as the sum of three stress components, namely thermal mismatch stress (rt), ageing stress (ra), and growth stress (rg). The thermal mismatch stress has been calculated and is shown [19, 18].

The thermal expansion coefficients for the CMC and EBC substrate are represented by aCMC and aEBC, correspondingly. Additionally, EEBC represents the Young's modulus of EBC, while mc denotes the Poisson's ratio of EBC. The Poisson's ratio of EBC is denoted by the symbol "mc". The thermal exposures of EBC result in alterations to its mechanical, physical, and chemical properties, which give rise to the phenomenon of ageing stress. This stress is a result of these modifications. Sintering, oxidation, chemical reactions, and phase transformations are some of the processes that contribute to these changes. Growth stress is a form of stress that becomes apparent during the deposition of EBC. Equation 5.1 does not take into account all of the stresses that can occur, such as the thermal shock stress that can be caused by a temperature differential.

A lucid and all-encompassing explication of every EBC stipulation pertaining to SiC / SiC CMCs is provided. Additionally, there is a discussion of the repercussions that each need has for EBC failure. The same line of thinking can be utilised when thinking about oxide/oxide CMCs. The silica volatility model established. It is able to express the reliance of EBC volatility on a number of different parameters, including the following: where $J\left(\mathrm{Si}\left(\mathrm{OH}\right)_4\right)$ represents the volatility of SiO_2, a(SiO_2) represents the silica activity of EBC, v represents the velocity of the gas, $P\left(H_2O\right)$ represents the pressure of the water vapour, and Ptotal represents the total pressure. Because of this, one should look for materials that have a reduced silica activity if they want water vapour stability [20].

Because of their natural propensity to catch fire, silica, mullite, and figured out with the assistance of the silica volatility model. The database of o withstand the impacts of water vapour the best, followed by mullite and then SiC. This statement pertains to the topics of kinetics, steam thermodynamics, and first principle calculations. The material known as BSAS was able tthermogravimetric analysis (TGA), which is also known as the thermogravimetric analysis, was utilised to ascertain the volatile nature of BSAS in addition to a number of different rare earth silicates [21].

The volatilities of rare earth disilicates such $Yb_2Si_2O_7$ and $Y_2Si_2O_7$ were comparable to those of BSAS, whereas the volatilities of rare earth monosilicates like

Yb_2SiO_5 and Y_2SiO_5 were significantly lower. In comparison to the silica activities exhibited by rare earth disilicates, rare earth monosilicates have significantly lower silica activities, which is approximately equivalent to a difference of two orders of magnitude. This is the reason why rare earth monosilicates have lower volatilities. The silicates with the highest volatility are mullite, BSAS, and rare earth disilicates. According to the following reactions, rare earth silicates are known to vapourise when exposed to water vapour.

When considering Yb as an illustration, Yb_2O_3 has the highest CTE (8.5 9 10–6/oC), followed by Yb_2SiO_5b (7–8 9 10–6/^{0}C) and $Yb_2Si_2O_7$ (* 4.6 9 10–6/^{0}C) (Ref 25). The creation of an RE monosilicate surface layer was corroborated by recession experiments conducted on RE disilicates in high-steam, high-velocity rig tests. Because of its high CTE, the RE monosilicate surface layer contributes to an increase in the EBC residual stress as described by Equation 2. The ability to resist oxidation Due to the fact that the permeability of H_2O in SiO_2 is approximately ten times higher than that of oxygen, water vapour is the principal oxidant for SiC and silicon in settings including burning.

Spallation is a commonly observed failure mode in EBC, primarily attributed to the expansion of TGO. This failure mode was discovered during laboratory testing and validated in rig and engine tests. The primary reason behind the failure of the EBC due to oxidation is believed to be the strain energy resulting from the residual stress generated by the TGO. CMAS Resistance The top use temperature of modern EBCs is limited by the degradation of EBCs caused by CMAS, which acts as a glass ceiling. CMAS deposits are formed when air-breathing turbine engines consume particles like sand, volcanic ash, and other siliceous material [5, 22]. This process takes place in the atmosphere. Ingestion of particles can lead to a variety of health problems. When temperatures rise over approximately 1230 °C (2246 °Fahrenheit), the melting and adhesion of CMAS onto the hot section components is observed. This leads to unfavourable chemical interactions with EBCs, which might produce an ageing stress. Furthermore, the liquid form of CMAS exhibits the ability to infiltrate porous EBCs, thereby elevating the modulus and consequently, the stress experienced by the EBC. The existing silicate-based EBCs do not offer adequate safeguard against the degradation of CMAS. The majority of the present EBC candidates, particularly oxides and silicates, demonstrates a significant inclination to engage in reciprocal reactions upon exposure to high temperatures. Chemical reactions may occur between layers of EBC in multilayer structures, resulting in alterations to their physical and mechanical properties. This can lead to an increase in residual stresses. Instances of interfaces that should be circumvented to prevent chemical reactions include the eutectic formation transpiring at the BSAS/SiO_2 interface (with a melting point of approximately 1300 °C), the $Y_2Si_2O_7$/-mullite interface (with a melting point of approximately 1400 °C), and the RE disilicate formation transpiring at the SiO_2/RE monosilicate interface. Adherence insufficient interfacial adherence between EBC and CMC results in accelerated growth of TGO, ultimately leading to a limited lifespan of EBC. The interfacial bonding between the EBC and CMC can be categorised into two distinct types: mechanical bonding and chemical bonding. The enhancement of mechanical bonding in CMC can be accomplished by employing various techniques to roughen the surface. These techniques include surface patterning, grit blasting, and chemical etching. The enhancement of chemical bonding can be attained by optimising the deposition process or selecting the coating chemistry with care. Various techniques, such as chemical vapour deposition and electron beam-physical vapour deposition (EB-PVD), have been proven to produce advantageous chemical bonding through the utilisation of elevated temperatures and/or chemical reactions [4]. A coating chemistry that inhibits CMC interdiffusion or reactivity may improve chemical bonding. Bind coatings bind refractory oxide or silicate EBCs to CMCs. Silicon is an ideal bond coat because it adheres well to EBC and CMC and matches CMC's thermal expansion qualities. Silicon also resists oxidation [23, 24].

When compared to other metals and metal alloys, silicon has the highest oxidation resistance in the temperature range most CMC applications require. CMC has a CTE between 4 and 5.3 times smaller than 10–6/0C, limiting the CTE-compatible candidate materials due to thermal expansion. Alumina, zirconia, and hafnia are inappropriate for EBCs since their CTEs are two times higher than SiCs. However, mullite, BSAS, and rare earth disilicates have a very tight CTE match with SiC, making them the most promising EBC options. The stability of phases in most cases, phase changes are accompanied by a change in volume, which, in turn, causes ageing stresses in the system. The mullite coating that is plasma-sprayed serves as a noteworthy illustration of an EBC that undergoes a detrimental phase transition. The conventional technique of plasma spraying mullite leads to the generation of a considerable quantity of amorphous phase. Subsequently, when subjected to a temperature of 1000 °C, the amorphous phase transforms into crystalline mullite [25].

The process of transition is concomitant with a notable reduction in volume, leading to significant fissuring and a limited lifespan of the EBC. In typical plasma spraying, BSAS and rare earth disilicates like $Y_2Si_2O_7$ and $Yb_2Si_2O_7$ create a glass in addition to metastable phases. In a rig test, it was reported that the phase change of $Y_2Si_2O_7$ had a negative effect. The section on EBC preparing by air plasma spray low modulus provides an overview of various techniques employed to prevent the formation of glass and metastable phases during the processing stage. EBC stresses are decreased when the modulus is low. The EBC modulus can be lowered by including pores in the structure. The formation of columnar pores is possible with the use of the processes of EB-PVD, PS-PVD (plasma spray physical vapour deposition), and SPS (suspension plasma spraying).

Plasma spraying enables the introduction of random pores in the material being processed by modifying the process parameters or applying a fugitive material that is subsequently eliminated through postprocess heat treatment. While plasma spraying with a fugitive substance is widely used in the fabrication of porous abradables, columnar structures produced by electron beam physical vapour deposition (EB-PVD) are frequently employed in the production of thermal barrier coatings for airfoils [4, 6].

However, interconnecting holes that make it easy for water vapour to diffuse through a material can reduce its oxidation resistance. For the EBC's performance to be at its very best, it is necessary to strike a precise balance between the management of stress and the mitigation of oxidation. The pursuit of an improved performing EBC remains a continuous endeavour for the next generation. Despite the recent accomplishments of rare earth silicates in safeguarding against environmental and molten corrosive species attacks, novel methods and pathways are constantly being explored. Notwithstanding the recent accomplishments that rare earth silicates have attained in safeguarding against environmental degradation and erosion by molten corrosive species, there persists a quest for enhanced performance [23, 26].

Figure 5.4 is a schematic representation of the construction of an EBC, which shares characteristics with TBCs. These hybrid TBC/EBC systems are designed to offer protection against the destructive effects of CMAS and steam.

Coatings are considered to be the next level of environmental protection. This section tries to present some of the most recent advancements in the subject of EBC, highlighting potential new possibilities that need to be further investigated by the industry before they can be adopted. In this part, different advancements are given, with the criterion of an EBC that displays a composition that goes beyond the single rare earth silicate that has already been explored serving as the unifying principle. As was mentioned earlier in this chapter, one of the fundamental roles of an effective EBC is to offer protection against the environment in which it will function while it is in operation. This is one of the primary tasks of an EBC. In light of the fact that the presence of steam has been shown to have a negative impact on SiC CMC components, this necessitates the application of a gas-tight coating that is able to restrict the passage of steam through to the substrate below [27, 28].

Cracks inside the coating are a preferred route for steam to travel to reach the substrate; as a result, it is abundantly evident that any cracks within the coating

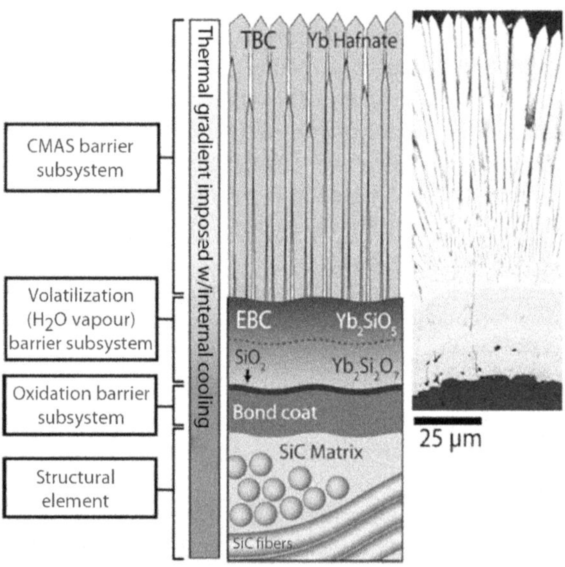

FIGURE 5.4 Schematic representation of the construction of an EBC.

should be avoided at all costs. However, it is possible that cracks will show up during operation as a result of the existence of a number of temperature cycles. This will lead to an accumulation of thermal strains, which will eventually lead to relaxation fractures appearing in vertical planes. Research has been done with the intention of incorporating self-healing capabilities into EBC systems in the hopes of lengthening the amount of time that these systems remain in service. The existence of patents pertaining to the topic and scholarly literature exploring diverse compositions and techniques, as indicated later, could be construed as indicative of the enthusiasm for a self-repairing EBC. Researchers provided one of the earliest examples of a self-healing EBC in their study.

The process accountable for self-healing in YbDS/YbMS +SiC systems is depicted schematically in Figure 5.5 The study includes visual representations that illustrate various aspects of the observed phenomena. Image (a) displays the cracked material, while image (b) demonstrates the formation of SiO_2 glass (indicated by dashed areas) resulting from the reaction between SiC nanoparticles (represented by black circles) and ingressed oxygen. In image (c), the crack is shown to be filled with viscous SiO_2 glass. Finally, image (d) depicts the healing of the crack through the reaction between SiO_2 glass and YbMS grains (depicted in light grey), leading to the formation of newly created YbDS (represented in dark grey). Redrawn from facilitates the ingress of oxygen from the ambient atmosphere into the material. During the annealing process, which occurs at a temperature of 1250 °C, the SiC nanofillers undergo a reaction with the oxygen present in the air, resulting in the formation of a dense SiO_2 glass.

The amorphous SiO_2 , possessing a high viscosity, exhibits the ability to occupy the fractures. Upon interaction with the pre-existing YbMS, the formation of YbDS takes place, which effectively seals the crack by virtue of the volume expansion that arises from the reaction. The phenomenon of crack self-healing was observed in

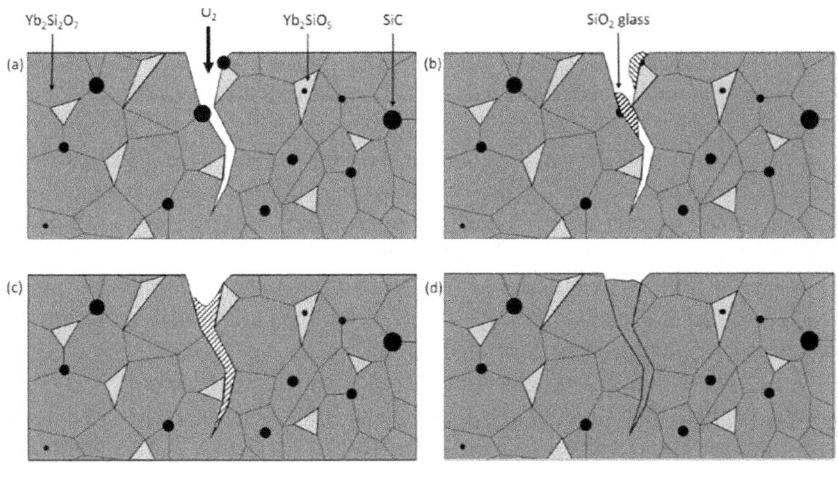

FIGURE 5.5 Process responsible for self-healing in YbDS/YbMS +SiC systems.

sintered bodies composed of yttrium monosilicate, disilicate, and a combination of both, along with 5% SiC nanoparticles. These bodies were subjected to temperatures ranging from 1000 °C to 1300 °C for durations of 1–24 hours in an air environment [4, 29]. This approach was recently documented in a previous research paper. The present investigation showcases the phenomenon of crack self-repair in sintered bodies. Their findings, which account for the investigation of this impact at a variety of temperatures and exposure times, contribute to a more complete comprehension of the self-healing process. Furthermore, the authors acknowledge the importance of ion diffusion as a potential explanation for the phenomenon of crack-healing. It is worth noting that this particular mechanism had not been previously investigated or addressed in existing literature. Notwithstanding the promising findings that have been published, there are several factors that warrant careful consideration. The researchers have observed that the self-healing capabilities are restricted to a single annealing operation due to the consumption of both the SiC nanofillers and the YbMS phase. This is because both of these components are needed for the self-healing process [30, 31].

Then there would not be a sufficient amount of SiC and YbMS present for the self-healing mechanism to activate in the event that a crack appeared. Second, sintered bodies were used to do the tests. It is a depiction of EBC that is less than perfect when compared to the processes that are now preferred by the industry. This is because the sintered body has a lower porosity. The predominant occurrence of the reaction between SiC nanofillers and oxygen primarily transpires at the cracks within the material, as opposed to being distributed uniformly throughout the entirety of the substance. Coatings applied using thermal spraying will have increased degrees of porosity, which may provide an extra source for reaction centres. This higher development of YbDS could generate excessive compressive levels in the coating, which could ultimately lead to failure. The volume expansion that is related with this increased synthesis of YbDS is also likely. Second, the annealing temperature that was chosen, which is 1250 °C, has a strong self-healing effect; nevertheless, this temperature is lower than the service temperature at which EBCs are anticipated to operate, which is approximately 1500 °C. It is important to analyse whether or not an increase in temperature would cause a change in the kinetics of the self-healing mechanism that has been described, as this is something that needs to be taken into consideration [29, 31].

To properly account for the dynamics of the situation, the next stage would need to involve the introduction of steam into the testing environment. Based on research findings, it has been observed that SiO_2 has the ability to undergo a reaction with steam at relatively low temperatures of 1200 °C. This reaction leads to the production of gaseous $Si(OH)_4$. The efficacy of the self-healing mechanism may be impeded in the event that the reaction kinetics at elevated temperatures surpass those of the reaction between amorphous SiO_2 and YbMS, which results in the production of YbDS and facilitates fracture healing. Researchers have developed YbDS APS coatings. Mullite, Al_2O_3, $Y_3Al_5O_{12}$ (YAG), and TiO_2 (less than 5 wt.%) improved these coatings. These additives slowed thermally grown oxide (TGO) growth at the YbDS-Si interface. This followed the plan of adding performance-proven additives to tried-and-true EBC formulas. The correlation between the TGO's growth rate and the shortcomings of EBC systems has rendered the prospect of a solution that mitigates the issue highly attractive. After conducting steam cyclic testing under specific conditions (90% H_2O /10% O_2 environment, flow velocity of 10 cm/s, 1 atm pressure, 20 minutes at 100 °C and 60 minutes at 1316 °C), it is noted that Al_2O_3 or Al_2O_3 -containing compounds reduced TGO thickness by 80% relative to the unmodified YbDS/Si baseline [9, 10].

Even though the author notes that there is insufficient trial evidence to properly obtain the nature of this occurrence, it is hypothesised that the additions cause a favourable outcome not by affecting the oxidiser permeability of the YbDS. This is despite the fact that the author makes this observation. On the other hand, the slower rate of TGO growth could be explained by a change in the SiO_2 network that exists inside the TGO itself.

This change would effectively obstruct the path that oxidisers take to reach the Si bond layer that lies beneath the TGO. This work gives a fresh and exciting path to pursue in the creation of the next generation of EBCs. The tests were performed on APS-applied coatings, and the baseline used was an unmodified EBC system that has been widely demonstrated. Moreover, owing to the minimal concentration of the recent inclusions, it is rational to expect that the influence on the characteristics of the EBC will be negligible. The steam cyclic test proved this theory, so this approach may be a reliable and easy way to make EBCs last longer [32]. The work that researchers have done on the study of something called thermal environmental barrier coatings (TEBCs) stands out. These coatings are a mix of the ideas behind TBCs and EBCs. To achieve the aforementioned objective, the fundamental prerequisites for the development of TEBCs were identified as follows: attainment of gas-tight shielding, which was realised by ensuring a robust coefficient of thermal expansion (CTE) alignment with the substrate; possession of elevated temperature capability or phase stability; possession of low thermal conductivity; and resistance to CMAS attack.

Both experimentation and theory-based measurement were carried out so that a comprehensive screening could be carried out to select the most promising possible candidate for usage as a TEBC. The selection of the initial rare earth silicates was based on two key criteria: coefficient of thermal expansion (CTE) match and high temperature phase stability. This led to the selection of -$Yb_2Si_2O_7$, $-Sc_2Si_2O_7$, and $-Lu_2Si_2O_7$. CTE match was selected as the first criterion to be used since it allowed for the selection of the initial rare earth silicates.

Prior to investigating the thermal conductivity and CMAS resistance of possible TEBCs, it was imperative to consider the feasibility of generating complete solid solutions that incorporate the targeted rare earth dopants, namely Y, Yb, Sc, and Lu. The study involved conducting comprehensive theoretical computations to determine the optimal blend of solid solution and rare earth silicates, as well as the optimal doping concentration for the material. The authors came to the conclusion that solid solution alloys of $YxYb(2-x)Si_2O_7$ would be able to satisfy the requirements that were outlined while simultaneously enhancing the existing thermal capabilities of EBCs. Conducting additional research to confirm the theoretical calculations for specific compositions in the field of EBCs has the potential to open up new avenues of investigation, leading to significant benefits.

5.3.3 EBC Processing and Manufacturing via Air Plasma Spray (APS)

First Attempts at Success Using the APS Method The deposition of EBCs has traditionally been accomplished through the use of the processing technique known as thermal spraying, more specifically air plasma spray (APS). Since the 1980s, TBCs have been successfully applied in gas turbines by means of APS, which is one of the primary reasons why this is the case. The market is aware that this method is dependable and that it may be carried out in a variety of locations across the globe. There is already in place a robust scientific community in addition to well-functioning industrial supply chains that include suppliers of equipment and feedstock as well as service centres [33, 34].

5.3.4 Processing and Manufacturing of EBCs Sprayed on SiC/SiC CMCs via APS

Although heat spraying has been the production process of choice to deposit EBCs, there is a significant obstacle that must be overcome whenever thermal spraying is performed on SiC/SiC CMC substrates. Typically, the stable phases of EBC materials for SiC/SiC CMC substrates, such as BSAS, mullite, and RE-silicates, are observed within limited regions (ranges) of their respective phase diagrams. The powdered form of these raw components, when subjected to thermal spraying, undergoes complete or partial melting during the duration that they are allowed to dwell in the thermal spray jet. These particles that have been thermally sprayed materials undergo high rates of solidification as soon as they reach the surface of the substrate. In addition to that, these silicates have a propensity to produce phases of a glassy consistency. In most cases, the combination of these two occurrences will result in the production of metastable phases that are either completely or partially amorphous. These amorphous and metastable phases (coatings) should be avoided since they crystallise when subjected to high temperatures, which is the temperature range in which they are intended to function [35].

At high temperatures, the process of crystallisation causes phase transformations, which in turn result in volume changes and the stresses that are associated with those changes. Furthermore, it is common for amorphous and metastable phases to display coefficients of thermal expansion (CTE) that are significantly different from those of their stable phases and CMC substrates. The coating and substrate system is subject to additional stress as a result of the mismatching of the CTE values. Because ceramic materials have a poor level of flexibility, they are unable to accommodate these levels of stress, which ultimately leads to the formation of crack networking. As per the findings of the research report, the crack networking advancements have resulted in the loss of gas-tight properties of EBCs, which can lead to potential spalling off of the substrate [36].

Before EBCs are put into service, they undergo a "robust" stabilisation process. Processing through APS in its conventional form, with post-spray heat treatment done in the regular way that used no aided heating when they deposited a YbDS-based powder on SiC substrates that had previously been covered with a Si bond. As a result of the temperature levels of the substrate being lower than those of the temperatures at which Yb-silicates crystallise, the as-sprayed coating was completely amorphous. Following the application of the coating, a standard post-spray heat treatment in air was carried out. This involved maintaining a temperature of 1200 °C for 25 hours [36].

After going through the heat treatment, the Yb-silicate EBC turned out to be extremely crystalline (a combination of YbDS and YbMS). The spallation of the coating does not seem to have been induced by the stresses resulting from phase transitions and CTE disparities. The controlled processing and Bespoke post-spray heat treatment approach was developed through the utilisation of a specific technique. The entirety of the processing can mostly be broken down into two primary components. Prior to spraying a $Yb_2Si_2O_7$ feedstock, it is recommended to optimise the chemistry and plasma power. This involves adjusting the combination of plasma gases and their respective flows to reduce the volatilisation of SiO_2. The objective is to enhance the occurrence of the $Yb_2Si_2O_7$ phase within the as-sprayed coating, which may exist in either the amorphous or crystalline form.

The coatings are applied through the use of APS in a manner that is consistent with the traditional process for spraying YSZ TBCs. This means that there is no requirement to subject the substrate to high temperatures within a furnace and that the sample temperatures during spraying are typical for standard APS processing. The substrate does not require high temperature treatment within a furnace, in essence. That is to say, there is no requirement to place the substrate into a furnace and heat it to a very high temperature. The second objective of this study is to illustrate the process of conducting a heat treatment procedure on the as-sprayed EBCs in an air environment. This procedure is aimed at gradually crystallising the EBCs before they are put into service. This is done to ensure that the EBCs are of the highest quality possible. The personalised thermal treatment is produced on the basis of the precise (i) CTE values and (ii) crystallisation temperatures of the various phases of the EBCs, with the state in which they are as-sprayed and continuing all the way through the state in which they are fully crystallised. It consists of five heating ramps and five plateau temperatures, each of which is greater than the one that came before it to achieve a maximum temperature of 1300 °C [37].

5.3.5 Processing and Manufacturing of EBCs Sprayed on Oxide-Oxide CMCs

Appropriate materials for EBC need to have a high water vapour stability and CTE values that are comparable to those of the CMC substrate. In addition to this, it is essential that the material be chemically compatible with the substrate, have resistance to CMAS assault, and maintain phase stability at high temperatures. Low heat conductivity is another feature that is much sought after.

Last but not least, strong erosion resistance is another criterion that must be met by any coating that is going to be used in the hot area of turbines. In light of these observations, it is suggested that EBCs may be constructed out of yttria (Y_2O_3). Gadolinium zirconate ($Gd_2Zr_2O_7$) and yttrium-aluminum garnet $(Y_3Al_5O_{12})$ were also among the materials that is suggested using in addition to yttrium oxide (Y_2O_3). It is proposed yttrium aluminum perovskite $(YAlO_3)$. It also took into account $Gd_2Zr_2O_7$ in their research. According to research hypothesis, yttrium-aluminum-garnet $(Y_3Al_5O_{12} - YAG)$ is still another possibility that could be considered.

Traditional APS was used to deposit each of these coatings; therefore, aided heating was not utilised in the process. To this point, it would appear that APS Y_2O_3 is one of the most "promising" candidates for use as an EBC for oxide-oxide CMCs. This is based on our current knowledge. The melting point of Y_2O_3 is 2425 °C, making it a material that is both chemically and thermodynamically stable [38].

It maintains its cubic structure (a-Y_2O_3) up until it reaches a temperature of 2300 °C, at which point it changes into a hexagonal structure. However, this temperature is higher than the threshold at which oxide-oxide CMCs can function properly. In addition, the CTE of Y_2O_3 is 99×10^{-6}/°C; this is a property that is comparable to that of Al_2O_3 of oxide-oxide CMCs (which is 89×10^{-6}/°C). Its resistance to an attack by water vapour is significantly higher than that of Al_2O_3 . Last but not least, unlike EBCs for SiC/SiC CMCs, it does not have a propensity to produce amorphous phases during the spraying process [39]. The presence of a metastable monoclinic Y_2O_3 (b-Y_2O_3) as a minor secondary phase in the as-sprayed coating was discovered by the researcher. This was found within a primary crystalline a-Y_2O_3 phase and was attributed to the rapid solidification rate of the sprayed particles. A rapid solidification rate of the sprayed particles was observed. After undergoing the heat treatment at 1200 °C for one hour, this phase totally changed into a-Y_2O_3 .

The deposition of Y_2O_3 EBC onto Al_2O_3 / Al_2O_3 CMCs, which are composed of thick alumina fibres embedded in a porous alumina matrix, is a crucial point to mention. To enhance the binding strength of the APS Y_2O_3 EBC on the porous CMC, a reaction-bonded Al_2O_3 (RBAO) bond coat was applied on top of the CMC. This was done due to the complex structure of the materials involved. The thickness of the RBAO bond coat was 100 lm. The CMC was coated with the RBAO bond coat using a slurry application method. The coated CMC was then heat treated in air at a temperature of 1300 °C for a duration of one hour. Subsequently, the EBC/CMC system underwent a boiler cycle test (FCT) in air to evaluate its performance. Each cycle included 50 minutes at 1200 °C, followed by 10 minutes of cooling to room temperature (RT) by pressurised air jets. After 500 cycles, the FCT valuation was halted since there was no macroscopic evidence of EBC spallation, as validated by the SEM analysis. This led to the decision to discontinue the FCT valuation. Because of the strong chemical bonding that was found between the Y_2O_3 EBC and the RBAO bond coat (which resulted in the creation of yttrium aluminates), this system was able to withstand high levels of thermal cycling without suffering significant damage. In spite of these encouraging findings, the protection afforded to the CMC by the EBC needs to be thermally cycled in a water vapour environment, as will be covered in the following section of this work. Only then can a more accurate assessment be made of the efficacy of this protection [40].

It should be emphasised that the volatility of Al_2O_3 via $Al(OH)_3$ gas generation in a high-temperature water vapour environment is not predicted to be an issue below a temperature level of 1300 °C, even for applications that last for a long period of time. This observation requires emphasis as it is deemed crucial. It can be inferred that the EBCs designed for oxide-oxide CMCs primarily serve to safeguard against CMAS attack, high temperatures, and erosion, given that porous oxide-oxide CMCs are restricted from operating beyond 1200 °C. Porous oxide-oxide CMCs are designed to operate within a temperature range not exceeding 1200 °C. In other words, the defence mechanism against water vapour attack is expected to have minimal significance. In point of fact, it is examined the performance of porous Al_2O_3 / Al_2O_3 CMCs in an environment characterised by a thermal gradient. During laser-rig cycling, surface fibre buckling and delamination were detected on the surface of the uncoated CMC. As a result of this, it has been hypothesised that an EBC would be able to shield the CMC from the damaging effects of high temperatures.

5.3.6 Processing and Manufacturing of EBCs by Other Thermal Spraying Techniques

The present chapter does not cover the description or elaboration of other thermal spraying techniques utilised for the deposition of EBCs, as it falls outside the scope of this work. In a nutshell, the production of EBCs was accomplished by the process of plasma spray physical vapour deposition (PS-PVD). PS-PVD is an intermediary technology that combines plasma spray and vapour phase deposition processes. This method of deposition does not require a direct line of sight, which is one of the many reasons why it has such promising potential. It is possible to construct coating microstructures that are dense or columnar. The preliminary findings appear to be encouraging; specifically, an isothermal exposure of PSPVD $Yb_2Si_2O_7$ EBCs to water vapour at a temperature of 1316 °C for 500 hours indicates negligible crystallographic change.

High velocity oxy-fuel, also known as HVOF, was utilised to engineer $Yb_2Si_2O_7$ EBCs. Despite the fact that the as-sprayed coatings generated by the HVOF were partially crystalline, one of the great potentials of the HVOF is related to the fact that it produced thick EBCs that were free of vertical cracks. The production of $Yb_2Si_2O_7$ -based EBCs is being pursued through the utilisation of suspension plasma spray (SPS) and very low-pressure plasma spraying (VLPPS) techniques [23]. The SPS EBC was amorphous for the most part, although XRD analysis revealed the existence of Yb_2SiO_5 in both its stable and metastable phases, in addition to Yb_2O_3 crystals. Crystalline structure was observed in the VLPPS-based EBCs, indicating the predominate presence of $Yb_2Si_2O_7$ as the major phase. The high temperatures that were reached during spraying (980 °C), as well as the regulated gradual cooling that occurred after deposition, were thought to have contributed to the VLPPS result. In research, it is also investigated a method known as the solution precursor plasma spray procedure (SPPS). This method resulted in the production of very crystalline EBCs in their as-sprayed state.

5.4 FOREIGN OBJECT DAMAGE (FOD) AND EROSION OF EBCS

Gas turbine engines, particularly those used in aviation, are frequently subjected to the intrusion of foreign materials such as dust, ash from volcanic eruptions, sand, hailstones, sleet pellets, debris from the runway and/or gravel, and even birds.

The force of these things colliding with the turbine components has the potential to cause damage to the EBC, which in turn can lead to the removal of all or almost all of the EBC. Eliminating the EBC can lead to two potential consequences. First, it may cause a reduction in the turbine's aerodynamic performance due to the presence of rough surfaces. Second, it may result in a catastrophic structural failure due to rapid recession or corrosion of unprotected components in the presence of water vapour. Both of these problems can be prevented. A study has been conducted on the foreign object damage (FOD) or impact resistance of SiC/SiC CMCs coated with APS EBCs. FOD can be caused by material that is internally generated or debris ingested from the outside. The APS EBC was made up of three layers: (i) an Si bond coat, (ii) a mullite-BSAS combination interlayer, and (iii) a BSAS top coat. Each of these layers had a specific function. (i) 2251m and (ii) 525 lm are two completely different EBC thickness levels that were created. In the course of conducting impact tests, a gas cannon was employed to launch a solitary, compact chrome-steel sphere (HRCC60) with a diameter of 1.59 mm and a density of 7.8 g/cm^3 onto the surface of the specimen at a perpendicular angle, with velocities spanning 110–400 metres per second. The density of the sphere was 7.8 g/cm^3. The impact tests were carried out in air at room temperature and 1316 °C. After being struck by an object, samples were afterwards subjected to an exposure at a high temperature (1316 °C for 500 hours) in the presence of water vapour (90% H_2O + 10% O_2). At the impact sites, EBC spallation was observed in addition to the formation of fractures and internal oxidation of the CMC underneath the impacted sites. On the other hand, there was no silica growth observed in the areas that were not harmed. This observation demonstrates how essential it is to protect the EBC from the damaging effects of big particles and to ensure that it remains intact. In addition to this, the findings imply that "thin" EBCs are not an efficient means of protecting substrates from damage caused by big particles. The levels of erosion resistance offered by APS EBCs that had been deposited on SiC-SiC CMCs were evaluated.

There was no disclosure regarding the make-up of the APS EBC levels. The erodent material was composed of silica particles that were both crushed and fused, with an average size of 50 micrometres. The erosion experiment was conducted within a wind tunnel, where a range of angles were tested under varying conditions of 1038 °C and 225 m/s. It is reported that upon comparing the erosion rate values of EBCs and TBCs under similar test conditions, the mean erosion rate of an atmospheric plasma sprayed (APS) EBC was notably lower than that of an APS TBC (approximately 1/6th) and was found to be similar to that of an electron beam-physical vapour deposition (EB-PVD) TBC. The deduction is the result of a comparative analysis between the erosion rate measurements of Erosion-Resistant Bond Coats (EBCs) and TBCs. The reliance on the impact angle is likewise comparable to that of TBCs; specifically, the erosion rate reduces for ceramic-based materials as the impingement angle moves further away from 90 degrees. The initial findings indicate that the use of APS EBCs on gas turbine parts has a favourable potential in mitigating erosion caused by small particles [42].

5.5 CMAS ATTACK ON EBCS

Apart from erosion, the elimination of EBC can also take place through the CMAS attack of calcium-magnesium alumino-silicate. The phenomenon known as CMAS arises from the ingestion of foreign particles, including dust, airborne or runway sand, and volcanic ash, with the intake air. These particles undergo complete or partial melting in the combustion chamber before being deposited onto the EBC-coated components of the engine. Foreign particulates refer to various types of particles that are not typically present in a given environment. Examples of foreign particulates include dust, airborne or runway sand, and volcanic ash. The acronym CMAS pertains to the typical make-up of these extraneous particles, which comprise a blend of CaO, MgO, Al_2O_3, and SiO_2, as stated in reference [43]. In technical terms, the combustion chamber functions as a flame-spray, while the foreign particles act as powder feedstocks. Additionally, the EBC-coated components of the engine serve as the substrate. When the combustion temperatures in the area surrounding the coated parts reach or surpass 1200 °C, which is the typical range for the melting point of CMAS materials, a crystalline crust will form on the EBCs. When temperatures drop below this degree, EBC erosion is more likely to take place.

The threat posed by CMAS should not be taken lightly by EBCs. As was said earlier, one of the primary driving causes behind the rise of EBCs and CMCs in the gas turbine sector is the continual push to raise the working temperature of gas turbines. This is one of the important driving forces behind the rise.

Damage to EBCs caused by CMAS can be caused either (i) mechanically or (ii) chemically. It is a mechanical process that happens when the molten CMAS seeps into the open pores and/or open vertical fissures in the EBC. Because of the densification of the EBC in the region that was penetrated (i.e., the pores and cracks were filled with CMAS), there was an increase in the zone's stiffness (also known as its elastic modulus). The heating and cooling stages result in the development of extra stresses in the area between the dense zone penetrated by CMAS and the underlying porous zone that remains non-penetrated. Spallation occurs when the stress levels exceed the fracture toughness of the EBC. The exposed non-penetrated porous zone results in increased CMAS penetration and EBC spallation.

The variation in Coefficient of Thermal Expansion (CTE) values between the Coating, Manufacturing, and Assembly System (CMAS) and the EBC can cause stress, which may result in the EBC cracking and spalling during thermal cycling. These discrepancies can be found during the cycles.

Manufacturing that is gas-tight, despite being dense or porous, vertically crack-free, and dense can be one viable solution to this type of challenge. The chemical damage caused by CMAS to EBCs is caused by chemical reactions that are thermally triggered and take place between the glassy CMAS deposits and the EBCs. These chemical processes can take place at high temperatures and cause the production of new phases on the EBC.

These new phases may have different levels of CTE and/or stiffness, and they may even generate phases with lower melting points than those of the original EBC. Because of all of these transformations and interactions, the C/CMC systems are subjected to extra stresses, which ultimately results in EBC spallation. Even by creating EBCs that are vertically crack-free, dense or porous yet gas-tight, it is not possible to simply solve this type of problem since the processes that are occurring are of a chemical nature and are initiated by heat. At a temperature of 1500 °C, for instance, it is investigated the MAS assault on RE-silicate ($Yb_2Si_2O_7$ and Yb_2SiO_5) based EBCs. The findings indicate that the CMAS exhibited a preferential penetration towards the grain boundaries of the EBCs, leading to the dissolution of the EBC material and the formation of pockets with lower melting points at the grain boundaries. This was executed in a visible manner. After being tested over a period of 200 hours, these low-melting were found to extend through the entirety of the samples' thicknesses. The CMAS attack at 1300 °C was investigated YbMS- and YbDS-based APS EBCs, which were created and analysed [44].

The samples were kept at this temperature for a range of intervals, anything from 53 hours to 200 hours. Both the YbMS and YbDS topcoats were dissolved when they were subjected to CMAS after being coated with any of the two coatings. The CMAS attack mechanism as a whole was described as the dissolution of the silicate and its subsequent re-precipitation as apatite. However, the reaction morphology was quite different for YbMS and YbDS, which was an important distinction. In situations where the reaction rate is not governed by the accessibility of silicate deposits, it has been determined that the interaction between CMAS and 100 lm-thick YbMS or YbDS top coats is considerably severe at 1300 °C, posing a significant threat to the intended operational lifespan of SiC composites, which ranges from 5,000 hours to 10,000 hours. The aforementioned pattern was identified in the study carried out. At a temperature of 1300 °C, the interaction between CMAS and $Yb_2Si_2O_7$-based EBCs caused alterations in the lattice spacing, fluctuations in the ambient temperature stresses, and notable fractures in the CMAS that penetrate the topcoat. Additionally, it can be suggested that the most suitable function for a substance derived from $Yb_2Si_2O_7$ is to act as an EBC which is subsequently layered with an oxide TBC. The proposed measure aims to restrict the CMAS interaction, thereby protecting the EBC from potential degradation resulting from such interactions. The notion of a thermal-environmental barrier coating (TEBC) may prove indispensable in addressing the issues linked to CMAS degradation, given the aforementioned factors. Specifically, an EBC system necessitates a coating of ceramic oxide that is resistant to CMAS to mitigate this problem. On the basis of this concept, it is investigated research into the possibility of employing a-alumina as a CMAS-resistance. The reaction between 99.9% pure a-alumina and volcanic ash was studied using differential thermal analysis (DTA) and x-ray diffraction (XRD) at temperatures ranging from room temperature (RT) to 1480 °C.

A comparative analysis of differential thermal analysis (DTA) was conducted using distinct specimens of (i) volcanic ash and (ii) a-alumina volcanic ash. According to the findings of the DTA conducted on the volcanic ash, it was observed that the temperature at which it melted was 1235 °C. Despite being subjected to a temperature of 1480 °C, the DTA pattern of the combination of a-alumina and volcanic ash did not display any significant endothermic valleys that could be associated with the mixture's melting temperature. Conversely, the onset of an endothermic region was observed when the temperature reached 1310 °C. Thus, despite the partial onset of melting at 1310 °C, the temperature was still higher than the initial melting point of volcanic ash, which was recorded at 1235 °C. According to the findings of XRD, the reaction between alumina and volcanic ash took place at a temperature of 1310 °C, producing the reactive phases anorthite $\left(CaAl_2Si_2O_8\right)$, magnetite $\left(Fe_3O_4\right)$, and spinel (Al1.75Mg0.889Mn0.351O4) at this temperature. Magnetite has a melting temperature of 1538 °C, while anorthite has a melting point of 1553 °C, and spinel has a melting point of 2135 °C. Based on research results, it has been observed that the current melting temperatures surpass the standard surface-operating temperature levels of turbine components. As a result, the penetration of volcanic ash and CMAS caused by melting can be considerably decreased. The reason for this is that the current melting temperatures exceed the standard surface-operating temperature levels of turbine components. It is hypothesised that a layer of thermally sprayed alumina might perhaps act as the CMAS-resistant top coat for TEBC systems, such as the ones developed to protect SiC/SiC and ox-ox CMCs. This is because of the reasons stated in the previous sentence. However, it is imperative to consider additional aspects. The challenges encountered in thermal spraying of EBC compositions are primarily attributed to three factors. First, there is a mismatch in the coefficient of thermal expansion (CTE) between alumina (8.9 × 10–6/oC) and EBC compositions. Second, there is a tendency for metastable alumina phases (c-Al_2O_3) to form during the thermal spraying process. Finally, the resistance of alumina, anorthite, magnetite, and spinel phases to water vapour attack is also a concern. Gadolinium zirconate ($Gd_2Zr_2O_7$) is a well-known material that has been specifically designed for use as a TBC. It has been found to exhibit exceptional resistance to the penetration of CMAS deposits. As a consequence of this, it may also be considered a possibility for use as a top coat that is resistant to CMAS. However, this material's high CTE level would make its employment in TEBC structures more difficult. CMAS assault resistance continues to be one of the most crucial issues in EBC research and development for all of the reasons that are highlighted in this chapter.

5.6 RECENT SUCCESSES OF EBC/CMC APPLICATION IN JET TURBINES

It is important to note that commercial jet turbines have begun using EBCs and CMCs in key components of the hot section (such as the combustion chamber, nozzles, and shrouds). This is something that should not be overlooked. The successful research and development efforts of EBC/CMC have enabled the incorporation of CMC high-pressure turbine shrouds into the LEAP engine, which is a collaborative project between GE Aviation and Safran Aircraft Engines. This development has paved the way for the commercial launch of the product. In 2019, the CMC shrouds have accumulated over four million hours of flight time in commercial LEAP engines that are utilised in aircraft produced by Airbus (A320neo), Boeing (737 MAX), and COMAC (C919). The General Electric (GE) GE9X engine incorporates CMC inner and outer combustor liners, high pressure torque (HPT) stage 1 and stage 2 nozzles, and high pressure torque (HPT) stage 1 shrouds. These components are safeguarded by EBCs. In 2017, the GE Aviation-designed GE9X turbine engine, intended for use in Boeing's new 777X aircraft, achieved a thrust of 60,000 kgf (134,000 lbf) during a test run.

EBC/CMC components for this engine are capable of operating at temperatures close to 1300 °C when they are at their most efficient. At the same time, they require less air for cooling and let more air into the mixer (i.e., the combustion chamber). By facilitating a cleaner burn, the increased air in the mixer contributes to a decrease in emissions. In addition to this, the GE9X demonstrates a 10% improvement in its fuel consumption when contrasted with its preceding engine, the GE90–115B.

5.7 CONCLUSIONS

Mullite was one of the first materials to gain attention owing to its favorable coefficient of thermal expansion (CTE) and chemical compatibility with the ceramic matrix of CMC. Despite its low silica activity, good CTE match with SiC, and low modulus, barium-strontium-aluminosilicate (BSAS) demonstrates enhanced heat cycle durability. RE denotes rare earth elements. As a result, Si/mullite/RE_2SiO_5 and RE_2SiO_7 were chosen as the materials that comprise the second generation of EBCs. A thermal barrier layer is included into the framework to increase temperature resistance and protect EBCs and CMCs in a steam environment up to 16500 °C.

Thermal barrier coatings (sometimes referred to as EBCs) are not the same as EBCs. TBCs most recent work has been focused on refining EBC technology and ensuring that the coatings function with the CMC base. To develop and optimise EBC coating materials and microstructures, damage mechanisms, driving force, and other parameters, physical and chemical qualities must be researched. Damage induced by mechanical forces, such as delamination and spalling, to the bonding layer.

Environmental and chemical processes, as well as thermal and thermomechanical processes, are examples of EBC deterioration. EBCs are designed to protect EBC/CMC systems against several degrading factors, hence extending the lifespan of EBC/CMC systems. As a result, if one wants water vapour stability, one should search for materials with low silica activity; the substance known as BSAS was capable of thermogravimetric analysis (TGA). Because the permeability of H_2O in SiO_2 is roughly ten times that of oxygen, water vapour is the primary oxidant for SiC and silicon under conditions such as burning. The traditional method of plasma-spraying mullite generates a significant amount of amorphous phase. After that, when heated to 1000 °C, the amorphous phase turns into crystalline mullite.

First Success Attempts APS Method Application EBC deposition has typically been performed with the use of the thermal spraying technology, especially air plasma spray (APS). To obtain a maximum temperature of 1300 °C, it consists of five heating ramps and five plateau temperatures, each of which is higher than the one before it.

EBC materials must have strong water vapour stability and CTE characteristics equivalent to the CMC substrate. As a result, it has been proposed that an EBC can protect the CMC from the harmful effects of high temperatures.

An approach known as the solution precursor plasma spray process (SPPS) is also being studied in research. As a result of this approach, highly crystalline EBCs were produced in their as-sprayed condition.

The first data suggests that using APS EBCs on gas turbine components has a high potential for reducing erosion due by tiny particles. These new phases may differ in CTE and/or stiffness, and they may even produce phases with lower melting points than the original EBC. All of these transformations and interactions put extra strain on the C/CMC systems, resulting in EBC spallation.

Finally, it is critical to note that the Boeing 777X, which was outfitted with GE9X engines, took to the skies for the first time on January 25, 2020 [45].

REFERENCES

[1] Kumar, S., and N. P. Padture. "Materials in the Aircraft Industry." *Metallurgical Design and Industry: Prehistory to the Space Age* (2018): 271–346.

[2] Gildersleeve, Edward J., and Robert Vaßen. "Thermally Sprayed Functional Coatings and Multilayers: A Selection of Historical Applications and Potential Pathways for Future Innovation." *Journal of Thermal Spray Technology* (2023): 1–40.

[3] Summers, William Daniel. "Degradation of Rare-Earth Silicate Environmental Barrier Coatings." PhD diss., University of California, Santa Barbara, 2019.

[4] Lee, Kang N., Dongming Zhu, and Rogerio S. Lima. "Perspectives on Environmental Barrier Coatings (EBCs) Manufactured via Air Plasma Spray (APS) on Ceramic Matrix Composites (CMCs): A Tutorial Paper." *Journal of Thermal Spray Technology* 30 (2021): 40–58.

[5] Fang, Guangwu, Xiguang Gao, and Yingdong Song. "A Review on Ceramic matrix Composites and Environmental Barrier Coatings for Aero-Engine: Material Development and Failure Analysis." *Coatings* 13, no. 2 (2023): 357.

[6] Suzuki, Masato, Mohammed Shahien, Kentaro Shinoda, and Jun Akedo. "The Current Status of Environmental Barrier Coatings and Future Direction of Thermal Spray Process." *Materials Transactions* 63, no. 8 (2022): 1101–1111.

[7] Binner, Jon, Matt Porter, Ben Baker, Ji Zou, Vinothini Venkatachalam, Virtudes Rubio Diaz, Andrea D'Angio, Prabhu Ramanujam, Tailin Zhang, and T. S. R. C. Murthy. "Selection, Processing, Properties and Applications of Ultra-High Temperature Ceramic Matrix Composites, UHTCMCs: A Review." *International Materials Reviews* 65, no. 7 (2020): 389–444.

[8] Tejero-Martin, D., C. Bennett, and T. Hussain. "A Review on Environmental Barrier Coatings: History, Current State of the Art and Future Developments." *Journal of the European Ceramic Society* 41, no. 3 (2021): 1747–1768.

[9] Lessing, Paul A. "A Review of Sealing Technologies Applicable to Solid Oxide Electrolysis Cells." *Journal of Materials Science* 42 (2007): 3465–3476.

[10] Perepezko, J. H. "Surface Engineering of Mo-Base Alloys for Elevated-Temperature Environmental Resistance." *Annual Review of Materials Research* 45 (2015): 519–542.

[11] Konegger, Thomas, Jessica Torrey, Octavio Flores, Tobias Fey, Bruno Ceron-Nicolat, Günter Motz, Franziska Scheffler, Michael Scheffler, Peter Greil, and Rajendra K. Bordia. "Ceramics for Sustainable Energy Technologies with a Focus on Polymer-Derived Ceramics." In *Novel Combustion Concepts for Sustainable Energy Development*, pp. 501–533. Springer India, 2014.

[12] Pasto, A. E., and B. J. Russell. *High Temperature Materials Laboratory, Eleventh Annual Report: October 1997 Through September 1998*. No. ORNL/TM-2000/2. Oak Ridge National Lab (ORNL), Oak Ridge, TN (United States), 2000.

[13] Wan, Caichao, Luyu Zhang, Ken-Tye Yong, Jian Li, and Yiqiang Wu. "Recent Progress in Flexible Nanocellulosic Structures for Wearable Piezoresistive Strain Sensors." *Journal of Materials Chemistry C* 9, no. 34 (2021): 11001–11029.

[14] Mariano, Anna, Claudia Lubrano, Ugo Bruno, Chiara Ausilio, Nikita Bhupesh Dinger, and Francesca Santoro. "Advances in Cell-Conductive Polymer Biointerfaces and Role of the Plasma Membrane." *Chemical Reviews* 122, no. 4 (2021): 4552–4580.

[15] Abdul-Aziz, Ali, and Adam C. Wroblewski. "Durability Analysis and Experimental Validation of Environmental Barrier Coating (EBC) Performance Using Combined Digital Image Correlation and NDE." *Coatings* 6, no. 4 (2016): 70.

[16] Alat, Ece. "Ceramic Coatings for Nuclear Fuel Cladding to Enhance Accident Tolerance." (2018).

[17] Mital, Subodh K., Trenton M. Ricks, Steven M. Arnold, and Bryan J. Harder. *Modeling of the Influence of Thermally Grown Oxide (TGO) Layers on the Driving Forces in Environmental Barrier Coating Systems*. No. E-19942, 2021.
[18] Di Iorio, Giulia. "Development of Novel Environmental Barrier Coatings for Next Generation Gas Turbine Engines." (2023).
[19] Webster, Rebekah I., and Elizabeth J. Opila. "Mixed Phase Ytterbium Silicate Environmental-Barrier Coating Materials for Improved Calcium-Magnesium-Alumino-Silicate Resistance." *Journal of Materials Research* 35, no. 17 (2020): 2358–2372.
[20] Badran, Aly A. "Relating Damage to Microstructure in SiC-SiC Ceramic Matrix Composites with μ-CT and Deep Learning Image Segmentation." PhD diss., University of Colorado, Boulder, 2021.
[21] Mubarik, Shamroza, Nawal Qureshi, Zainab Sattar, Aqeela Shaheen, Ambreen Kalsoom, Marryam Imran, and Farzana Hanif. "Synthetic Approach to Rice Waste-Derived Carbon-Based Nanomaterials and Their Applications." *Nanomanufacturing* 1, no. 3 (2021): 109–159.
[22] Presby, Michael J. "Foreign Object Damage and Solid Particle Erosion Behavior of Ceramic Matrix Composites." PhD diss., The University of Akron, 2019.
[23] Paksoy, Ahmet Hilmi, and Ping Xiao. "Review of Processing and Design Methodologies of Environmental Barrier Coatings for Next Generation Gas Turbine Applications." *Advances in Applied Ceramics* 122, no. 1 (2023): 36–56.
[24] Shvydyuk, Kateryna O., João Nunes-Pereira, Frederico F. Rodrigues, and Abílio P. Silva. "Review of Ceramic Composites in Aeronautical and Aerospace: A Multifunctional Approach for TPS, TBC and DBD Applications." *Ceramics* 6, no. 1 (2023): 195–230.
[25] Waddar, Sunil Shankar. "Buckling and Dynamic Characteristics of Cenosphere/Epoxy Syntactic Foam Composites and Sandwiches under Mechanical and Thermal Loads." PhD diss., National Institute of Technology Karnataka, Surathkal, 2019.
[26] Dong, Lin, Mei-Jun Liu, Xiao-Feng Zhang, Guan-Jun Yang, and Ke-Song Zhou. "Improved Water Vapor Resistance of Environmental Barrier Coatings Densified by Aluminum Infiltration." *Ceramics International* 48, no. 16 (2022): 23638–23648.
[27] Alsharif, Rashed, Mehrdad Arashpour, Victor Chang, and Jenny Zhou. "A Review of Building Parameters' Roles in Conserving Energy Versus Maintaining Comfort." *Journal of Building Engineering* 35 (2021): 102087.
[28] Bernardo-Bermejo, Samuel, Elena Sanchez-Lopez, Maria Castro-Puyana, and María Luisa Marina. "Chiral Capillary Electrophoresis." *TrAC Trends in Analytical Chemistry* 124 (2020): 115807.
[29] Tejero-Martin, Daniel, Chris Bennett, and Tanvir Hussain. "A Review on Environmental Barrier Coatings: History, Current State of the Art and Future Developments." *Journal of the European Ceramic Society* 41, no. 3 (2021): 1747–1768.
[30] Roque, Bruno Augusto Cabral, Pedro Pinto Ferreira Brasileiro, Yana Batista Brandão, Alessandro Alberto Casazza, Attilio Converti, Mohand Benachour, and Leonie Asfora Sarubbo. "Self-Healing Concrete: Concepts, Energy Saving and Sustainability." *Energies* 16, no. 4 (2023): 1650.
[31] Sinha, Sudip Kumar. "Additive Manufacturing (AM) of Medical Devices and Scaffolds for Tissue Engineering Based on 3D and 4D Printing." In *3D and 4D Printing of Polymer Nanocomposite Materials*, pp. 119–160. Elsevier, 2020.
[32] Dahl, Amy L. "Comparison of Direct and Operational Methods for Probing Metal Bioavailability and Speciation in Aquatic Systems." PhD diss., Northwestern University, 2005.
[33] Vaßen, Robert, Emine Bakan, Daniel Emil Mack, and Olivier Guillon. "A Perspective on Thermally Sprayed Thermal Barrier Coatings: Current Status and Trends." *Journal of Thermal Spray Technology* 31, no. 4 (2022): 685–698.

[34] Lee, Kang N. "Environmental Barrier Coatings for SiCf/SiC." *Ceramic Matrix Composites: Materials, Modeling and Technology* (2014): 430–451.
[35] Lambert, Paul, Janet McGovern, Dean Putnam, Nick Olah, John Gallagher, John Kieffer, Tracy Frost, and Summer Jones. "Navy SBIR fy09 1 Proposal Submission Instructions."
[36] Seidl, Markus, Alice Fayter, Josef N. Stern, Katrin Amann-Winkel, Marion Bauer, and Thomas Loerting. "High-Performance Dilatometry Under Extreme Conditions." *Proceedings of the 6th Zwick Academia Day 2015* 76 (2015).
[37] Gledhill, Andrew. *Thermal Barrier Coatings Chemically and Mechanically Resistant to High Temperature Attack by Molten Ashes*. The Ohio State University, 2011.
[38] Paksoy, Ahmet Hilmi, and Ping Xiao. "Study on Sintering Behaviour of Ytterbium Disilicate and Ytterbium Monosilicate for Environmental Barrier Coating Applications." *Ceramics International* 48, no. 8 (2022): 11238–11250.
[39] Sun, Dongwei. *Finite Element Analysis of Machining Damages in Manufacturing Ceramic Implants*. The University of Alabama, 2015.
[40] Porter, Matthew Thomas. "High-Temperature Ceramic Matrix Composites Prepared via Microwave Energy Enhanced Chemical Vapour Infiltration." PhD diss., University of Birmingham, 2020.
[41] Kumar, Sharvan, and Nitin P. Padture. "Materials in the Aircraft Industry." *Metallurgical Design and Industry: Prehistory to the Space Age* (2018): 271–346.
[42] Nieto, Andy, Richa Agrawal, Luis Bravo, Clara Hofmeister-Mock, Marc Pepi, and Anindya Ghoshal. "Calcia-Magnesia-Alumina-Silicate (CMAS) Attack Mechanisms and Roadmap Towards Sandphobic Thermal and Environmental Barrier Coatings." *International Materials Reviews* 66, no. 7 (2021): 451–492.
[43] Lee, K. N., D. Zhu, and R. S. Lima. "Perspectives on Environmental Barrier Coatings (EBCs) Manufactured via Air Plasma Spray (APS) on Ceramic Matrix Composites (CMCs): A Tutorial Paper." *Journal of Thermal Spray Technology* 30 (2021): 40–58.
[44] Harder, Bryan J. "Phase Transformations and Residual Stresses in Environmental Barrier Coatings." PhD diss., Northwestern University, 2009.
[45] Akinc, Mufit. "Processing and Oxidation Behavior of ZrB2-SiC-AlN Composites." *ECerS2017* (2017): 172.

6 EIA of Engineered Ceramics and Their Sustainable Management

K. Ihjas and Neethu Bhaskar

6.1 INTRODUCTION

The advancement of science and technology has significantly influenced various sectors, including ceramic technologies, which now play a crucial role in addressing the multifaceted demands of a growing global population. As an enabling technology, ceramics have expanded their scope to encompass multidisciplinary fields within academia and industrial applications. This chapter delves into the environmental impact assessment of ceramic-based products and their waste management, providing a comprehensive overview of how these materials interact with and affect social, environmental, and economic aspects.

Ceramics have been a cornerstone of human civilization, from ancient pottery to modern high-tech applications. The evolution of ceramic technologies has led to their widespread use in diverse fields such as electronics and communications, energy and environmental management, and mechanical engineering. The properties of ceramics – such as high thermal resistance, durability, and biocompatibility – make them invaluable in both scientific research and practical applications (Rödel et al., 2009). The burgeoning growth opportunities in ceramics underscore their importance in innovative technological advancements and sustainable development.

In the realm of electronics and communications, ceramics are essential components in the manufacture of capacitors, insulators, and substrates. Their unique electrical properties, including high dielectric strength and low loss, make them ideal for these applications. Similarly, in the energy sector, ceramics contribute to the development of fuel cells, batteries, and solar panels, enhancing energy efficiency and sustainability. The environmental benefits of these applications are significant, as they help reduce reliance on fossil fuels and lower greenhouse gas emissions.

Mechanical engineering also benefits from ceramic technologies, particularly in applications requiring high strength, wear resistance, and thermal stability. Advanced ceramics are used in cutting tools, bearings, and engine components, where they outperform traditional materials like metals and polymers under extreme conditions. These properties not only improve the performance and longevity of

DOI: 10.1201/9781003731672-6

mechanical systems but also contribute to environmental sustainability by reducing the frequency of replacements and maintenance, thereby lowering material and energy consumption.

However, the production and disposal of ceramic-based products pose environmental challenges that need to be addressed through effective waste management strategies. The manufacturing process of ceramics involves high-temperature treatments, which consume substantial amounts of energy and contribute to carbon emissions. Additionally, the extraction of raw materials for ceramics, such as clay and other minerals, can lead to habitat destruction and resource depletion. Therefore, an environmental impact assessment is crucial to identify and mitigate these negative effects.

Waste management of ceramic products is another critical aspect that requires attention. Ceramic waste, including defective products, manufacturing residues, and end-of-life ceramics, needs to be managed efficiently to minimize environmental harm. Recycling and reusing ceramic waste can significantly reduce the burden on landfills and the demand for raw materials. For instance, waste ceramics can be crushed and used as aggregates in construction, reducing the need for new raw materials and promoting circular economy principles. Furthermore, advancements in recycling technologies are essential to improve the sustainability of ceramic industries. Innovative processes, such as thermal and mechanical recycling, can convert ceramic waste into valuable secondary raw materials. These processes not only help in waste reduction but also conserve energy and resources, contributing to the overall sustainability of ceramic technologies.

Environmental impact assessment of ceramic-based products and their waste management is a critical area of study that intersects with multiple disciplines and industries. By understanding the environmental implications of ceramic technologies and developing effective waste management strategies, we can harness the full potential of ceramics while minimizing their ecological footprint. The continued advancement and innovation in ceramic technologies hold promise for a sustainable future, where environmental, social, and economic goals are harmoniously achieved.

6.2 CERAMIC-BASED INDUSTRIES AND THEIR OVERVIEW

Ceramic industries are probably the best- and old-known industries with its wide range of applications from earthware pottery making to the application of ceramics in aerospace turbine blades and space shuttle tiles (O'Keeffe, 2008). Ceramic industries are consistently advancing and innovatively developing in their wide range of applications, from construction to consumer goods, industrial processes, and cutting-edge technologies for science and industries. Applications of ceramic-based products range as diversely as EV batteries, fiber optics, tiles, artificial joints, dental crowns, membranes, chemical sensors, micro positioners, and so on. Other examples of advanced ceramics applications include nuclear fuel rods, bearings in food processing equipment, thermal barriers, cutting tools, abrasives, and so on. The highlight of the boom in ceramic industries/materials is their significant dominance over metals and polymers at high temperatures and in corrosive environments (Dean, 2019).

Advanced ceramics are one of the major promising technologies which have a wide range of industrial applications as well as market beneficiaries. It is considered as one of the capable technologies that have the potential to deliver high-value contributions to meet the challenges of the future. The advanced ceramics sector comprises the following ceramics, and its applications are listed in Table 6.1 (*Powder Matrix Technology Roadmaps 2004, Advanced Ceramics Report No. RM/01/04/PX, 2004.*).

Table 6.1 summarizes various advanced ceramics and their corresponding sectoral applications, illustrating the versatility and wide-ranging uses of these materials across different industries.

6.3 ENVIRONMENTAL IMPACT ASSESSMENT

Environmental impact assessment (EIA) has become a widely accepted tool in environmental management for all kinds of developmental works. EIA is nothing but a procedure for assessment of the existing as well as potential impacts of projects or activities to implement policies, to enact legislation and guidelines, or to initiate/start some developmental projects so that the associated impacts can be controlled, reduced, or minimized during the project planning stage itself (Mareddy, 2017). The processes involved in industries manufacturing ceramic-based products, from the

TABLE 6.1
Various Advanced Ceramics and Their Corresponding Sectoral Applications

Advanced Ceramics	Sectoral Applications
Structural ceramics	Monoliths and composites, for example, oxides, nitrides, carbides, borides, and composite materials based on these materials
Functional ceramics	Electrical and magnetic ceramics (i.e., dielectrics, piezoelectric, ferromagnetics), ionic conductors, and superconductive ceramics
Ceramic coatings	Oxides, nitrides, carbides, borides, cermets, and diamond-like coatings, deposited by technologies such as spraying, vapor deposition, and sol-gel coating
Bioceramics	For example, hydroxyapatite and alumina
Special glasses	Processed flat glass, fire-resistant glazing, and glasses for optoelectronics
Alumina (Aluminum oxide)	Electrical insulators, wear-resistant parts, cutting tools
Zirconia (Zirconium oxide)	Dental crowns, fuel cell membranes, thermal barrier coatings
Silicon carbide	Abrasives, high-temperature semiconductors, mechanical seals
Barium titanate	Capacitors, piezoelectric devices, thermistors
Silicon nitride	Engine components, bearings, cutting tools
Lithium disilicate	Dental restorations, high-strength glass-ceramics
Lead zirconate titanate (PZT)	Sensors, actuators, ultrasound transducers
Boron carbide	Armor plating, neutron absorbers in nuclear reactors
Magnesium diboride	Superconductors, magnetic shielding materials

procurement of raw materials to the dispatch of finished products pose immense threats and hazards to the social well-being of human beings as well as the environment. The literature suggests the manufacturing process itself has a direct impact on the environment and humans by generating an enormous amount of harmful and hazardous gases. It has been reported that a plant manufacturing the widely used ceramic tiles generates 15.89 kg CO_2-eq. per square meter of the ceramic tiles, which leads to 1.14×10^{-5} kg CFC 11-eq in depletion of the ozone layer. The workers as well as society are further affected by the ultrafine and nanosized airborne particles produced by the production plants (Muthukannan & Ganesh, 2019).

Environmental pollution is one of the major challenges the world is facing today. Rapid urbanization and industrialization prompted the increasing requirement of energy, and also the world around struggles with the waste generated. The waste generated in any form – solid, liquid or gas – causes multiple hazards to environment, acute health problems, ozone layer depletion, greenhouse gas emissions, and water, air, and land pollution (Muthukannan & Ganesh, 2019). The growing demands of the world resulted in industrial boom around. The industries started producing multiple products and by-products, thus generating waste throughout its product lifecycle. The effluents produced from all industries are causing severe toxic metal pollution or some type of physico-chemical pollution (Asif Hanif et al., 2005). The activity cycle of industries includes from procurement of raw materials to processing, production, and dispatch of the finished good, that is, cradle-to-grave life cycle approach including the stages of mining the raw materials, transportation to the facility, the production process, the product check, the transport to the construction site, the application and the final disposal of the product as well as the waste generated (Audretsch et al., 1996) All these processes have direct as well as indirect impact on the environment, and environmental sustainability has to be the fundamental urgency for all industrial activities. The market should grow concerns on the environment, and the industries have to be developed based on the impact assessment studies. The carbon and environmental footprint of the products has to be assessed by identifying the carbon maps of the supply chains, thus emphasizing the additional benefits of higher production efficiency in lower production costs (Gaussin et al., 2013).

Multiple studies have explored the impact of ceramic industries on human health as well as environment. The ceramic industry generates environmental impacts throughout its lifecycle through its interactions with multiple environmental aspects including air, water, energy, noise during its activities of consumption of resources to processing and waste generation, the most significant is its impact on the air environment (Almeida et al., 2013). The risk of particle emissions to workers as well as environment raises serious concerns, especially the risk of exposure to ultrafine and nanoparticles above the nano reference value (NRV, 4×10^4 cm^{-3}) to the workforce. Evidence of environmental (outdoor air) exposure despite having effective filtration system was also recorded. The context justifies the significance of impact assessment studies, as the emission patterns are strongly related to the temperature, chemical composition, and raw material processing technologies adopted by the industries (Fonseca et al., 2015). Long-term exposure of air emissions from the ceramic industries could cause a reduction in lung function of the workers exposed, and a synergistic effect with smoking habits could further worsen the impact (Jaakkola et al., 2011). Most of the research work on the impact of ceramic industries highlights the emissions, energy consumption, etc. as critical and thus emphasizes on the need of proper environmental impact assessment studies during the planning and operational phases of a ceramic industry. Major techniques are adopted for assessing the impact of pollution and environmental footprint due to the ceramic industry across the globe.

6.4 ENVIRONMENT IMPACT ASSESSMENT AND PERFORMANCE METRICS

Different environmental performance metrics are employed to assess quantitatively the impacts of industrial products, processes, and activities on the environment. Each method is based on different methodologies, assumptions, and weighting techniques; the performance evaluation based on the magnitude of output cannot be scientific while the relative examination of the equivalent functionality distinguishes the different methods available (Luo et al., 2001). Life cycle assessment (LCA), Eco-indicator 95, Eco-indicator 99, BAT, etc., are various impact assessment methods and metrics discussed here.

6.4.1 LCA Methodology

The LCA approach is one of the major environmental impact assessment tools and is identified as the only internationally standardized tool of ecological impact assessment (Tikul & Srichandr, 2010). The LCA tool was developed in the 1970s and 1980s from modest seeds, and its decisive transition happened with the intervention of Society of Environmental Toxicology and Chemistry (SETAC) and International Organization for Standardization (ISO) in harmonizing the largely adopted diverse older techniques. The International Standards, IS 14040 and IS 14044, serving as a framework defines LCA as a method to analyze the aspects and impacts of products on environment (Hunkeler, 2016).

LCA is a tool/method employed to appraise a product/industry's impact on the environment through its whole life cycle on all sides of its development, from the extraction of the raw materials to manufacturing, distribution, use, recycling, and final disposal into the environment (Ilgin & Gupta, 2010). In LCA approach, the environmental impact is assessed on all stages of industrial operations and production from cradle to grave (Olmez et al., 2016). The industrial production process has to be analyzed for a certain period of operation time in years, and various aspects on different stages of processes are to be identified for assessing the impacts such as global warming, ozone layer depletion, human toxicity, acidification, photochemical oxidation, and eutrophication (Nitschelm et al., 2016). All input as well as output data in each stage of the life cycle of the product, including raw materials, energy, water, airborne emissions, emissions into water and land, solid as well as liquid waste produced, are collected and presented in Table 6.2. The collected data has to be related to the unit processes in various functional stages (Su et al., 2016).

TABLE 6.2
General Inventory Data Format on Various Stages of Ceramic Industrial Life Cycle (Ibáñez-Forés et al., 2011; Muthukannan & Ganesh, 2019).

Stages / Aspect	Raw Materials (Source: Mines)	Processing of RM	Prepartory works	Manufacturing/ Production	Post Production works	Distribution/ Transportation	Waste Generated	Application & Installation	Product Fate
Non-Renewable Energy									
Electricity kWh/m^2									
Diesel L/m^2									
Natural Gas kWh/m^2									
Air borne emissions (Kg/m^2)									
PM10									
NOx									
Sox									
CO									
HF									
Pb									
As									
Hg									
Cu									
Cr									
Hazardous and Non-Hazardous waste (kg/m^2)									
HW									
NHW									
Water inputs and Outputs (m^3/m^2)									
Incoming Water									
Outgoing water									
Other Specific Impacts									
Impact Assessment									

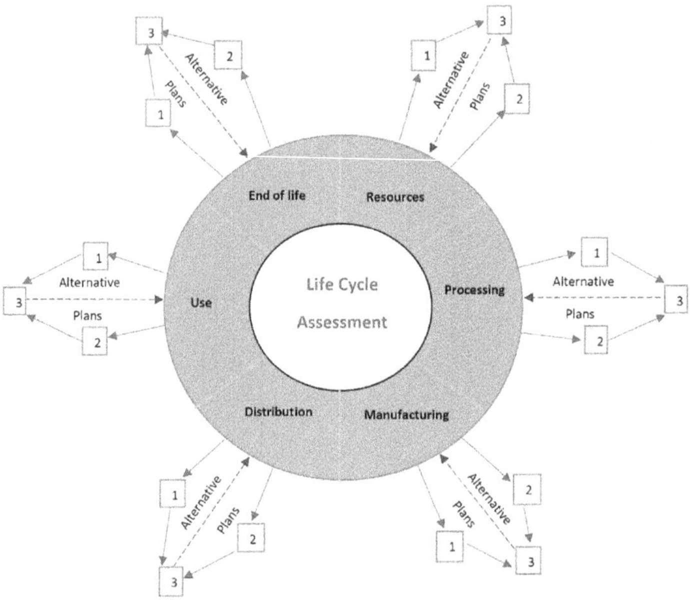

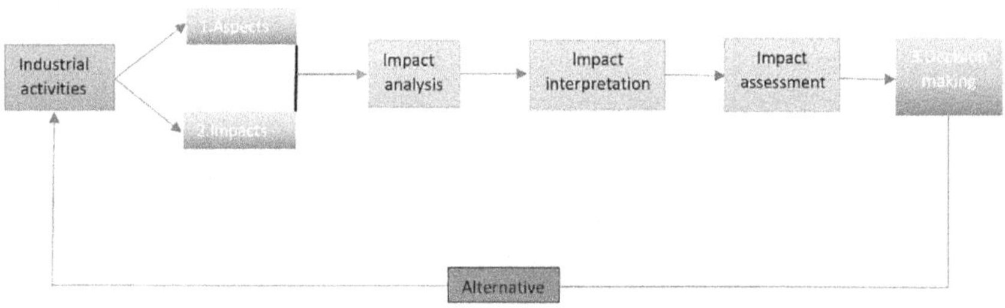

FIGURE 6.1 **a** Schematic representation of the process life of ceramic industries and b schematic representation of the life cycle assessment for ceramic industries.

6.5 EDIP METHODOLOGY

EDIP refers to environmental development of industrial products, which is an impact assessment methodology measured based on the impact of different aspects in terms of global warming and human toxicity. The methodology involves assessment of midpoint indicators for the identified environmental impacts, and the impact values are represented as CO_2 equivalents. The calculations for global warming are denoted as CO_2-eq, where CO_2 is taken as the reference substance indicating the impact on global warming. Similarly, the reference item CFC-11 indicates ozone depletion, where SO_2 stands for the impact on acidification. In this way, impact values are calculated for all known aspects of industrial products, and hence impacts are assessed (Tikul & Srichandr, 2010).

In EDIP, the impact values are calculated as (Wenzel & Alting, 1999; Pommer et al., 2003):

$$EP_j = \Sigma\ (Q_i \times EF_{ij}),$$

where,

EP_j: Potential environmental for an environmental problem. (j)

Q_i: Substance quantity (i)

EF_{ij}: Substance for equivalency factor on the impact of the environmental problem.

6.5.1 Eco-Indicator 95

Eco-indicator 95 is a weighting method which is normally used to present the LCA results in all possible levels such as normalization, characterization, and weighting so as to assess the environmental impacts of industrial products or activities (Tikul, 2014). This is also an indicator tool in which the adopted methodology helps in determining the impacts in terms of harmful gases like CFC , CH_4 , N_2O , and CO_2 that affect the environment (Luo et al., 2001).

6.5.2 Eco-Indicator 99

Eco-indicator 99 is an LCA tool that helps industries to conduct an environmental impact assessment for a product by calculating eco-indicator scores for raw materials as well as the processes employed. The main advantage of this method is that it provides a single score as an indication of areas for product improvements (*High Temperature Thermal Storage Systems Using PCMs.PDF*, 2017).

Eco-indicator 99 was developed as a pragmatic science-based eco-design model for environmental impact assessment using the LCA approach that provides us with a single score, which specifically indicates the damage to resources. However, the recently developed new damage models provide the damages in more categories, including damage to human health, damage to ecosystem, and damage to resources (*Lees' Loss Prevention in the Process Industries*, 2012).

Eco-indicator 99, being a damage-oriented impact assessment methodology measures the end point indicators of the environmental impacts on human health, quality of the ecosystem, as well as resources (Goedkoop, 2007). The impact values are measured as points or scores. Using the relationships detailed in the following, the impact values and damage points (DP) of various categories can be calculated (Tikul & Srichandr, 2010).

$$DP_{ij} = Q_i \times DF_{ij} ,$$

where,

DP_{ij} : Damage scores for substance i and categories j

Q_i : Substance quantity i (kg)

DF_{ij} : Impacts of the damage categories j and substance damage factor i

Using the following equation, the damage scores are normalized.

$$NP_{ij} = DP_{ij} \times NF_{ij},$$

where,

NP_{ij} : Normalized scores for substance i and damage categories j
NF_{ij} : Normalization impacts of the damage categories j and substance damage factor i

Using the following equation, normalized scores are weighted, and weighted scores are arrived at.

$$WP_j = \sum \left(WF_{ij} \times NP_{ij} \right),$$

where,

WP_j : Damage categories j, weighted scores of all substances
WF_{ij} : Damage categories j and substance i, Weighted damage factor.

6.5.3 Best Available Technique (BAT)

Best available technique (BAT) is the best possible method to assess the environmental impacts of ceramic industries in which different methods are employed to arrive at the best feasible alternative to reduce the anticipated impacts. Prior to application of the BAT tool, a baseline measurement of the impacts has to be determined using the LCA method. After establishing the baseline conditions, all available sustainable alternatives should be selected based on the guidelines of the industrial emission standards as well as by comparing the relevant indicators, including environmental, economic, technical, and social aspects. Once the alternatives are selected, an initial screening has to be done by adopting BAT alternatives to evaluate the economic and environmental performance of the production process. Based on the evaluated performance, the best and efficient alternative can be selected considering the economic as well as environment commitment of the industry (Ibáñez-Forés et al., 2013; Muthukannan & Ganesh, 2019). The BAT method becomes pertinent especially for the ceramic industries considering the option itself serves as a reference document for the industries targeting reduction in environmental footprint as well as sustainability, subject to the assumption that all indicators are given equal importance even though it is based on the preference of the reviewer.

6.6 LEGISLATION

The first International Conference on Ceramics (ICC), held in Toronto in 2006 aimed at developing a future road map of advanced ceramics, defined them as nonmetallic inorganic materials and identified the impacts of ceramics in addressing the future scope of developing solutions in the major markets of environment, energy, industries, health and safety, etc. (Freiman et al., 2007). The EIA regulations and the development of BAT emerged as key factors in ensuring regulatory stringency and enforcing rigorous standards for ceramic industries. The legislative changes in the sector improved by adopting BAT standards in raw materials and waste management, strengthening monitoring, management, and licensing. Another remarkable achievement was in improving the energy efficiency by decarbonizing and fuel switching (OECD, 2022).

BAT based permitting, stringent emission standards, and enhanced monitoring with the EU CER BREF materials and revisions have tightened the international guidelines on ceramic industry legislations. BAT conclusions were recorded in Sevilla process in 2024 and proposed for binding BAT-AELs and stringent regulations for energy use as well as emissions, including particulates, fluorides, and SO_2 (Karlis et al., 2024). In parallel, NESHAP and EPA finalized amendments, and the international organizations as well as major producers pushed for BAT harmonization, thus ensuring EIAs are mandatory for the ceramic-based industries and projects to unambiguously direct BAT enforcement, emission standards and monitoring, waste management, hazardous-waste controls, and decarbonization/energy efficiency (Stationary Sources of Air Pollution | US EPA, 2025).

6.7 SUMMARY

This chapter highlights the importance of engineered ceramics across the technology sector alongside providing a comprehensive overview of its environmental implications. Exceptional thermal stability, durability, and functional versatility of ceramics have widespread applications in industries including electronics, energy, healthcare, aerospace, and so on. Even though, ceramics contribute highly towards the sustainable goals via their application as sensors, batteries, coatings, etc., their production process remains energy intensive and contributes significantly to environmental pollution. The process cycle involving sintering and raw material extraction gives rise to emissions of particulate matter and greenhouse gases, causing serious environmental and occupational health risks.

To address these challenges, the chapter details key EIA tools such as Life Cycle Assessment (LCA), EDIP, Eco-indicator 95, Eco-indicator 99, and Best Available Techniques (BAT). In addition, a briefing on the environmental footprint calculation, hotspot identification, and informed decision making using these tools is furnished in this chapter. The role of international legislation, including EU BAT documents and EPA standards, is highlighted in strengthening regulatory compliance and promoting sustainable manufacturing practices. Overall, the chapter emphasizes the significance of approaches towards circular economy, advanced waste management strategies, energy-efficient technologies, and better impact assessment frameworks to ensure the sustainable growth of the ceramic industry.

REFERENCES

Almeida, M., Demertzi, M., Dias, A., & Arooja, L. (2013). Environmental product declaration for ceramic tile. *Energy for Sustainability 2013*.

Audretsch, D. B., & Feldman, M. P. (1996). Innovative clusters and the industry life cycle. *Review of Industrial Organization, 11*(2), 253–273.

Asif Hanif, M., Nadeem, R., Rashid, U., & Zafar, M. N. (2005). Assessing pollution levels in effluents of industries in city zone of Faisalabad, Pakistan. *Journal of Applied Sciences*, *5*(10), 1713–1717. https://doi.org/10.3923/jas.2005.1713.1717

Best Available Techniques (BAT) for preventing and controlling industrial pollution, activity 5: Value chain approaches to determining BAT for industrial installations. Environment, Health and Safety, Environment Directorate, OECD.

Dean, Y. (2019). Applications of ceramic materials. *Finishes*, 56–116. https://doi.org/10.4324/9781315842837-5

Fonseca, A. S., Maragkidou, A., Viana, M., Querol, X., Hämeri, K., de Francisco, I., Estepa, C., Borrell, C., Lennikov, V., & de la Fuente, G. F. (2015). Process-generated nanoparticles from ceramic tile sintering: Emissions, exposure and environmental release. *Science of the Total Environment*, *565*, 922–932. https://doi.org/10.1016/j.scitotenv.2016.01.106

Freiman, S., Singh, M., Fischman, G. S., & Cook, R. (2007). *Global roadmap for ceramics and glass technology*, 966. http://books.google.co.jp/books?id=jm9hGqECbXcC

Gaussin, M., Hu, G., Abolghasem, S., Basu, S., Shankar, M. R., & Bidanda, B. (2013). Assessing the environmental footprint of manufactured products: A survey of current literature. *International Journal of Production Economics*, *146*(2), 515–523. https://doi.org/10.1016/j.ijpe.2011.12.002

Goedkoop, M. (2007). The eco-indicator 99 methodology. *Journal of Life Cycle Assessment, Japan*, *3*(1), 32–38. https://doi.org/10.3370/lca.3.32

High Temperature Thermal Storage Systems Using PCMs.PDF. (2017).

Hunkeler, D. (2016). Life cycle assessment (LCA): A guide to best practice. *The International Journal of Life Cycle Assessment*, *21*(7), 1063–1066. https://doi.org/10.1007/s11367-016-1083-z

Ibáñez-Forés, V., Bovea, M. D., & Azapagic, A. (2013). Assessing the sustainability of best available techniques (BAT): Methodology and application in the ceramic tiles industry. *Journal of Cleaner Production*, *51*, 162–176. https://doi.org/10.1016/j.jclepro.2013.01.020

Ibáñez-Forés, V., Bovea, M. D., & Simó, A. (2011). Life cycle assessment of ceramic tiles: Environmental and statistical analysis. *International Journal of Life Cycle Assessment*, *16*(9), 916–928. https://doi.org/10.1007/s11367-011-0322-6

Ilgin, M. A., & Gupta, S. M. (2010). Environmentally conscious manufacturing and product recovery (ECMPRO): A review of the state of the art. *Journal of Environmental Management*, *91*(3), 563–591. https://doi.org/10.1016/j.jenvman.2009.09.037

Jaakkola, M. S., Sripaiboonkij, P., & Jaakkola, J. J. K. (2011). Effects of occupational exposures and smoking on lung function in tile factory workers. *International Archives of Occupational and Environmental Health*, *84*(2), 151–158. https://doi.org/10.1007/s00420-010-0603-6

Karlis, P., Presicce, F., Giner-Santonja, G., Brinkmann, T., & Roudier, S. (2024). *Best available techniques (BAT) reference document for the slaughterhouses, animal by-products and/or edible co-products industries.*

Lees' Loss Prevention in the Process Industries. (2012).

Luo, Y., Wirojanagud, P., & Caudill, R. J. (2001). Comparison of major environmental performance metrics and their application to typical electronic products. *IEEE International Symposium on Electronics and the Environment*, 94–99. https://doi.org/10.1109/isee.2001.924508

Mareddy, A. R. (2017). Environmental impact assessment: Theory and practice. *Environmental Impact Assessment: Theory and Practice*, 1–616. https://doi.org/10.1016/C2015-0-06055-5

Muthukannan, M., & Ganesh, A. S. C. (2019). The environmental impact caused by the ceramic industries and assessment methodologies. *International Journal for Quality Research*, *13*(2), 315–334. https://doi.org/10.24874/IJQR13.02–05

Nitschelm, L., Aubin, J., Corson, M. S., Viaud, V., & Walter, C. (2016). Spatial differentiation in life cycle assessment LCA applied to an agricultural territory: Current practices and method development. *Journal of Cleaner Production*, *112*, 2472–2484. https://doi.org/10.1016/j.jclepro.2015.09.138

O'Keeffe, T. (2008). Europe, northern and western: Medieval. *Encyclopedia of Archaeology*, 1240–1249. https://doi.org/10.1016/B978-012373962-9.00181-3

Olmez, G. M., Dilek, F. B., Karanfil, T., & Yetis, U. (2016). The environmental impacts of iron and steel industry: A life cycle assessment study. *Journal of Cleaner Production*, *130*, 195–201. https://doi.org/10.1016/j.jclepro.2015.09.139

Pommer, K., Bech, P., Wenzel, H., Caspersen, N., & Olsen, S. I. (2003). *Handbook on environmental assessment of products. Environmental Project No. 813, 813, 181*. https://www2.mst.dk/udgiv/publications/2003/87-7972-683-6/pdf/87-7972-684-4.pdf

Powder Matrix Technology Roadmaps 2004, Advanced Ceramics Report No. RM/01/04/PX. (2004).

Rödel, J., Kounga, A. B. N., Weissenberger-Eibl, M., Koch, D., Bierwisch, A., Rossner, W., Hoffmann, M. J., Danzer, R., & Schneider, G. (2009). Development of a roadmap for advanced ceramics: 2010–2025. *Journal of the European Ceramic Society*, *29*(9), 1549–1560. https://doi.org/10.1016/j.jeurceramsoc.2008.10.015

Stationary Sources of Air Pollution | US EPA. (2025, November 19). https://www.epa.gov/stationary-sources-air-pollution

Su, M., Chen, C., & Yang, Z. (2016). Urban energy structure optimization at the sector scale: Considering environmental impact based on life cycle assessment. *Journal of Cleaner Production*, *112*, 1464–1474. https://doi.org/10.1016/j.jclepro.2015.01.059

Tikul, N. (2014). Assessing environmental impact of small and medium ceramic tile manufacturing enterprises in Thailand. *Journal of Manufacturing Systems*, *33*(1), 1–6. https://doi.org/10.1016/j.jmsy.2013.12.002

Tikul, N., & Srichandr, P. (2010). Assessing the environmental impact of ceramic tile production in Thailand. *Journal of the Ceramic Society of Japan*, *118*(1382), 887–894. https://doi.org/10.2109/jcersj2.118.887

Wenzel, H., & Alting, L. (1999). Danish experience with the EDIP tool for environmental design of industrial products. *Proceedings – 1st International Symposium on Environmentally Conscious Design and Inverse Manufacturing, EcoDesign, 1999*, 370–379. https://doi.org/10.1109/ECODIM.1999.747640

7 Advanced Ceramic-Polymer Composite Electrolyte Materials and Technologies for Lithium-Based Electrochemical Energy Storage Systems

Megha Sara Jacob, Raman Vedarajan, and Sahana B. Moodakare

7.1 INTRODUCTION

The commercialization of lithium-ion batteries (LIBs) occurred in 1991 by Sony and thenceforward, laboratories around the world have been greatly involved in developing this energy storage system [1, 2]. In comparison with several commercially available electrochemical secondary batteries, LIBs have taken the center stage for their application in numerous electronic devices [3, 4]. This trend can be attributed to their higher gravimetric and volumetric energy density along with notable cycling performance [5]. Despite the significant progress of LIBs, there are several challenges in materials science, processing science, and the fabrication of practical full cells that curb its use for practical applications [6, 7]. Identification of electrolytes with high ion conductivity, manufacturing of effective electrodes, and optimization of the electrode–electrolyte interface without compromising on cycle life, cost, and safety remain as major objectives of research [8–10].

The concept of all-solid-state batteries (ASSBs) using solid state electrolytes (SSE) has been introduced to address the safety problems associated with traditional LIBs that make use of flammable liquid electrolytes [11, 12]. SSE for LIBs can be ceramic-based, or polymer-based, or a composite of ceramic and polymer [13, 14]. Ceramic/inorganic electrolytes such as NaSICON-type Lithium Aluminium Titanium Phosphate (LATP) with general formula $Li_{1+x}Al_xTi_{2-x}(PO_4)_3$ provide high ionic conductivity values in the range of 10^{-4}–10^{-3} S/cm along with high thermal and electrochemical stability [15–17]. However, they are associated with poor mechanical properties and induce greater interfacial resistance upon contact with

 DOI: 10.1201/9781003731672-7

the electrodes [18]. In comparison, polymer electrolytes such as polyethylene oxide (PEO) and polyvinylidene fluoride (PVDF) offer exceptional mechanical properties such as flexibility as well as low interfacial resistance with the electrodes [19]. Nevertheless, their ionic conductivity is found to be comparatively lower [20].

A third set of SSEs is the composite solid electrolytes (CSEs), which is a combination of the former-mentioned ceramic and polymer electrolytes [21, 22]. The latter one is found to be more advantageous by combining the pros of both electrolytes thereby suppressing the cons [23]. CSEs possess higher ionic conductivity, enhanced mechanical properties, and better interfacial stability in comparison with its individual components [24]. In addition to the inorganic component, the CSEs contain a lithium salt as well. The polymer components along with the anionic part of the lithium salts are considered as Lewis bases, while the inorganic component and the Li^+ cations are referred to as Lewis acids [25, 26]. The acid–base interactions between these components facilitate salt dissociation and promote the transport of free Li^+ ions thereby enhancing the ionic conductivity [27]. Instead of an inorganic component, electrolytes formed by the addition of alternatives such as ionic liquids and metal organic frameworks to the polymer matrix are also termed as composite electrolytes [28].

In this chapter, we aim to provide a detailed analysis on CSEs. We begin by explaining the inorganic and polymer solid electrolyte systems, where the different types of inorganic electrolytes and the components of the polymer electrolyte system are discussed in detail. Further, by discussing the failures of the latter ones, we move on to the composite electrolyte systems. An overview of the traditional methods adopted in fabricating composite electrolytes is provided. Three major electrochemical properties of a composite electrolyte – ionic conductivity, transference number, and electrochemical stability window, are further explained. Finally, we have attempted to throw light on the numerous modifications incorporated by different research groups at the material level and the processing level in fabricating composite electrolytes with enhanced properties and performance.

7.2 PROPERTIES OF AN IDEAL SOLID ELECTROLYTE

To be suitable for application in electrochemical devices, the solid electrolyte materials should inherently possess the following properties.

(1) **Ionic conductivity**: The room temperature ionic conductivity for solid electrolytes is expected to be greater than 10^{-3} S/cm [29].
(2) **Transference Number**: In general, the transference number is expected to be close to unity [30]. Also, it is essential that the electrolyte is a single-ion conductor. A larger cation transference number can reduce the concentration polarization of the electrolytes thereby improving the power density [31].
(3) **Electrochemical Stability:** A wide electrochemical stability window of 0–5 V versus Li/Li^+ is desirable for SSEs. In an ASSB, the reduction potential limit of the electrolyte must be lower than that of the electrochemical potential of lithium in anode while the oxidation potential limit should be

higher than that of the cathode's potential [32]. A wider electrochemical window ensures longer cycle life and reversibility of the battery.

(4) **Chemical and thermal stability**: To avoid undesirable side reactions that cause safety issues, it is essential that the electrolyte is chemically stable when in contact with the electrodes. At the same time, the SSE must possess a strong thermal stability to facilitate the operation of LIBs under a wide range of temperatures [24].

(5) **Mechanical Strength**: Upon cycling, there arises volume changes that may lead to microcracks on the electrolytes thereby harnessing its long-term performance. Thus, for the practical application of ASSBs, the electrolyte used in the system must be mechanically strong and flexible.

The relative values of the aforementioned variables of polymer, inorganic, and hybrid electrolytes are given in Figure 7.1.

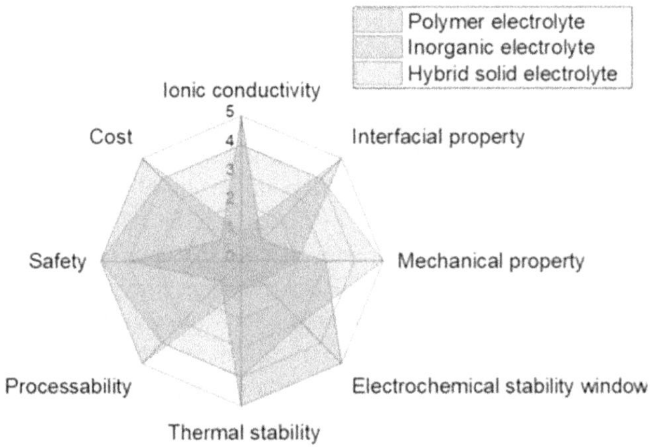

FIGURE 7.1 Properties of solid electrolytes (5 = best, 1 = worst).

Source: Reprinted with permission from [33].

7.3 INORGANIC/CERAMIC SOLID ELECTROLYTES

Inorganic solid electrolytes, also termed as 'ceramic electrolytes', are crystalline or amorphous glasses that exhibit the property of Li-ion conduction [34]. In comparison to liquid electrolytes that conduct via ion motion in solvents, inorganic solid electrolytes contain defects in the form of vacancies and interstitial sites that serve as ion conduction pathways [35]. Perovskites, NaSICON, Garnet, Thio-LiSICON, Argyrodites, and Sulfide glasses are some common examples of ceramic electrolytes that can be grouped as oxides or sulfides [36].

The crystal structure of a few of the important classes of inorganic electrolytes are given in Figure 7.2. The representative conductivity values of various inorganic electrolytes are given in Table 7.1.

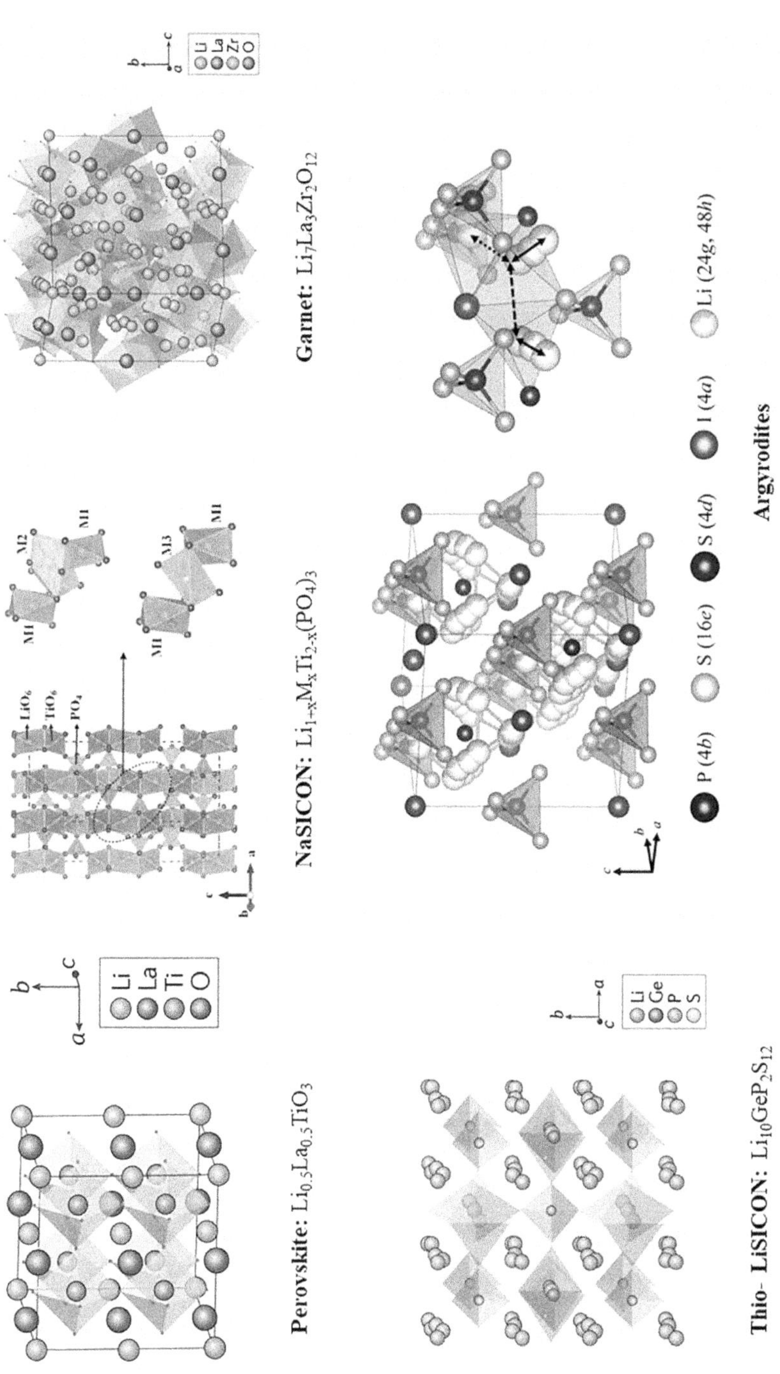

FIGURE 7.2 Crystal structure of inorganic electrolytes.

Source: Reprinted with permission from [37–39].

TABLE 7.1
Classification of Inorganic Electrolytes

Type	Example	Synthesis Routes	Ionic Conductivity (S cm^{-1})	Ref.
Perovskites	$Li_{0.34}La_{0.56}TiO_3$ (LLTO)	High-temperature solid-phase method	9.6×10^{-6}	[56]
	$Li0_{.38}Sr_{0.44}Ta_{0.7}Hf_{0.3}O_{2.95}F_{0.05}$ ($LSTHF_5$)	Solid state reaction	4.8×10^{-4}	[57]
	$Li_{0.33}La_{0.55}TiO_3$	Molten salt method	2.70×10^{-5}	[58]
NaSICON	$Li_{1.3}Al_{0.3}Ti_{1.7}(PO_4)_3$ LATP	Pechini synthesis	6×10^{-4}	[59]
	$Li_{1.3}Al_{0.3}Ti_{1.7}(PO_4)_3$ (LATP)	Co-precipitation	2.19×10^{-4}	[60]
	$Li_{1.7}Al_{0.3}Ti_{1.7}Si_{0.4}P_{2.6}O_{12}$LATSP	Solution-based method	1.33×10^{-3}	[61]
	$Li_{1.5}Al_{0.5}Ti_{1.5}(PO_4)_3$	Sol-gel	3.4×10^{-3}	[62]
Garnet	$Li_{6.4}La_3Zr_{1.4}Ta_{0.6}O_{12}$ (LLZTO)	Ball milling and fast oxygen-assisted sintering	6.13×10^{-4}	[63]
	$Li_7La_3Zr_2O_{12}$ (LLZO)	Co-precipitation	2.0×10^{-4}	[64]
	$Li_{6.4}La_3Zr_{1.4}Ta_{0.6}O_{12}$	Oscillatory pressure sintering	4.43×10^{-4}	[65]
	$Li_{6.75}La_3Zr_{1.75}Nb_{0.25}O_{12}$ (LLZNb)	Co-precipitation	1.0×10^{-4}	[66]
LISICON	$Li_{3.75}Ge_{0.75}P_{0.25}O_4$ (LGPO)	Solid state reaction	2×10^{-5}	[67]
	$Li_{3.68}(Ge_{0.6}V_{0.36}Ga_{0.04})O_4$	Solid state reaction	1.5×10^{-4}	[68]
	$Li_{10}SiP_2O_{12}$	Melt quenching	3.5×10^{-6}	[69]
Thio-LiSICON	$Li_{10}GeP_2S_{12}$ (LGPS)	Mechanical ball milling	3.27×10^{-3}	[70]
	$Li_{10}SiP_2S_{12}$ (LSiPS)	Mechanochemical milling	2.3×10^{-3}	[71]
	$Li_{11}AlP_2S_{12}$	Mechanical milling	8.0×10^{-4}	[72]
Argyrodites	Li_6PS_5Cl	Microwave irradiation	2.1×10^{-3}	[73]
	Li_6PS_5Br	Ball milling	6.8×10^{-3}	[74]
	$Li_{6.55}P_{0.45}Si_{0.55}S_5I$	Ball milling	1.1×10^{-3}	[75]

7.3.1 Oxide Solid Electrolytes

Perovskites: Perovskites are of interest to the scientific community owing to their optimal bulk ionic conductivity as high as 10^{-3} S/cm and appropriate electrochemical window [40]. Lithium lanthanum titanate (LLTO) is a famous example for perovskite-type electrolyte with the general formula $Li_3La_{(2/3-x)}TiO_3$. LLTO contains A sites formed by TiO_6 octahedra groups that are responsible for Li^+ migration leading to high ionic conductivity [41]. However, the large number of grain boundaries remains a major drawback in practical applications. In addition, the Ti^{4+} reduces to Ti^{3+} causing electrolyte decomposition and cell short-circuiting during cycling with lithium metal anodes [42].

NaSICON Type: The NaSICON-type electrolytes such as LATP with general formula $Li_{1+x}M_xTi_{2-x}(PO_4)_3$ are low-cost electrolytes with low toxicity and high electrochemical stability [43]. The ionic conductivity values in the range of 10^{-4}–10^{-3} S/cm have further attracted researchers toward this type of electrolytes [44]. LATP is made up of TiO_6 octahedra connected to PO_4 tetrahedra forming 3D structures [45]. The incorporation of additional lithium ions at M3 sites for the charge compensation during Al substitution at Ti sites and creation of vacancies at M1 sites in the structure are responsible for Li-ion passage [46].

Garnet: Garnet-type electrolytes exhibit notable chemical stability against the electrodes with a wide electrochemical window. $Li_7La_3Zr_2O_{12}$ (LLZO) is a garnet-type electrolyte with high stability and suitable ionic conductivity [47]. Though LLZO exists in both cubic and tetragonal phases, the cubic phase has a desirable bulk ionic conductivity and is stable in air in comparison with the tetragonal phase. However, a higher synthesis temperature is required to obtain the cubic phase [48]. Over the years, several research have been carried out in the elemental doping of LLZO to enhance its performance and properties [49].

7.3.2 Sulfide Solid Electrolytes

Sulfide-based electrolytes are shown to possess an exceptional ionic conductivity similar to liquid electrolytes. This property can be attributed to the lower electronegativity of S atoms leading to weaker S-Li bonds in comparison with O [50]. However, the handling of sulfide-based electrolytes seems to be a tedious task due to its immediate reactivity with moisture and air [51].

Thio-LiSICON: $Li_{10}GeP_2S_{12}$ (LGPS) falls under the category of thio-LiSICONs and possess exceptional ionic conductivity and a wide potential window [52]. LGPS has a three-dimensional structural framework, where LiS_6 octahedra and the $(Ge_{0.5}P_{0.5})$ S4 tetrahedral units are connected by a common edge to form the 1D chain and are further connected with the PS_4 tetrahedral units at the corners [53].

Argyrodites: Argyrodites characterized by good ionic conductivities and electrochemical stabilities are potential candidates for solid electrolytes [54]. However, it is to be noted that argyrodites are air and moisture sensitive, and their structures are metastable leading to a limitation in synthesis. Research on argyrodites has brought in the concept of aliovalent substitution for ionic conductivity and structural enhancements [55].

7.4 LIMITATIONS OF CERAMIC ELECTROLYTES

Some major challenges include electrolyte decomposition reactions, interfacial contact issues, and dendrite growth (Figure 7.3).

7.4.1 Electrolyte Reactions and Interfacial Layers

The formation of ion-resistant and electronically conducting decomposition products at the electrolyte–electrode interface restricts the practical application of SSEs. Such reactions are a result of the volume changes during repeated cycling as well as contact issues. When cathodic limit of the electrolyte falls below the electrochemical potential of the anode, an interfacial reaction takes place between the electrode and the electrolyte [37]. Such reactions lead to the creation of an interfacial layer that is detrimental to the functioning of the system. In NASICON-type electrolytes containing Ti^{4+} ions, a rapid reduction of the ion to Ti^{3+} occurs by the Li metal [76]. Hence, it becomes essential to ensure that batteries are operated within the electrochemical potential window of the electrolytes. Stability of electrolytes in the ambient atmosphere also poses a serious threat to ASSBs. Most of the sulfide-based electrolytes are highly unstable to oxygen or moisture in atmosphere thereby limiting its use for practical purposes [77].

7.4.2 Contact Issues

The ion movement in solid state systems depends on the contact between the particles. An interfacial barrier between the electrolyte and electrode particles reduces the overall ionic conduction [78]. In case of ceramic electrolytes, though they have a large bulk ionic conductivity, the grain boundaries create an interfacial barrier leading to a decline in the total ionic conductivity of the electrolyte [50]. These gaps also affect the cycling abilities of the batteries [79]. The contact particles experience greater stress upon cycling thereby leading to cracks in the system.

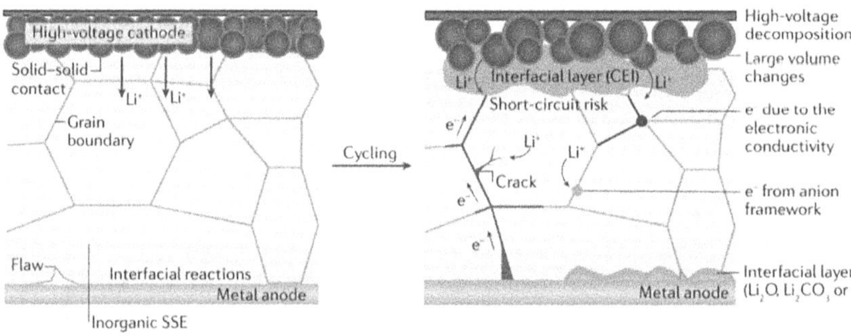

FIGURE 7.3 Major demerits of inorganic solid electrolyte-based batteries.

Source: Reprinted with permission from [37].

7.4.3 Dendrite Growth

Void spaces of the electrodes, electrolytes, and their interfaces not only reduce the lithium-ion motion but also act as suitable spaces for dendrite generation [80]. Lithium dendrite growth arises at such voids leading to short circuits of the system [81]. The grain boundaries of the electrolytes are nucleation sites with higher current density that cause dendrite formation [82]. The initiation and growth of dendrites occurs upon charging during the plating of Li-metal anode. Upon stripping of Li metals from the anode during the discharge process, void spaces are generated in the anode. This further generates an increase in current concentration at the contact areas and finally forms dendrites on charging [83].

7.5 POLYMER ELECTROLYTES

The study on polymer electrolytes started after Fenton reported, in the year of 1973, about the ionic conductivity properties of polymer salts that are formed with poly ethylene oxide and alkali metal salts [84]. Since then, a large number of polymer materials such as poly (ethylene oxide) (PEO), poly (carbonate) (PEC), poly (vinylidene fluoride) (PVDF), poly (acrylonitrile) (PAN), poly (methyl methacrylate) (PMMA), and their derivatives are being studied for practical applications [85–87].

The conductivity of polymer electrolytes occurs via the migration of Li ions from one coordination point to the other along the length of the polymer chain (Figure 7.4)

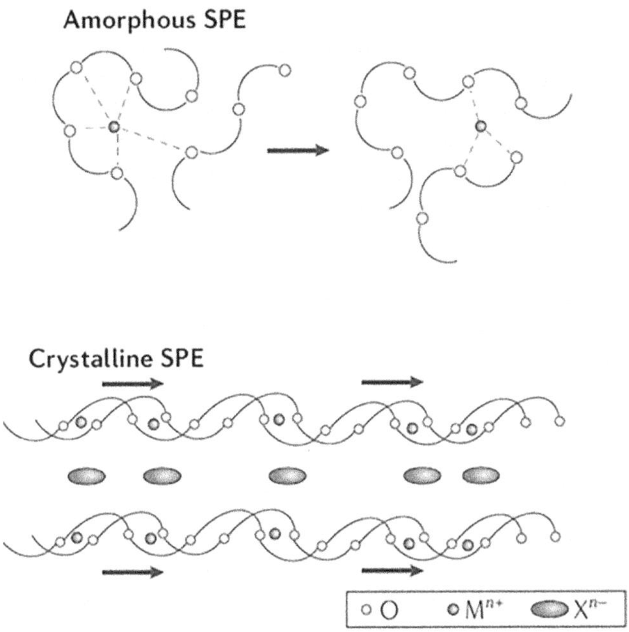

FIGURE 7.4 Ion conduction in solid polymer electrolytes.

Source: Reprinted with permission from [37].

[88]. The migration of ions is a result of the segmental motion of the polymer chains [89, 90]. In amorphous polymeric systems, the lithium ion transport is a result of sub diffusive motion along the polymer chains, intersegmental hopping, and the overall motion of the polymer chain with the coordinated ions [91, 92]. While, for high-molecular-weight polymers, intersegmental hopping is the predominant form of ion motion, diffusion in the solvated form is the conduction pathway for low-molecular-weight polymers [37]. In case of crystalline polymer systems, a folding of the polymer chains into cylindrical tunnels occurs where Li^+ ions migrate via hopping through the tunnels [93, 94].

7.5.1 Lithium Salts

To facilitate high conductivity, the lithium-based salts should be dissolved in the polymer and coupled with lithium ions [95]. The ionic conductivity of a few of the important polymer-salt matrices is given in Table 7.2. Use of salts with larger anionic groups such as lithium bis(trifluoromethylsulfonyl)-imide (LiTFSI), lithium bis(trifluoromethylsulfonyl)-methide (LiTFSM), and lithium bis(perfluoroethylsulfonyl)imide are found to provide better cationic transference number as a result of minimized anionic migration [96]. However, these salts are found to corrode the current collectors [97]. The number of migratable lithium ions depend on the salt dissociation ability of the polymer host. The good solubility of alkali salts is associated with the presence of oligoethers due to their metal complexing abilities [98]. The linear oligoether chains contain -O- atom groups with lone pairs of electrons that can function as ligating groups for the Li metal ion [99]. This causes the dissociation of cations and anions of the lithium salt [100]. Further, dissociation of salt in polymer requires a salt with low lattice energy and a polymer with high dielectric constant [101]. However, an increased Li+–salt concentration leads to a decrease in ionic conductivity as well as transference number [102]. This occurs due to a reduction in the number of charge carriers as a result of the formation of ion aggregates. There is also a hindrance to the polymer chain motion due to reduction in the free spaces within the sample [103].

TABLE 7.2
Ionic Conductivity Values of Polymer Electrolyte Systems

Polymer/Salt Matrix	Conductivity (S/cm)	Reference
p(TMC)/$LiBF_4$	~ 3×10^{-4} (95 °C)	[104]
PVDF/$LiClO_4$	1.0×10^{-4}	[105]
MEEP/$NaCF_3SO_3$	1×10^{-5}	[106]
PEC-LiTFSI	~0.5×10^{-5}	[107]
PAN/LiTFSI	8.5×10^{-5}	[108]

7.5.2 Organic Solvent

The structure and electrochemical performance of the electrolyte depends on the solvent used for polymer and salt dissolution. Solvents such as dimethyl formamide (DMF), dimethylacetamide (DMAc), and acetonitrile (ACN) are utilized to dissolve the polymer and lithium salt at the time of electrolyte preparation [109–111]. The solvent gets trapped within the polymer matrix and performs the function of a plasticizer thereby increasing conductivity [112]. Lahiri et al. studied the effect of acetonitrile solvent on the ionic conductivity of a hybrid polymer-gel electrolyte based on polyvinylidene fluoride-co-hexafluoropropylene (PVDF-HFP), 1-butyl-1-methylpyrrolidinium bis(fluorosulfonyl)imide ([Py1,4]FSI), and lithium bis(fluorosulfonyl) imide (LiFSI) [113]. A combination of two different solvents has also been reported to aid in preparing polymer electrolyte membranes with better properties. Tian et al. reported the use of a combination of DMF with acetone as solvent generated gel polymer electrolytes that exhibited a high porosity of 42% along with superior electrochemical performance [114].

In general, polymer electrolytes are flexible and less prone to decomposition reactions [115–117]. Also, they provide the added advantage of blocking the dendrite growth on Li metal anodes [118]. However, their room temperature ionic conductivities are found to be much lower than the required conductivity of 10^{-3} S cm^{-1} for practical applications [119, 120].

7.6 COMPOSITE SOLID ELECTROLYTES

During the synthesis of CSEs, polymer and the Li-based salt dissolve in the organic solvent while the inorganic electrolyte component disperses into the solvent forming a homogeneous solution [121]. The inorganic component can either be inert/passive ceramic fillers that simply play the role of plasticizers or active ceramic fillers that provide a second path for Li-ion transport in addition to polymer electrolytes [122]. The concept of inert ceramic fillers was first proposed by Croce et al., who reported the addition of inorganic ceramic nanoparticles such as SiO_2, Al_2O_3, and TiO_2 into polymer electrolytes as a means to obtaining enhanced ionic conductivity and electrode–electrolyte stability [27]. Ceramic fillers dispersed in polymer matrix prevent polymer crystallization, improve mechanical stabilities, as well as electrode–electrolyte stability [123, 124]. Moreover, they speed up the segmental motion of the polymer chains thereby enhancing the ionic motion.

Several studies have reported the conductivity enhancement to depend on particle size and concentration of the filler. CSE composing of ionically conducting Li_3N in PEO matrix was first introduced in 1988 by Skaarup et al. [125]. They have reported an increase in ionic conductivity with the increase in Li_3N content. Rajendran et al. have reported an increase in conductivity value by an order of two magnitudes upon dispersing inert TiO_2 particles into polyvinyl chloride (PVC)/PAN composite polymer electrolyte system [126]. Better electrochemical performance by nanosized filler-incorporated polymer electrolytes than micro-sized filler-based electrolytes are reported by Manuel et al. [127]. Further modifications of the filler, such as silane modification of SiO_2 fillers, were experimented by Fan et al., and an increase in ionic conductivity along with improvement of mechanical properties have been noted [128].

7.7 PREPARATION OF COMPOSITE ELECTROLYTES

The preparation traditionally involves a solution casting technique (Figure 7.5) [129]. It begins with dissolution of the polymer constituent in a suitable solvent followed by addition of the inorganic component. This mixture is further made into a thick slurry via magnetic stirring, poured onto a dish or mold, and finally air/vacuum dried or placed on a hot plate to generate thin films [130–132]. If the thick slurry is coated on a glass plate with a doctor blade, the preparation technique is sometimes termed as tape casting [133]. Based on the appearance of the films obtained, the solvent used, weight ratio of the components, and the process of slurry preparation, coating and drying have been optimized differently in different works (Table 7.3).

Zha et al., on the other hand, dissolved PEO and LiTFSI salt in acetonitrile with an EO: Li molar ratio of 18:1. The doctor blade-coated slurry was dried at room temperature for 24 h followed by vacuum drying at 45 °C [140]. Instead of doctor blade coating, Laura C. et al. have standardized the film thickness by sandwiching the slurry in between two glass plates [141].

On the other hand, Kim et al. have produced phase inversion membranes by dipping the slurry-coated glass plate into a flowing water bath [142]. This might help in better removal of the solvent and in efficient penetration of water molecules in one

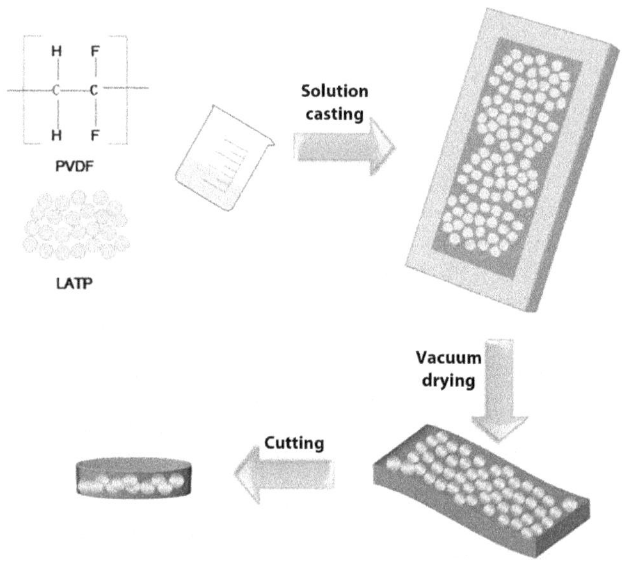

FIGURE 7.5 Schematic representation of the preparation of composite electrolyte membranes. Reprinted with permission from [134].

TABLE 7.3
Preparation Conditions of Composite Membranes via Solution Casting

System	Polymer: Salt	Solvent	Stirring Duration	Drying Method	Ref.
PVDF-HFP/$LiClO_4$	90:10	Dimethylformamide	12 h	Air drying	[135]
PEO/$LiClO_4$/LATP	8:1	Acetonitrile	24 h at 60–70 °C	Slow evaporation followed by high vacuum drying at 50 °C for 24 h	[136]
PVDF/$LiClO_4$/LLZO	3:1	N, N-dimethylformamide	4 h at 50 °C	Vacuum drying at 60 °C for 16 h	[137]
PEO/LiTFSI/LLTO	18:1	Acetonitrile	24 h	Drying at room temperature for 36 h followed by heating at 60 °C for 12 h. Finally, vacuum drying at 60° C for 12 h	[138]
PEO/LiTFSI/silica nanotubes	16:1	Acetonitrile	16 h	Slow evaporation at room temperature followed by vacuum drying for 48 h	[139]

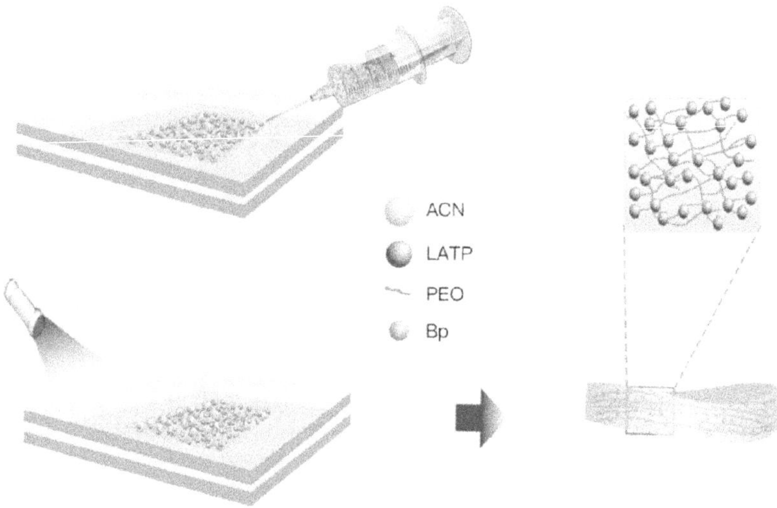

FIGURE 7.6 Composite membrane preparation via UV irradiation.

Source: Reprinted with permission from [145].

direction. The prepared phase inversion membranes were found to have enhanced physical and electrochemical properties in comparison to the conventionally casted membranes. In contrast, Bonizzoni et al. mixed together LATP, PEO, and LiTFSI in acetonitrile and applied a pressure to the mixture after solvent evaporation. A load of 500 MPa was applied uniaxially at room temperature to form the membranes [143]. Pandian et al. have reported the use of spray coating followed by hot pressing to fabricate composite membranes with high ceramic loading of over 60 volume % [144]. They have observed an enhanced conductivity of one order of magnitude than those prepared via solution casting.

To achieve high mechanical strength and improved electrochemical performance, Siyal et al. fabricated composite membranes of PEO and LATP via UV irradiation with Benzophenone as the photo initiator (Figure 7.6). A magnetically stirred solution of LATP and PEO with Benzophenone was injected into a silicon mold and cured under UV light to obtain the membranes [145]. Similarly, a composite electrolyte composed of Al_2O_3 inert filler in liquid electrolyte-swollen acrylate macromer matrix is fabricated by UV curing of the precursor solution containing 2-hydroxy-2-methyl-1-phenyl-1-propanon as the photoinitiator [146].

7.8 ELECTROCHEMICAL PROPERTIES OF THE COMPOSITE MEMBRANES

7.8.1 Ionic Conductivity

Composite electrolytes make use of electrochemical impedance spectroscopy (EIS) to identify the ionic conductivity [147, 148]. Upon application of an AC potential to a cell containing the composite electrolyte sandwiched between two blocking electrodes, a Nyquist plot is obtained [149]. The plot is usually composed of one or two semicircles from high frequency to intermediate frequency and a sloped line at low frequency. The real axis intercept at the high-frequency point or the first semicircle represents the bulk resistance and x intercept at the intermediate frequency or a second semicircle corresponds to grain boundary resistance [150, 151]. The sloped line at low frequency was associated with the Warburg diffusion of lithium ions at the blocking electrodes [152].

The ionic conductivity (σ) is calculated based on the equation [153],

$$\sigma = \frac{D}{RA}$$

where,

D is the thickness of the membrane;
R is the resistance of the electrolyte;
A is the area of the electrolyte.

EIS measurements are also carried out at various temperatures to calculate the activation energy of the material based on the Arrhenius equation [154]. For those materials that obey the Arrhenius concept of ionic conductivity, ion transport occurs via hopping of cations to vacant sites and not from the relaxation of the polymer chains [155].

The Arrhenius type equation can be represented as [156]

$$\sigma = \sigma_0 e^{\frac{-E_a}{K_b T}}$$

where

σ is the ionic conductivity;
σ_0 is the pre-exponential factor;
K_b is the Boltzmann constant;
T is the absolute temperature;
E_a is the activation energy (computed from slope of the linear plot of log σ versus 1/T).

However, certain polymer-based materials are reported to exhibit a nonlinear Arrhenius plot [157–159]. Such systems follow the Vogal–Tamman–Fulcher (VTF) type relationship [160, 161] given by

$$\sigma = \sigma_0 e^{\frac{-E_a}{K_b (T - T_0)}}$$

where

σ is the ionic conductivity;
σ_0 is the pre-exponential factor;
E_a is the activation energy (computed from the log σ versus 1/T plots);
K_b is the Boltzmann constant;
T_o is the equilibrium glass transition temperature.

The conductivity mechanism based on the VTF equation involves the hopping motion of ions coupled with the relational or segmental motion of the polymer chains [162].

However, the absence of a standardized procedure for EIS measurements creates an ambiguity in the conductivity values of composite electrolytes. Different research groups use different frequency ranges, voltage amplitudes, blocking electrodes, and membrane compositions for even the same material (Table 7.4). For instance, Liang et al. reported the optimum ratio of LATP:PVDF for the composite matrix to be 9:1. This composition exhibited the highest ionic conductivity of 6.26×10^{-4} S cm^{-1} at 30 °C [134]. On the other hand, from the various mass ratios of LATP:PVDF, Shi et al. have reported a ratio of 2:1 to provide the highest ionic conductivity of 0.96×10^{-3} S cm^{-1} at the room temperature [168].

Also, to calculate the bulk ionic conductivity of an electrolyte from an EIS spectra, the resistance value corresponding to the first point of the semicircle at a high frequency region is considered in the literature. However, at such higher frequencies, a phase segregation is observed at the grain boundaries which give rise to fluctuations in the ionic conductivity values [169]. On comparing the LATP/PVDF-HFP composite electrolyte membranes of different compositions, we have identified the role of parameters such as frequency and stack pressure in ionic conductivity measurements (Figure 7.7) [170]. Impedance measurement for the determination of the ionic conductivity is to be carried out under pressure and at a fixed middle frequency perturbation.

TABLE 7.4
Parameters for EIS Measurements

System	Blocking Electrode	Frequency Range	Voltage Amplitude	Ionic Conductivity (S cm^{-1})	Ref.
Ga-LLZO–PEO (LiTFSI)	Lithium	5 MHz–1Hz	10 mV	1.2×10^{-4}	[163]
LATP/PVDF–LiTFSI (30 wt.% LATP)	Stainless steel	1 MHz–1 Hz	10 mV	1.64×10^{-3}	[164]
$LiSnZr(PO_4)_3$/PVDF–LiTFSI (15 wt.% $LiSnZr(PO_4)_3$)	Stainless steel	5 MHz–1 Hz	100 mV	5.76×10^{-5}	[165]
$BaTiO_3$ nanoparticles in PVDF–HFP–PVAc–LiTFSI (7.5 wt% $BaTiO_3$)	Stainless steel	1 MHz–1 Hz	10 mV	2.3×10^{-3}	[166]
LGPS/PEO/LiTFSI (90 wt.% LGPS)	Stainless steel	1 MHz–1 Hz	10 mV	1.22×10^{-5}	[167]

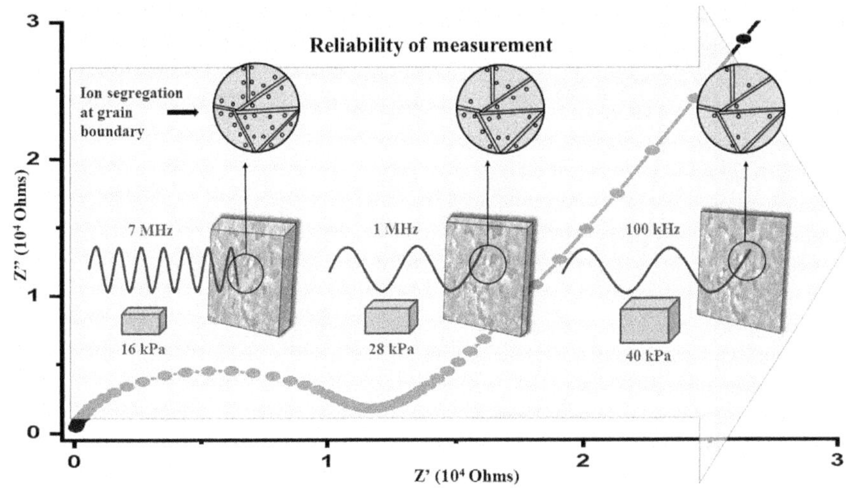

FIGURE 7.7 Representation of parameters affecting EIS measurements.

7.8.2 Transference Number (t_{Li+})

Transference number is a dimensionless parameter that relates the transport of a particular charged species present in the electrolyte to the total charge transport contribution by all the ions present in the system [171]. In case of Li-ion batteries, at high currents, the movement of anions to the anode would cause a concentration polarization thereby leading to cell failures [172, 173]. Hence, it is only the Li+ ions that are expected to be transported through the system while the anions remain stationary, and this would give rise to a high cation transference number. A high t_{Li}^+ close to unity corresponds to a higher fraction of transported ions composed of only cations [174]. However, most composite electrolytes are reported to exhibit a t_{Li}^+ less than 0.5 [175]. Though the cationic transference number is generally thought to exist between 0 and 1, the presence of several neutral and charged species as well as the method adopted for transference number determination could lead to a negative value as well [176, 177].

Bruce and Vincent have combined AC impedance and DC polarization measurements on a symmetrical cell of Li/SPE/Li to calculate the lithium-ion transference number [178, 179].

The Bruce-Vincent equation is given by [180, 181],

$$t_{Li^+} = \frac{I_{ss}\left(\Delta V - I_0 R_0\right)}{I_0\left(\Delta V - I_{ss} R_{ss}\right)}$$

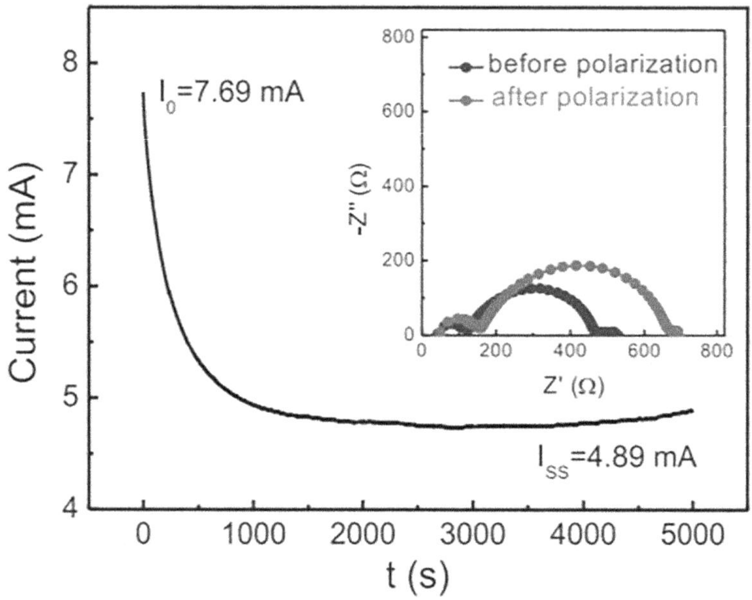

FIGURE 7.8 Chronoamperometric plot for transference number studies (Inset: EIS measurements before and after polarization).

Source: Reprinted with permission from [182].

where,

ΔV=DC voltage applied for cell polarization
I_0 = Initial current through cell
I_{ss} = Steady current through cell
R_0 = Resistance before perturbation of the system obtained from EIS spectra
R_{ss} = Resistance after perturbation of the system obtained from EIS spectra

The chronoamperometric plot used for the determination of transference number in electrolytes, showing the current response over time for a step change in the applied potential along with initial and steady state currents, is provided in Figure 7.8.

Based on the Bruce-Vincent equation, Zhao et al. have calculated the transference number of the PEO–LiTFSI system with 1% LGSP to be 0.26 [183]. On the other hand, LGPS/polyethylene oxide (PEO) composite electrolyte obtained via hot press is reported to have a transference number of 0.87 by Li et al. [184]. The higher transference number is attributed to the high content and dense packing of LGPS within the composite. Also, a decline in transference number is observed with decrease in molding pressure. Xie et al. have reported the transference number of PVDF–HFP with 10% sub-micro-sized alumina to be 0.46 [185]. As reported by Li et al., the incorporation of 1% $Li_{10}SnP_2S_{12}$ into the PEO–LiTFSI polymer matrix is observed to increase the transference number of the system from 0.25 to 0.38. The sulfhydryl groups of $Li_{10}SnP_2S_{12}$ are found to enhance Li^+ conduction via favorable bonding with TFSI anions of the salt PEO–$Li_{10}SnP_2S_{12}$ [186].

7.8.3 Electrochemical Stability Window (ESW)

A wide electrochemical stability window of 0–5 V is desirable for electrolytes to ensure longer cycle life and higher energy density of the battery involving high-voltage cathodes and lithium anodes [187, 188]. Composite electrolytes make use of the electrochemical tools linear sweep voltammetry (LSV) and cyclic voltammetry (CV) to interpret the stability window of the material [189–192]. In LSV, as the name suggests, a potential is linearly swept from low to high value with time through a lithium| composite electrolyte| inert electrode (platinum, stainless steel) cell configuration. Consequently, the current is measured, and the potential at which the electrolyte gets oxidized or reduced is detected (Figure 7.9). Oxidation of the material is represented by an increase in the current value at a particular potential [193].

Peng et al. have reported a very high electrochemical window up to 6 V versus Li/Li + for LLZTO/PEO-LiTFSI composite electrolytes in comparison to a value of 2.5 V versus Li/Li + for PEO-LiTFSI system [194]. Similarly, PEO-LiTFSI incorporated with nanosized LATP exhibited an electrochemical window of 5 V versus Li/Li+ [195]. From the LSV curve, Liang et al. have identified the prepared LLZO/PVDF-HFP membrane to remain undecomposed up to 4.65 V (versus Li/Li +) [182]. PVDF-5% LLZO and the PVDF-10% LLZO composite membranes have the highest stable oxidation potential up to 4.5 V, as reported by Chen et al. [196].

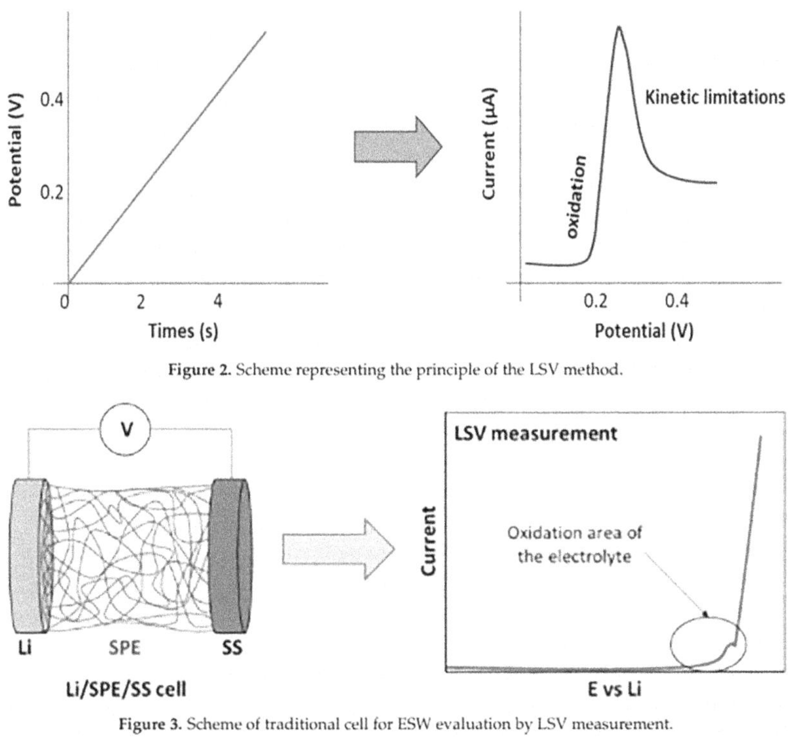

FIGURE 7.9 ESW evaluation by LSV measurements.

Source: Reprinted with permission from [193].

In case of CV, similar to the LSV, the potential is initially swept with time up to a maximum set potential (forward scan) after which the potential reverts back in the opposite direction to the initial potential (backward scan). The intense peaks observed along the positive and negative current correspond to oxidation and reduction occurring with the material [197, 198]. The corresponding potential range between the peaks is identified to be the stable potential window within which the material can be operated. Also, during the experimentation, the scan rate at which the potential is applied can alter the voltammograms obtained. A faster scan rate leads to higher currents [199].

7.9 STRATEGIES TO MITIGATE ISSUES RELATED TO COMPOSITE ELECTROLYTES

Though composite electrolytes are advantageous in comparison to pure inorganic or polymer electrolytes, it is essential to identify suitable strategies to modify the electrolyte membranes fabricated via the traditional methods described earlier. These modifications should have the potential to improve the mechanical strength, ionic conductivity, and compatibility of composite electrolytes with electrode materials, thereby advancing the development of high-performance solid-state batteries.

One approach to modifying composite electrolytes is by incorporating nanostructures such as nanofibers and nanowires [200, 201]. This method enhances the mechanical strength and ionic conductivity of the composite electrolytes (Figure 7.10). The nanostructures act as a scaffold for the polymer electrolyte, providing a high surface area and high aspect ratio, which improves the interface. Moreover, the presence of nanofibers and nanowires in the composite electrolyte promotes the formation of a

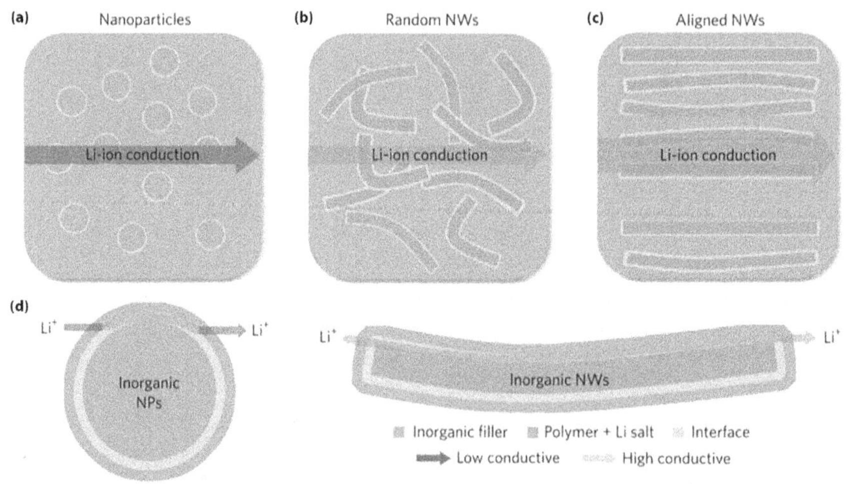

FIGURE 7.10 Li-ion conduction pathways in nanofiber/nanowire-based composite electrolytes.

Source: Reprinted with permission from [203].

percolation network, which enables the efficient transport of ions across the electrolyte. In addition to providing longer ion pathways, nanofibers also prevent polymer crystallization and enhance segmental motion of the polymer chains [202].

In comparison to spherical or zero-dimensional ceramic particles, one-dimensional inorganic nanofiber with high surface area-to-volume ratio ensures good filler–polymer contact with minimized particle–particle junctions [204]. Aligning the nanowires in one direction provides faster ion conduction without having to cross junctions [203]. Ion conduction also depends on the angle of alignment where a placement of the nanowires in the direction of the electrode and perpendicular to the electrolyte surface provides the maximum conductivity [205].

Another method for modifying composite electrolytes is through the addition of organic plasticizers [206, 207]. These plasticizers act as a softening agent for the polymer electrolyte, which reduces the glass transition temperature (Tg) of the polymer electrolyte and enhances the ionic conductivity. It further improves the electrode–electrolyte interfacial contact, and its highly polar nature helps in the lithium salt dissolution [208].

Though polymer electrolytes crosslinked via traditional methods such as UV curing provide high ionic conductivity, the material remains tough and brittle. Hence, polymerization and grafting are other methods for modifying composite electrolytes [209, 210]. The polymer network provides mechanical stability to the electrolyte, while the grafting of functional groups onto the polymer backbone improves its compatibility with the electrode materials.

The surface of active fillers becomes contaminated when the composite electrolytes come in contact with moist air. This further affects the polymer-ceramic interfacial contact and reduces lithium ion conduction within the material [211, 212]. Hence, the design of composite electrolytes via in situ interactions with the help of electrolyte surface contaminants has been proposed [213]. In situ interactions involve the formation of interfacial layers between the composite electrolyte and the electrode material during the battery operation. These layers can be formed through the electrochemical reduction or oxidation of the electrode material or through the reaction of the electrode material with the electrolyte components. The in situ formed layers can improve the adhesion between the electrode and the electrolyte and reduce the interfacial resistance, which results in better electrochemical performance.

Electrolytes composed of certain inorganic components such as the NaSICON like LATP are not compatible with the lithium metal and exhibits titanium reduction over cycling [214]. Also, for some of the polymer electrolytes such as PEO, the ether groups have a low oxidative stability thereby making them unsuitable for use with high-voltage cathodes [215]. Hence, a bilayer composite electrolyte composed of a cathode-contacting layer that is stable at high voltage and another low-voltage stable heterogeneous layer contacting the anode is proposed (Figure 7.11) [216, 217].

Yu et al. presented a laminated dual-polymer/polymer–ceramic composite electrolyte composed of a PEO-SN layer at anode and a PAN-LATP layer at the cathode [218]. Liang et al. engineered Janus interfaces of composite electrolytes with antioxidative PAN and reduction-reaction-inhibited PEO to the $Li_{1.4}Al_{0.4}Ti_{1.6}(PO_4)_3$ (LATP) surface at the cathode and anode, respectively [219].

And finally, to overcome the predicament of high cost, and toxic solvent traces in the final composite electrolyte, a solvent-free method was adopted to fabricate

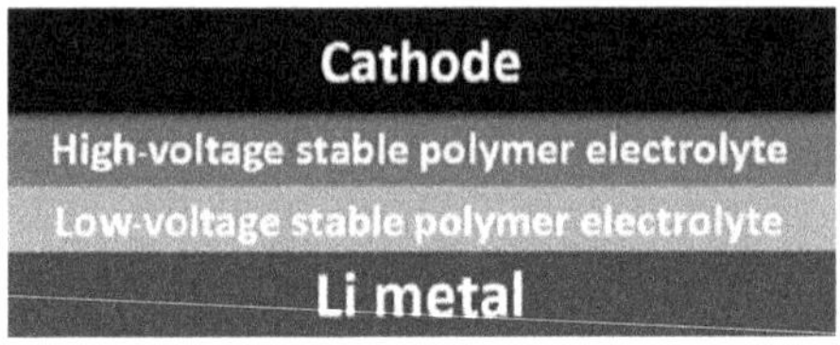

FIGURE 7.11 Stacking of bilayer composites between electrodes.

Source: Reprinted with permission from [216].

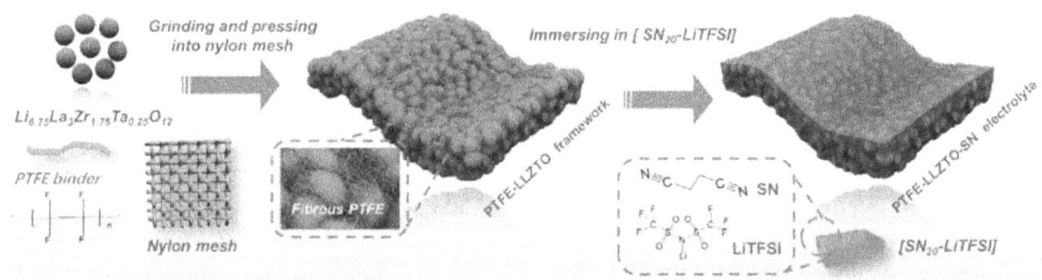

FIGURE 7.12 Solvent-free fabrication of LLZTO-based garnet framework.

Source: Reprinted with permission from [220].

composite electrolyte [220]. A 3D $Li_{6.75}La_3Zr_{1.75}Ta_{0.25}O_{12}$ (LLZTO) framework interconnected by polytetrafluoroethylene (PTFE) binder was combined with LiTFSI dissolved SN organic fillers to form thin, flexible, and nonflammable composite solid electrolyte films (Figure 7.12).

7.10 SOLID STATE ELECTROLYTES FOR COMMERCIAL APPLICATIONS

Batteries with solid electrolytes have been seen as a promising alternative for traditional batteries that pose safety concerns [221]. The number of patents registered in the field of ASSB over the past two decades has skyrocketed between 2010 and 2020 in comparison to the previous decade indicating an increased interest in SSB research [222]. However, owing to the numerous challenges associated with shifting to large-scale production, the developments in this field are still at the research level in research and development centers. Fraunhofer, in an ASSB roadmap, has identified the share of ASSBs in the global demand for LIB to be currently less than 0.5%. The need for interfacial contact between the electrode and electrolyte layers demands better coating technologies and upgradation of the current production line [223]. Though deposition techniques such as photolithography and 3D printing have succeeded in fabricating 3D structures with better interfaces, they still possess several issues and are under active studies. Yet words of many manufacturing companies give us the hope of seeing the SSB technology turn into a reality on a large scale.

7.11 CONCLUSION

This chapter provides an overall glimpse on composite solid electrolytes in terms of the materials being used and the methodologies adopted for the fabrication and modification of composite membranes. On comparing the shortcomings of using either polymer or inorganic electrolytes individually, composite electrolytes seem to be a promising alternative for all-solid-state lithium-ion batteries. Over the past years, researchers have been tremendously involved in proposing suitable modifications on composite electrolytes that could counteract the existing challenges. While every modification is successful in bringing about some positive impact on the property of the material and performance of the cell, there still exists many facets to be worked on. For instance, a modified composite electrolyte may exhibit good electrochemical properties but may not be successful in preventing Li dendrites. Another material may possess high ionic conductivity, yet could have low cyclic stability. Hence, identifying a composite material that possesses all the desirable properties of an ideal electrolyte for practical application is still a challenge. However, the increased interest of researchers and battery manufacturers in the field of SSB is indeed a positive sign of development.

REFERENCES

[1] Chen X, Vereecken PM (2019) Solid and Solid-Like Composite Electrolyte for Lithium Ion Batteries: Engineering the Ion Conductivity at Interfaces. *Adv Mater Interfaces* 6:1–31. https://doi.org/10.1002/admi.201800899

[2] Armand M, Tarascon J-M (2008) Building Better Batteries. *Nature* 451:652–657

[3] Kazyak E, Garcia-Mendez R, LePage WS, et al (2020) Li Penetration in Ceramic Solid Electrolytes: Operando Microscopy Analysis of Morphology, Propagation, and Reversibility. *Matter* 2:1025–1048. https://doi.org/10.1016/j.matt.2020.02.008

[4] Wang W, Yue X, Meng J, et al (2019) Comparative Study of Water-Based LA133 and CMC/SBR Binders for Sulfur Cathode in Advanced Lithium-Sulfur Batteries. *J Phys Chem C* 123:250–257. https://doi.org/10.1021/acs.jpcc.8b10736

[5] Yu X, Manthiram A (2020) A Long Cycle Life, All-Solid-State Lithium Battery with a Ceramic-Polymer Composite Electrolyte. *ACS Appl. Energy Mater* 3:2916–2924. https://doi.org/10.1021/acsaem.9b02547

[6] Albertus P, Anandan V, Ban C, et al (2021) Challenges for and Pathways Toward Li-Metal-Based All-Solid-State Batteries. *ACS Energy Lett* 6:1399–1404. https://doi.org/10.1021/acsenergylett.1c00445

[7] Yi E, Shen H, Heywood S, et al (2020) All-Solid-State Batteries Using Rationally Designed Garnet Electrolyte Frameworks. *ACS Appl Energy Mater* 3:170–175. https://doi.org/10.1021/acsaem.9b02101

[8] Ding Z, Li J, Li J, An C (2020) Review – Interfaces: Key Issue to Be Solved for All Solid-State Lithium Battery Technologies. *J Electrochem Soc* 167:070541. https://doi.org/10.1149/1945-7111/ab7f84

[9] Natarajan S, Moodakare SB, Haridoss P, Gopalan R (2020) Concentration Gradient-Driven Aluminum Diffusion in a Single-Step Coprecipitation of a Compositionally Graded Precursor for LiNi0.8Co0.135Al0.065O2with Mitigated Irreversibility of H2 â H3 Phase Transition. *ACS Appl Mater Interfaces* 12:34959–34970. https://doi.org/10.1021/acsami.0c08502

[10] Zhang Z, Zhang L, Liu Y, et al (2020) One-Step Solution Process Toward Formation of Li6PS5Cl Argyrodite Solid Electrolyte for All-Solid-State Lithium-Ion Batteries. *J Alloys Compd* 812. https://doi.org/10.1016/j.jallcom.2019.152103

[11] Yu S, Schmidt RD, Garcia-Mendez R, et al (2016) Elastic Properties of the Solid Electrolyte Li7La3Zr2O12 (LLZO). *Chem Mater* 28:197–206. https://doi.org/10.1021/acs.chemmater.5b03854
[12] Yu S, Schmohl S, Liu Z, et al (2019) Insights into a Layered Hybrid Solid Electrolyte and Its Application in Long Lifespan High-Voltage All-Solid-State Lithium Batteries. *J Mater Chem A Mater* 7:3882–3894. https://doi.org/10.1039/c8ta11259b
[13] Gao J, Shao Q, Chen J (2020) Lithiated Nafion-Garnet Ceramic Composite Electrolyte Membrane for Solid-State Lithium Metal Battery. *J Energy Chem* 46:237–247. https://doi.org/10.1016/j.jechem.2019.11.012
[14] Xi G, Xiao M, Wang S, et al (2021) Polymer-Based Solid Electrolytes: Material Selection, Design, and Application. *Adv Funct Mater* 31:1–28. https://doi.org/10.1002/adfm.202007598
[15] DeWees R, Wang H (2019) Synthesis and Properties of NaSICON-Type LATP and LAGP Solid Electrolytes. *ChemSusChem* 12:3713–3725. https://doi.org/10.1002/cssc.201900725
[16] Liu J, Liu T, Pu Y, et al (2017) Facile Synthesis of NASICON-Type Li1.3Al0.3Ti1.7(PO4)3 Solid Electrolyte and Its Application for Enhanced Cyclic Performance in Lithium Ion Batteries Through the Introduction of an Artificial Li3PO4 SEI Layer. *RSC Adv* 7:46545–46552. https://doi.org/10.1039/c7ra09335g
[17] Kosova NV, Devyatkina ET, Stepanov AP, Buzlukov AL (2008) Lithium Conductivity and Lithium Diffusion in NASICON-Type Li 1+xTi2-xAlx(PO4)3 (x=∊0; 0.3) Prepared by Mechanical Activation. *Ionics (Kiel)* 14:303–311. https://doi.org/10.1007/s11581-007-0197-5
[18] Zheng Y, Yao Y, Ou J, et al (2020) A Review of Composite Solid-State Electrolytes for Lithium Batteries: Fundamentals, Key Materials and Advanced Structures. *Chem Soc Rev* 49:8790–8839. https://doi.org/10.1039/d0cs00305k
[19] Yue L, Ma J, Zhang J, et al (2016) All Solid-State Polymer Electrolytes for High-Performance Lithium Ion Batteries. *Energy Storage Mater* 5:139–164. https://doi.org/10.1016/j.ensm.2016.07.003
[20] Cheng X, Pan J, Zhao Y, et al (2018) Gel Polymer Electrolytes for Electrochemical Energy Storage. *Adv Energy Mater* 8:1–16. https://doi.org/10.1002/aenm.201702184
[21] Fu X, Yu D, Zhou J, et al (2016) Inorganic and Organic Hybrid Solid Electrolytes for Lithium-Ion Batteries. *CrystEngComm* 18:4236–4258. https://doi.org/10.1039/c6ce00171h
[22] Yang F, Liu Q, Xie W, et al (2022) High-Content Lithium Aluminum Titanium Phosphate-Based Composite Solid Electrolyte with Poly(Ionic Liquid) Binder. *Polymers* 14. https://doi.org/10.3390/polym14071274
[23] Zheng J, Wang P, Liu H, Hu YY (2019) Interface-Enabled Ion Conduction in Li10GeP2S12-Poly(Ethylene Oxide) Hybrid Electrolytes. *ACS Appl Energy Mater* 2:1452–1459. https://doi.org/10.1021/acsaem.8b02008
[24] Agrawal RC, Pandey GP (2008) Solid Polymer Electrolytes: Materials Designing and All-Solid-State Battery Applications: An Overview. *J Phys D Appl Phys* 41:223001. https://doi.org/10.1088/0022-3727/41/22/223001
[25] Fu J, Li Z, Zhou X, Guo X (2022) Ion Transport in Composite Polymer Electrolytes. *Mater Adv* 3809–3819. https://doi.org/10.1039/d2ma00215a
[26] Manuel Stephan A, Nahm KS (2006) Review on Composite Polymer Electrolytes for Lithium Batteries. *Polymer* 47:5952–5964. https://doi.org/10.1016/j.polymer.2006.05.069
[27] Croce F, Persi LL, Scrosati B, et al (2001) Role of the Ceramic Fillers in Enhancing the Transport Properties of Composite Polymer Electrolytes. *Electrochim Acta* 46:2457–2461. https://doi.org/10.1016/S0013-4686(01)00458-3
[28] Tang S, Guo W, Fu Y (2021) Advances in Composite Polymer Electrolytes for Lithium Batteries and Beyond. *Adv Energy Mater* 11:1–29. https://doi.org/10.1002/aenm.202000802

[29] Gulino V, Brighi M, Murgia F, et al (2021) Room-Temperature Solid-State Lithium-Ion Battery Using a LiBH4-MgO Composite Electrolyte. *ACS Appl Energy Mater* 4:1228–1236. https://doi.org/10.1021/acsaem.0c02525
[30] Kim JG, Son B, Mukherjee S, et al (2015) A Review of Lithium and Non-Lithium Based Solid State Batteries. *J Power Sources* 282:299–322. https://doi.org/10.1016/j.jpowsour.2015.02.054
[31] Song JY, Wang YY, Wan CC (1999) Review of Gel-Type Polymer Electrolytes for Lithium-Ion Batteries. *J Power Sources* 77:183–197. https://doi.org/10.1016/S0378-7753(98)00193-1
[32] Huo S, Sheng L, Xue W, et al (2023) Challenges of Polymer Electrolyte with Wide Electrochemical Window for High Energy Solid-State Lithium Batteries. *InfoMat* 5:e12394–12425. https://doi.org/10.1002/inf2.12394
[33] Han L, Lehmann ML, Zhu J, et al (2020) Recent Developments and Challenges in Hybrid Solid Electrolytes for Lithium-Ion Batteries. *Front Energy Res* 8:1–14. https://doi.org/10.3389/fenrg.2020.00202
[34] Das A, Sahu S, Mohapatra M, et al (2022) Lithium-Ion Conductive Glass-Ceramic Electrolytes Enable Safe and Practical Li Batteries. *Mater Today Energy* 29:101118. https://doi.org/10.1016/j.mtener.2022.101118
[35] Yang H, Wu N (2022) Ionic Conductivity and Ion Transport Mechanisms of Solid-State Lithium-Ion Battery Electrolytes: A Review. *Energy Sci Eng* 10:1643–1671. https://doi.org/10.1002/ese3.1163
[36] Chen X, Guan Z, Chu F, et al (2022) Air-Stable Inorganic Solid-State Electrolytes for High Energy Density Lithium Batteries: Challenges, Strategies, and Prospects. *InfoMat* 4:e12248. https://doi.org/10.1002/inf2.12248
[37] Zhao Q, Stalin S, Zhao CZ, Archer LA (2020) Designing Solid-State Electrolytes for Safe, Energy-Dense Batteries. *Nat Rev Mater* 5:229–252. https://doi.org/10.1038/s41578-019-0165-5
[38] Xiao W, Wang J, Fan L, et al (2019) Recent Advances in Li1+xAlxTi2–x(PO4)3 Solid-State Electrolyte for Safe Lithium Batteries. *Energy Storage Mater* 19:379–400
[39] Kraft MA, Ohno S, Zinkevich T, et al (2018) Inducing High Ionic Conductivity in the Lithium Superionic Argyrodites Li6+xP1-xGexS5I for All-Solid-State Batteries. *J Am Chem Soc* 140:16330–16339. https://doi.org/10.1021/jacs.8b10282
[40] Yu T, Yang X, Yang R, et al (2021) Progress and Perspectives on Typical Inorganic Solid-State Electrolytes. *J Alloys Compd* 885:161013. https://doi.org/10.1016/j.jallcom.2021.161013
[41] Bohnke O, Bohnke C, Fourquet J (1996) Mechanism of Ionic Conduction and Electrochemical Intercalation of Lithium into the Perovskite Lanthanum Lithium Titanate. *Solid State Ion* 91:21–31. https://doi.org/10.1016/S0167-2738(96)00434-1
[42] Le HTT, Kalubarme RS, Ngo DT, et al (2015) Bi-Layer Lithium Phosphorous Oxynitride/Aluminium Substituted Lithium Lanthanum Titanate as a Promising Solid Electrolyte for Long-Life Rechargeable Lithium-Oxygen Batteries. *J Mater Chem A Mater* 3:22421–22431. https://doi.org/10.1039/c5ta06374d
[43] Morimoto H, Awano H, Terashima J, et al (2013) Preparation of Lithium Ion Conducting Solid Electrolyte of NASICON-Type Li1+xAlxTi2-x(PO4)3 (x = 0.3) Obtained by Using the Mechanochemical Method and Its Application as Surface Modification Materials of LiCoO2 Cathode for Lithium Cell. *J Power Sources* 240:636–643. https://doi.org/10.1016/j.jpowsour.2013.05.039
[44] Wu P, Zhou W, Su X, et al (2023) Recent Advances in Conduction Mechanisms, Synthesis Methods, and Improvement Strategies for Li1+xAlxTi2–x(PO4)3 Solid Electrolyte for All-Solid-State Lithium Batteries. *Adv Energy Mater* 13:2203440. https://doi.org/10.1002/aenm.202203440

[45] Kwatek K, Ślubowska W, Nowiński JL, et al (2021) Electrical and Structural Properties of Li1.3Al0.3Ti1.7(PO4)3-Based Ceramics Prepared with the Addition of Li4SiO4. *Materials* 14:5729. https://doi.org/10.3390/ma14195729

[46] Feng J, Gao Z, Sheng L, et al (2021) Progress and Perspective of Interface Design in Garnet Electrolyte-Based All-Solid-State Batteries. *Carbon Energy* 3:385–409. https://doi.org/10.1002/cey2.100

[47] Raju MM, Altayran F, Johnson M, et al (2021) Crystal Structure and Preparation of Li7La3Zr2O12 (LLZO) Solid-State Electrolyte and Doping Impacts on the Conductivity: An Overview. *Electrochem* 2:390–414. https://doi.org/10.3390/electrochem2030026

[48] Parascos K, Watts JL, Alarco JA, et al (2022) Compositional and Structural Control in LLZO Solid Electrolytes. *RSC Adv* 12:23466–23480. https://doi.org/10.1039/d2ra03303h

[49] Yang H, Wu N (2022) Ionic Conductivity and Ion Transport Mechanisms of Solid-State Lithium-Ion Battery Electrolytes: A Review. *Energy Sci Eng* 10:1643–1671. https://doi.org/10.1002/ese3.1163

[50] Kang SG, Kim DH, Kim BJ, Yoon CB (2023) Sn-Substituted Argyrodite Li6PS5Cl Solid Electrolyte for Improving Interfacial and Atmospheric Stability. *Materials* 16:2751. https://doi.org/10.3390/ma16072751

[51] Zhang Q, Hu J, Chu Y, et al (2019) Electrochemical Performance of Sulfide Solid Electrolyte Li 10 GeP 2 S 12 Synthesized by a New Method. *Mater Lett* 248:153–156. https://doi.org/10.1016/j.matlet.2019.04.022

[52] Liang F, Sun Y, Yuan Y, et al (2021) Designing Inorganic Electrolytes for Solid-State Li-Ion Batteries: A Perspective of LGPS and Garnet. *Mater Today* 50:418–441. https://doi.org/10.1016/j.mattod.2021.03.013

[53] Schneider H, Du H, Kelley T, et al (2017) A Novel Class of Halogen-Free, Super-Conductive Lithium Argyrodites: Synthesis and Characterization. *J Power Sources* 366:151–160. https://doi.org/10.1016/j.jpowsour.2017.09.012

[54] Ohno S, Helm B, Fuchs T, et al (2019) Further Evidence for Energy Landscape Flattening in the Superionic Argyrodites Li6+ xP1-xMxS5I (M = Si, Ge, Sn). *Chem Mater* 31:4936–4944. https://doi.org/10.1021/acs.chemmater.9b01857

[55] Jiang Z, Wang S, Chen X, et al (2020) Tape-Casting Li0.34La0.56TiO3 Ceramic Electrolyte Films Permit High Energy Density of Lithium-Metal Batteries. *Adv Mater* 32:1–7. https://doi.org/10.1002/adma.201906221

[56] Li Y, Xu H, Chien PH, et al (2018) A Perovskite Electrolyte That Is Stable in Moist Air for Lithium-Ion Batteries. *Angew Chem Int Ed* 57:8587–8591. https://doi.org/10.1002/anie.201804114

[57] Huang Y, Liu X, Jiang Y, Zhu X (2021) Synthesis of Textured Li0.33La0.55TiO3 Solid Electrolytes by Molten Salt Method. *Ceram Int* 47:11654–11661. https://doi.org/10.1016/j.ceramint.2021.01.003

[58] Zhao E, Ma F, Jin Y, Kanamura K (2016) Pechini Synthesis of High Ionic Conductivity Li1.3Al0.3Ti1.7 (PO4)3 Solid Electrolytes: The Effect of Dispersant. *J Alloys Compd* 680:646–653. https://doi.org/10.1016/j.jallcom.2016.04.173

[59] Yao Z, Zhu K, Zhang J, et al (2021) Co-Precipitation Synthesis and Electrochemical Properties of NASICON-Type Li1.3Al0.3Ti1.7(PO4)3 Solid Electrolytes. *J Mater Sci Mater Electron* 32:24834–24844. https://doi.org/10.1007/s10854-021-06943-x

[60] Liu M, Li X, Wang X, et al (2018) Facile Synthesis and Electrochemical Properties of High Lithium Ionic Conductivity Li1.7Al0.3Ti1.7Si0.4P2.6O12 Ceramic Solid Electrolyte. *J Alloys Compd* 756:103–110. https://doi.org/10.1016/j.jallcom.2018.04.333

[61] Breuer S, Prutsch D, Ma Q, et al (2015) Separating Bulk from Grain Boundary Li Ion Conductivity in the Sol-Gel Prepared Solid Electrolyte Li1.5Al0.5Ti1.5(PO4)3. *J Mater Chem A Mater* 3:21343–21350. https://doi.org/10.1039/c5ta06379e

[62] Shen F, Guo W, Zeng D, et al (2020) A Simple and Highly Efficient Method Toward High-Density Garnet-Type LLZTO Solid-State Electrolyte. *ACS Appl Mater Interfaces* 12:30313–30319. https://doi.org/10.1021/acsami.0c04850
[63] Shao C, Liu H, Yu Z, et al (2016) Structure and Ionic Conductivity of Cubic Li7La3Zr2O12 Solid Electrolyte Prepared by Chemical Co-Precipitation Method. *Solid State Ion* 287:13–16. https://doi.org/10.1016/j.ssi.2016.01.042
[64] Li HY, Huang B, Huang Z, Wang CA (2019) Enhanced Mechanical Strength and Ionic Conductivity of LLZO Solid Electrolytes by Oscillatory Pressure Sintering. *Ceram Int* 45:18115–18118. https://doi.org/10.1016/j.ceramint.2019.05.241
[65] Imagawa H, Ohta S, Kihira Y, Asaoka T (2014) Garnet-Type Li6.75La3Zr1.75Nb 0.25O12 Synthesized by Coprecipitation Method and Its Lithium Ion Conductivity. *Solid State Ion* 262:609–612. https://doi.org/10.1016/j.ssi.2013.10.059
[66] Okumura T, Taminato S, Miyazaki Y, et al (2020) LISICON-Based Amorphous Oxide for Bulk-Type All-Solid-State Lithium-Ion Battery. *ACS Appl Energy Mater* 3:3220–3229. https://doi.org/10.1021/acsaem.9b01949
[67] Zhao G, Suzuki K, Okumura T, et al (2021) Extending the Frontiers of Lithium-Ion Conducting Oxides: Development of Multicomponent Materials with γ-Li3PO4-Type Structures. *Chem Mater* 34:3948–3959. https://doi.org/10.1021/acs.chemmater.1c04335
[68] Song S, Dong Z, Deng F, Hu N (2018) Lithium Superionic Conductors Li10MP2O12 (M = Ge, Si). *Funct Mater Lett* 11:1850039. https://doi.org/10.1142/S179360471850039X
[69] Zhang Q, Hu J, Chu Y, et al (2019) Electrochemical Performance of Sulfide Solid Electrolyte Li 10 GeP 2 S 12 Synthesized by a New Method. *Mater Lett* 248:153–156. https://doi.org/10.1016/j.matlet.2019.04.022
[70] Whiteley JM, Woo JH, Hu E, et al (2014) Empowering the Lithium Metal Battery Through a Silicon-Based Superionic Conductor. *J Electrochem Soc* 161:A1812–A1817. https://doi.org/10.1149/2.0501412jes
[71] Zhou P, Wang J, Cheng F, et al (2016) A Solid Lithium Superionic Conductor Li11AlP2S12 with a Thio-LISICON Analogous Structure. *Chem Commun* 52:6091–6094. https://doi.org/10.1039/c6cc02131j
[72] Maniwa R, Calpa M, Rosero-Navarro NC, et al (2021) Synthesis of Sulfide Solid Electrolytes from Li2S and P2S5in Anisole. *J Mater Chem A Mater* 9:400–405. https://doi.org/10.1039/d0ta08658d
[73] Rao RP, Adams S (2011) Studies of Lithium Argyrodite Solid Electrolytes for All-Solid-State Batteries. *Phys Status Solidi (A) Appl Mater Sci* 208:1804–1807. https://doi.org/10.1002/pssa.201001117
[74] Zhang J, Li L, Zheng C, et al (2020) Silicon-Doped Argyrodite Solid Electrolyte Li6PS5I with Improved Ionic Conductivity and Interfacial Compatibility for High-Performance All-Solid-State Lithium Batteries. *ACS Appl Mater Interfaces* 12:41538–41545. https://doi.org/10.1021/acsami.0c11683
[75] Yang K, Chen L, Ma J, et al (2021) Progress and Perspective of Li 1 + x Al x Ti 2 -x (PO 4) 3 Ceramic Electrolyte in Lithium Batteries. *InfoMat* 3:1195–1217. https://doi.org/10.1002/inf2.12222
[76] Deshmukh A, Thripuranthaka M, Chaturvedi V, et al (2022) A Review on Recent Advancements in Solid State Lithium-Sulfur Batteries: Fundamentals, Challenges, and Perspectives. *Prog Energy* 4:042001. https://doi.org/10.1088/2516-1083/ac78bd
[77] Sun Z, Liu M, Zhu Y, et al (2022) Issues Concerning Interfaces with Inorganic Solid Electrolytes in All-Solid-State Lithium Metal Batteries. *Sustainability (Switzerland)* 14:9090. https://doi.org/10.3390/su14159090
[78] Horowitz Y, Lifshitz M, Greenbaum A, et al (2020) Review – Polymer/Ceramic Interface Barriers: The Fundamental Challenge for Advancing Composite Solid Electrolytes for Li-Ion Batteries. *J Electrochem Soc* 167:160514. https://doi.org/10.1149/1945-7111/abcd12

[79] Lu Y, Zhao C-Z, Hu J-K, et al (2022) The Void Formation Behaviors in Working Solid-State Li Metal Batteries. *Sci Adv* 8:eadd0510. https://doi.org/10.1126/sciadv.add0510
[80] Yuan C, Lu W, Xu J (2021) Unlocking the Electrochemical–Mechanical Coupling Behaviors of Dendrite Growth and Crack Propagation in All-Solid-State Batteries. *Adv Energy Mater* 11:2101807. https://doi.org/10.1002/aenm.202101807
[81] Raj R, Wolfenstine J (2017) Current Limit Diagrams for Dendrite Formation in Solid-State Electrolytes for Li-Ion Batteries. *J Power Sources* 343:119–126. https://doi.org/10.1016/j.jpowsour.2017.01.037
[82] Cheng Y, Chen J, Chen Y, et al (2021) Lithium Host: Advanced Architecture Components for Lithium Metal Anode. *Energy Storage Mater* 38:276–298. https://doi.org/10.1016/j.ensm.2021.03.008
[83] Nazir K, Aziz AF, Yahya MZA, Ali AMM (2017) Ionic Conductivity Studies of Epoxidized Poly (Methyl Methacrylate)-Grafted Natural Rubber Based Gel Polymer Electrolyte for Dye Sensitized Polymer Solar Cell. *AIP Conf Proc* 1877. https://doi.org/10.1063/1.4999869
[84] Ban X, Zhang W, Chen N, Sun C (2018) A High-Performance and Durable Poly(Ethylene Oxide)-Based Composite Solid Electrolyte for All Solid-State Lithium Battery. *J Phys Chem C* 122:9852–9858. https://doi.org/10.1021/acs.jpcc.8b02556
[85] Costa CM, Lizundia E, Lanceros-Méndez S (2020) Polymers for Advanced Lithium-Ion Batteries: State of the Art and Future Needs on Polymers for the Different Battery Components. *Prog Energy Combust Sci* 79:100846. https://doi.org/10.1016/j.pecs.2020.100846
[86] Cong L, Li Y, Lu W, et al (2020) Unlocking the Poly(Vinylidene Fluoride-Co-Hexafluoropropylene)/Li10GeP2S12 Composite Solid-State Electrolytes for Dendrite-Free Li Metal Batteries Assisting with Perfluoropolyethers as Bifunctional Adjuvant. *J Power Sources* 446:227365. https://doi.org/10.1016/j.jpowsour.2019.227365
[87] Arya A, Sharma AL (2020) A Glimpse on All-Solid-State Li-Ion Battery (ASSLIB) Performance Based on Novel Solid Polymer Electrolytes: A Topical Review. *J Mater Sci* 55:6242–6304. https://doi.org/10.1007/s10853-020-04434-8
[88] Gray F, Armand M (2011) Polymer Electrolytes. *Handbook of Battery Materials: Second Edition*, 627–656. https://doi.org/10.1002/9783527637188.ch18
[89] Golodnitsky D, Strauss E, Peled E, Greenbaum S (2015) Review – On Order and Disorder in Polymer Electrolytes. *J Electrochem Soc* 162:A2551–A2566. https://doi.org/10.1149/2.0161514jes
[90] Diddens D, Heuer A, Borodin O (2010) Understanding the Lithium Transport Within a Rouse-Based Model for a PEO/LiTFSI Polymer Electrolyte. *Macromolecules* 43:2028–2036. https://doi.org/10.1021/ma901893h
[91] Diddens D, Paillard E, Heuer A (2017) Improving the Lithium Ion Transport in Polymer Electrolytes by Functionalized Ionic-Liquid Additives: Simulations and Modeling. *J Electrochem Soc* 164:E3225–E3231. https://doi.org/10.1149/2.0271711jes
[92] MacGlashan GS, Andreev YG (1999) Structure of the Polymer Electrolyte Poly (Ethylene Oxide) 6: LiAsF6. *Nature* 398:792–794. https://doi.org/10.1038/19730
[93] Gadjourova Z, Andreev YG, Tunstall DP, Bruce PG (2001) Ionic Conductivity in Crystalline Polymer Electrolytes. *Nature* 412:520–523. https://doi.org/10.1038/35087538
[94] Arya A, Sharma AL (2018) Effect of Salt Concentration on Dielectric Properties of Li-Ion Conducting Blend Polymer Electrolytes. *J Mater Sci Mater Electron* 29:17903–17920. https://doi.org/10.1007/s10854-018-9905-3
[95] Moskwiak M, Giska I, Borkowska R, et al (2006) Physico- and Electrochemistry of Composite Electrolytes Based on PEODME-LiTFSI with TiO2. *J Power Sources* 159:443–448. https://doi.org/10.1016/j.jpowsour.2006.02.040
[96] Zhang Q, Liu K, Ding F, Liu X (2017) Recent Advances in Solid Polymer Electrolytes for Lithium Batteries. *Nano Res* 10:4139–4174. https://doi.org/10.1007/s12274-017-1763-4

[97] Young WS, Kuan WF, Epps TH (2014) Block Copolymer Electrolytes for Rechargeable Lithium Batteries. *J Polym Sci B Polym Phys* 52:1–16. https://doi.org/10.1002/polb.23404
[98] Badi N, Theodore AM, Alghamdi SA, et al (2022) The Impact of Polymer Electrolyte Properties on Lithium-Ion Batteries. *Polymers* 14:3101–3122. https://doi.org/10.3390/polym14153101
[99] Volkov VI, Yarmolenko O V., Chernyak A V., et al (2022) Polymer Electrolytes for Lithium-Ion Batteries Studied by NMR Techniques. *Membranes* 12:416. https://doi.org/10.3390/membranes12040416
[100] Meyer WH (1998) Polymer Electrolytes for Lithium-Ion Batteries. *Adv Mater* 10:439–448. https://doi.org/10.7536/PC190627
[101] Hooper R, Lyons LJ, Mapes MK, et al (2001) Highly Conductive Siloxane Polymers. *Macromolecules* 34:931–936. https://doi.org/10.1021/ma0018446
[102] Osman Z, Mohd Ghazali MI, Othman L, Md Isa KB (2012) AC Ionic Conductivity and DC Polarization Method of Lithium Ion Transport in PMMA-LiBF 4 Gel Polymer Electrolytes. *Results Phys* 2:1–4. https://doi.org/10.1016/j.rinp.2011.12.001
[103] Silva MM, Barros SC, Smith MJ, MacCallum JR (2004) Characterization of Solid Polymer Electrolytes Based on Poly(trimethylenecarbonate) and Lithium Tetrafluoroborate. *Electrochim Acta* 49:1887–1891. https://doi.org/10.1016/j.electacta.2003.12.017
[104] Shen YJ, Reddy MJ, Chu PP (2004) Porous PVDF with LiClO4 Complex as "Solid" and "Wet" Polymer Electrolyte. *Solid State Ion* 175:747–750. https://doi.org/10.1016/j.ssi.2003.10.020
[105] Greenbaum SG, Adamić KJ, Pak YS, et al (1988) NMR, DSC and Electrical Conductivity Studies of MEEP Complexed with NaCF3SO3. *Solid State Ion* 28–30:1042–1046. https://doi.org/10.1016/0167-2738(88)90328–1
[106] Tominaga Y, Nanthana V, Tohyama D (2012) Ionic Conduction in Poly(ethylene carbonate)-Based Rubbery Electrolytes Including Lithium Salts. *Polym J* 44:1155–1158. https://doi.org/10.1038/pj.2012.97
[107] Gao H, Grundish NS, Zhao Y, et al (2021) Formation of Stable Interphase of Polymer-in-Salt Electrolyte in All-Solid-State Lithium Batteries. *Energy Mater Adv* 2021:1–10. https://doi.org/10.34133/2021/1932952
[108] Jin L, Jang G, Lim H, et al (2022) Improving the Ionic Conductivity of PEGDMA-Based Polymer Electrolytes by Reducing the Interfacial Resistance for LIBs. *Polymers* 14:3443–3455. https://doi.org/10.3390/polym14173443
[109] Nakazawa S, Matsuda Y, Ochiai M, et al (2021) Enhancing Lithium Ion Conductivity and All-Solid-State Secondary Battery Performance in Polymer Composite Electrolyte Membranes with β-Crystalline-rich Poly(vinylidene fluoride) Nanofibers. *Electrochim Acta* 394:139114. https://doi.org/10.1016/j.electacta.2021.139114
[110] He KQ, Zha JW, Du P, et al (2019) Tailored High Cycling Performance in a Solid Polymer Electrolyte with Perovskite-type Li0.33La0.557TiO3 Nanofibers for All-Solid-State Lithium Ion Batteries. *Dalton Trans* 48:3263–3269. https://doi.org/10.1039/c9dt00074g
[111] Zhou C, Bag S, Lv B, Thangadurai V (2020) Understanding the Role of Solvents on the Morphological Structure and Li-Ion Conductivity of Poly(vinylidene fluoride)-Based Polymer Electrolytes. *J Electrochem Soc* 167:070552. https://doi.org/10.1149/1945-7111/ab7c3a
[112] Lahiri A, Pulletikurthi G, Shapouri Ghazvini M, et al (2018) Ionic Liquid-Organic Solvent Mixture-Based Polymer Gel Electrolyte with High Lithium Concentration for Li-Ion Batteries. *J Phys Chem C* 122:24788–24800. https://doi.org/10.1021/acs.jpcc.8b07745
[113] Tian L, Wang M, Xiong L, et al (2018) The Effect of Different Mixed Organic Solvents on the Properties of p(OPal-MMA) Gel Electrolyte Membrane for Lithium Ion Batteries. *Applied Sciences* 8:2587. https://doi.org/10.3390/app8122587

[114] Borodin O, Smith GD (2006) Mechanism of Ion Transport in Amorphous Poly(ethylene oxide)/LiTFSI from Molecular Dynamics Simulations. *Macromolecules* 39:1620–1629. https://doi.org/10.1021/ma052277v

[115] Mindemark J, Lacey MJ, Bowden T, Brandell D (2018) Beyond PEO – Alternative Host Materials for Li+-Conducting Solid Polymer Electrolytes. *Prog Polym Sci* 81:114–143. https://doi.org/10.1016/j.progpolymsci.2017.12.004

[116] Rollo-Walker G, Malic N, Wang X, et al (2021) Development and Progression of Polymer Electrolytes for Batteries: Influence of Structure and Chemistry. *Polymers (Basel)* 13:1–20. https://doi.org/10.3390/polym13234127

[117] Zhou D, Shanmukaraj D, Tkacheva A, et al (2019) Polymer Electrolytes for Lithium-Based Batteries: Advances and Prospects. *Chem* 5:2326–2352. https://doi.org/10.1016/j.chempr.2019.05.009

[118] Liu L, Chu L, Jiang B, Li M (2019) Li 1.4 Al 0.4 Ti 1.6 (PO 4) 3 Nanoparticle-Reinforced Solid Polymer Electrolytes for All-Solid-State Lithium Batteries. *Solid State Ion* 331:89–95. https://doi.org/10.1016/j.ssi.2019.01.007

[119] Najjar R, Passerini S, Nematdoust S, Bresser D (2020) Understanding the Role of Nanoparticles in PEO-Based Hybrid Polymer Electrolytes for Solid-State Lithium–Polymer Batteries. *J Phys Chemi C* 124:27907–27915. https://doi.org/10.1021/acs.jpcc.0c08749

[120] Feng J, Wang L, Chen Y, et al (2021) PEO Based Polymer-Ceramic Hybrid Solid Electrolytes: A Review. *Nano Converg* 8. https://doi.org/10.1186/s40580-020-00252-5

[121] Fan P, Liu H, Marosz V, et al (2021) High Performance Composite Polymer Electrolytes for Lithium-Ion Batteries. *Adv Funct Mater* 31:1–29. https://doi.org/10.1002/adfm.202101380

[122] Jagadeesan A, Sasikumar M, Krishna RH, et al (2019) High Electrochemical Performance of Nano TiO2 Ceramic Filler Incorporated PVC-PEMA Composite Gel Polymer Electrolyte for Liion Battery Applications. *Mater Res Express* 6:105524. https://doi.org/10.1088/2053-1591/ab3cb8

[123] Wang W, Yi E, Fici AJ, et al (2017) Lithium Ion Conducting Poly(ethylene oxide)-Based Solid Electrolytes Containing Active or Passive Ceramic Nanoparticles. *J Phys Chem C* 121:2563–2573. https://doi.org/10.1021/acs.jpcc.6b11136

[124] Skaarup S, West K, Zachau-Christiansen B (1988) Mixed Phase Solid Electrolytes. *Solid State Ion* 28–30:975–978. https://doi.org/10.1016/0167-2738(88)90314–1

[125] Rajendran S, Babu R shanker, Sivakumar P (2008) Investigations on PVC/PAN composite Polymer Electrolytes. *J Memb Sci* 315:67–73. https://doi.org/10.1016/j.memsci.2008.02.007

[126] Manuel Stephan A, Nahm KS, Prem Kumar T, et al (2006) Nanofiller Incorporated Poly(vinylidene fluoride-hexafluoropropylene) (PVdF-HFP) Composite Electrolytes for Lithium Batteries. *J Power Sources* 159:1316–1321. https://doi.org/10.1016/j.jpowsour.2005.11.055

[127] Fan L, Nan CW, Zhao S (2003) Effect of Modified SiO2 on the Properties of PEO-Based Polymer Electrolytes. *Solid State Ion* 164:81–86. https://doi.org/10.1016/j.ssi.2003.08.004

[128] Zheng J, Hu YY (2018) New Insights into the Compositional Dependence of Li-Ion Transport in Polymer-Ceramic Composite Electrolytes. *ACS Appl Mater Interfaces* 10:4113–4120. https://doi.org/10.1021/acsami.7b17301

[129] Stephan AM, Nahm KS, Anbu Kulandainathan M, et al (2006) Poly(vinylidene fluoride-hexafluoropropylene) (PVdF-HFP) Based Composite Electrolytes for Lithium Batteries. *Eur Polym J* 42:1728–1734. https://doi.org/10.1016/j.eurpolymj.2006.02.006

[130] Varishetty MM, Qiu W, Gao Y, Chen W (2009) Structure, Electrical and Optical Properties of (PVA/LiAsF6) Polymer Composite Electrolyte Films. *Polym Eng Sci* 50:878–884. https://doi.org/10.1002/pen

[131] Liang B, Tang S, Jiang Q, et al (2015) Preparation and Characterization of PEO-PMMA Polymer Composite Electrolytes Doped with Nano-Al2O3. *Electrochim Acta* 169:334–341. https://doi.org/10.1016/j.electacta.2015.04.039
[132] Duan H, Fan M, Chen WP, et al (2019) Extended Electrochemical Window of Solid Electrolytes via Heterogeneous Multilayered Structure for High-Voltage Lithium Metal Batteries. *Adv Mater* 31:e1807789. https://doi.org/10.1002/adma.201807789
[133] Liang X, Han D, Wang Y, et al (2018) Preparation and Performance Study of a PVDF-LATP Ceramic Composite Polymer Electrolyte Membrane for Solid-State Batteries. *RSC Adv* 8:40498–40504. https://doi.org/10.1039/C8RA08436J
[134] Yesappa L, Niranjana M, Ashokkumar SP, et al (2018) Optical Properties and Ionic Conductivity Studies of an 8 MeV Electron Beam Irradiated Poly(vinylidene fluoride-: co-hexafluoropropylene)/LiClO4 Electrolyte Film for Opto-electronic Applications. *RSC Adv* 8:15297–15309. https://doi.org/10.1039/c8ra00970h
[135] Wang YJ, Pan Y, Kim D (2006) Conductivity Studies on Ceramic Li1.3Al0.3Ti1.7(PO4) 3-Filled PEO-Based Solid Composite Polymer Electrolytes. *J Power Sources* 159:690–701. https://doi.org/10.1016/j.jpowsour.2005.10.104
[136] Sun Y, Zhan X, Hu J, et al (2019) Improving Ionic Conductivity with Bimodal-Sized Li 7 La 3 Zr 2 O 12 Fillers for Composite Polymer Electrolytes. *ACS Appl Mater Interfaces* 11:12467–12475. https://doi.org/10.1021/acsami.8b21770
[137] Zhu L, Zhu P, Fang Q, et al (2018) A Novel Solid PEO/LLTO-nanowires Polymer Composite Electrolyte for Solid-State Lithium-Ion Battery. *Electrochim Acta* 292:718–726. https://doi.org/10.1016/j.electacta.2018.10.005
[138] Hu J, Wang W, Zhu X, et al (2021) Composite Polymer Electrolytes Reinforced by Hollow Silica Nanotubes for Lithium Metal Batteries. *J Memb Sci* 618:118697. https://doi.org/10.1016/j.memsci.2020.118697
[139] Zha W, Chen F, Yang D, et al (2018) High-Performance Li6.4La3Zr1.4Ta0.6O12/Poly(ethylene oxide)/Succinonitrile Composite Electrolyte for Solid-State Lithium Batteries. *J Power Sources* 397:87–94. https://doi.org/10.1016/j.jpowsour.2018.07.005
[140] Merrill LC, Chen XC, Zhang Y, et al (2020) Polymer-Ceramic Composite Electrolytes for Lithium Batteries: A Comparison Between the Single-Ion-Conducting Polymer Matrix and Its Counterpart. *ACS Appl Energy Mater* 3:8871–8881. https://doi.org/10.1021/acsaem.0c01358
[141] Kim KM, Park NG, Ryu KS, Chang SH (2006) Characteristics of PVdF-HFP/TiO2 Composite Membrane Electrolytes Prepared by Phase Inversion and Conventional Casting Methods. *Electrochim Acta* 51:5636–5644. https://doi.org/10.1016/j.electacta.2006.02.038
[142] Bonizzoni S, Ferrara C, Berbenni V, et al (2019) NASICON-type Polymer-in-ceramic Composite Electrolytes for Lithium Batteries. *Phys Chem Chem Phys* 21:6142–6149. https://doi.org/10.1039/c9cp00405j
[143] Pandian AS, Chen XC, Chen J, et al (2018) Facile and Scalable Fabrication of Polymer-ceramic Composite Electrolyte with High Ceramic Loadings. *J Power Sources* 390:153–164. https://doi.org/10.1016/j.jpowsour.2018.04.006
[144] Siyal SH, Li M, Li H, et al (2019) Ultraviolet Irradiated PEO/LATP Composite Gel Polymer Electrolytes for Lithium-metallic Batteries (LMBs). *Appl Surf Sci* 494:1119–1126. https://doi.org/10.1016/j.apsusc.2019.07.179
[145] Kim SH, Choi KH, Cho SJ, et al (2013) Mechanically Compliant and Lithium Dendrite Growth-Suppressing Composite Polymer Electrolytes for Flexible Lithium-Ion Batteries. *J Mater Chem A Mater* 1:4949–4955. https://doi.org/10.1039/c3ta10612h
[146] Zhang X, Liu T, Zhang S, et al (2017) Synergistic Coupling Between Li6.75La3Zr-175Ta0.25O12 and Poly(vinylidene fluoride) Induces High Ionic Conductivity, Mechanical Strength, and Thermal Stability of Solid Composite Electrolytes. *J Am Chem Soc* 139:13779–13785. https://doi.org/10.1021/jacs.7b06364

[147] Vadhva P, Hu J, Johnson MJ, et al (2021) Electrochemical Impedance Spectroscopy for All-Solid-State Batteries: Theory, Methods and Future Outlook. *ChemElectroChem* 8:1930–1947. https://doi.org/10.1002/celc.202100108
[148] Zhang W, Nie J, Li F, et al (2018) A Durable and Safe Solid-State Lithium Battery with a Hybrid Electrolyte Membrane. *Nano Energy* 45:413–419. https://doi.org/10.1016/j.nanoen.2018.01.028
[149] Pan X, Sun J, Jin C, et al (2020) A Flexible Composite Electrolyte Membrane with Ultrahigh LLZTO Garnet Content for Quasi Solid State Li-air Batteries. *Solid State Ion* 351:115340. https://doi.org/10.1016/j.ssi.2020.115340
[150] Dutta S, Choudhary RNP, Sinha PK, Thakur AK (2004) Microstructural Studies of (PbLa)(ZrTi)O3 Ceramics Using Complex Impedance Spectroscopy. *J Appl Phys* 96:1607–1613. https://doi.org/10.1063/1.1765869
[151] Kotobuki M, Kanamura K, Sato Y, Yoshida T (2011) Fabrication of All-Solid-State Lithium Battery with Lithium Metal Anode Using A12O3-Added Li7La3Zr 2O12 Solid Electrolyte. *J Power Sources* 196:7750–7754. https://doi.org/10.1016/j.jpowsour.2011.04.047
[152] Liu L, Sun C (2020) Flexible Quasi-Solid-State Composite Electrolyte Membrane Derived from a Metal-Organic Framework for Lithium-Metal Batteries. *ChemElectroChem* 7:707–715. https://doi.org/10.1002/celc.201902032
[153] Zhao E, Ma F, Guo Y, Jin Y (2016) Stable LATP/LAGP Double-Layer Solid Electrolyte Prepared: Via a Simple Dry-Pressing Method For Solid State Lithium Ion Batteries. *RSC Adv* 6:92579–92585. https://doi.org/10.1039/c6ra19415j
[154] Aziz SB, Woo TJ, Kadir MFZ, Ahmed HM (2018) A Conceptual Review on Polymer Electrolytes and ion Transport Models. *J Sci Adv Mater Dev* 3:1–17. https://doi.org/10.1016/j.jsamd.2018.01.002
[155] Ahmed HT, Abdullah OG (2020) Structural and Ionic Conductivity Characterization of PEO:MC-NH4I Proton-Conducting Polymer Blend Electrolytes Based Films. *Results Phys* 16:102861. https://doi.org/10.1016/j.rinp.2019.102861
[156] Wan Z, Lei D, Yang W, et al (2019) Low Resistance–Integrated All-Solid-State Battery Achieved by Li7La3Zr2O12 Nanowire Upgrading Polyethylene Oxide (PEO) Composite Electrolyte and PEO Cathode Binder. *Adv Funct Mater* 29:1–10. https://doi.org/10.1002/adfm.201805301
[157] Bobade RS, Pakade (Yawale) S V., Yawale SP (2009) Electrical Investigation of Polythiophene-Poly(vinyl acetate) Composite Films via VTF and Impedance Spectroscopy. *J Non Cryst Solids* 355:2410–2414. https://doi.org/10.1016/j.jnoncrysol.2009.08.041
[158] Mertens IJA, Wübbenhorst M, Oosterbaan WD, et al (1999) Novel Polymer Electrolytes Based on Amorphous Poly(ether-ester)s Containing 1,4,7-trioxanonyl Main Chain Units. Ionic Conductivity Versus Polymer Chain Mobility. *Macromolecules* 32:3314–3324. https://doi.org/10.1021/ma981901j
[159] Sangeetha M, Mallikarjun A, Jaipal Reddy M, Siva Kumar J (2017) FTIR Spectroscopic and DC Ionic Conductivity Studies of PVDF-HFP: LiBF4: EC Plasticized Polymer Electrolyte Membrane. *IOP Conf Ser Mater Sci Eng* 225. https://doi.org/10.1088/1757-899X/225/1/012049
[160] Diederichsen KM, Buss HG, McCloskey BD (2017) The Compensation Effect in the Vogel-Tammann-Fulcher (VTF) Equation for Polymer-Based Electrolytes. *Macromolecules* 50:3831–3840. https://doi.org/10.1021/acs.macromol.7b00423
[161] Bashiri P, Rao TP, Naik VM, et al (2019) AC Conductivity Studies of Polyethylene Oxide-garnet-type Li7La3Zr2O12 Hybrid Composite Solid Polymer Electrolyte Films. *Solid State Ion* 343:115089. https://doi.org/10.1016/j.ssi.2019.115089
[162] Li Z, Sha WX, Guo X (2019) Three-Dimensional Garnet Framework-Reinforced Solid Composite Electrolytes with High Lithium-Ion Conductivity and Excellent Stability. *ACS Appl Mater Interfaces* 11:26920–26927. https://doi.org/10.1021/acsami.9b07830

[163] Jin Y, Liu C, Jia Z, et al (2021) Building a Highly Functional Li1.3Al0.3Ti1.7(PO4)3/Poly(vinylidene fluoride) Composite Electrolyte for All-Solid-State Lithium Batteries. *J Alloys Compd* 874:159890. https://doi.org/10.1016/j.jallcom.2021.159890
[164] Pareek T, Dwivedi S, Ahmad SA, et al (2020) Effect of NASICON-type LiSnZr(PO4)3 Ceramic Filler on the Ionic Conductivity and Electrochemical Behavior of PVDF Based Composite Electrolyte. *J Alloys Compd* 824:153991. https://doi.org/10.1016/j.jallcom.2020.153991
[165] Sasikumar M, Raja M, Krishna RH, et al (2018) Influence of Hydrothermally Synthesized Cubic-Structured BaTiO3 Ceramic Fillers on Ionic Conductivity, Mechanical Integrity, and Thermal Behavior of P(VDF-HFP)/PVAc-Based Composite Solid Polymer Electrolytes for Lithium-Ion Batteries. *J Phys Chem C* 122:25741–25752. https://doi.org/10.1021/acs.jpcc.8b03952
[166] Li M, Kolek M, Frerichs JE, et al (2021) Investigation of Polymer/Ceramic Composite Solid Electrolyte System: The Case of PEO/LGPS Composite Electrolytes. *ACS Sustain Chem Eng* 9:11314–11322. https://doi.org/10.1021/acssuschemeng.1c00904
[167] Shi X, Ma N, Wu Y, et al (2018) Fabrication and Electrochemical Properties of LATP/PVDF Composite Electrolytes for Rechargeable Lithium-ion Battery. *Solid State Ion* 325:112–119. https://doi.org/10.1016/j.ssi.2018.08.010
[168] Uddin MJ, Cho SJ (2018) Reassessing the Bulk Ionic Conductivity of Solid-state Electrolytes. *Sustain Energy Fuels* 2:1458–1462. https://doi.org/10.1039/c8se00139a
[169] Jacob MS, Nikil D, Shanmugam V, et al (2022) Standardization of Ionic Conductivity Measurements in. *Mater Sci Eng B* 286:116049–116057. https://doi.org/10.1016/j.mseb.2022.116049
[170] Liu K, Li X, Cai J, et al (2021) Design of High-Voltage Stable Hybrid Electrolyte with an Ultrahigh Li Transference Number. *ACS Energy Lett* 6:1315–1323. https://doi.org/10.1021/acsenergylett.0c02559
[171] Jorne J (2006) Transference Number Approaching Unity in Nanocomposite Electrolytes. *Nano Lett* 6:2973–2976. https://doi.org/10.1021/nl062182m
[172] Zugmann S, Gores HJ (2014) Transference Numbers of Ions in Electrolytes. In *Encyclopedia of Applied Electrochemistry* 2086–2091. https://doi.org/10.1007/978-1-4419-6996-5_501
[173] Deivanayagam R, Shahbazian-Yassar R (2021) Electrochemical Methods and Protocols for Characterization of Ceramic and Polymer Electrolytes for Rechargeable Batteries. *Batter Supercaps* 4:596–606. https://doi.org/10.1002/batt.202000221
[174] Ye L, Feng Z (2010) Polymer electrolytes as solid solvents and their applications. Polymer Electrolytes 550–582. https://doi.org/10.1533/9781845699772.2.550
[175] Siekierski M, Bukat M, Ciosek M, et al (2021) Transference Number Determination in Poor-dissociated Low Dielectric Constant Lithium and Protonic Electrolytes. *Polymers* 13. https://doi.org/10.3390/polym13060895
[176] Pesko DM, Timachova K, Bhattacharya R, et al (2017) Negative Transference Numbers in Poly(ethylene oxide)-Based Electrolytes. *J Electrochem Soc* 164:E3569–E3575. https://doi.org/10.1149/2.0581711jes
[177] Pesko DM, Sawhney S, Newman J, Balsara NP (2018) Comparing Two Electrochemical Approaches for Measuring Transference Numbers in Concentrated Electrolytes. *J Electrochem Soc* 165:A3014–A3021. https://doi.org/10.1149/2.0231813jes
[178] Bruce PG, Evans J, Vincent CA (1988) Conductivity and transference number measurements on polymer electrolytes. *Solid State Ion* 28–30:918–922. https://doi.org/10.1016/0167-2738(88)90304-9
[179] Chen L, Li W, Fan LZ, et al (2019) Intercalated Electrolyte with High Transference Number for Dendrite-Free Solid-State Lithium Batteries. *Adv Funct Mater* 29:1–11. https://doi.org/10.1002/adfm.201901047

[180] Liu X, Liu J, Lin B, et al (2022) PVDF-HFP-Based Composite Electrolyte Membranes having High Conductivity and Lithium-Ion Transference Number for Lithium Metal Batteries. *ACS Appl Energy Mater* 5:1031–1040. https://doi.org/10.1021/acsaem.1c03417
[181] Liang YF, Deng SJ, Xia Y, et al (2018) A Superior Composite Gel Polymer Electrolyte of Li7La3Zr2O12-Poly(vinylidene fluoride-hexafluoropropylene) (PVDF-HFP) for Rechargeable Solid-state Lithium Ion Batteries. *Mater Res Bull* 102:412–417. https://doi.org/10.1016/j.materresbull.2018.02.051
[182] Zhao Y, Wu C, Peng G, et al (2016) A New Solid Polymer Electrolyte Incorporating Li10GeP2S12 into a Polyethylene Oxide Matrix for All-Solid-State Lithium Batteries. *J Power Sources* 301:47–53. https://doi.org/10.1016/j.jpowsour.2015.09.111
[183] Li M, Frerichs JE, Kolek M, et al (2020) Solid-State Lithium–Sulfur Battery Enabled by Thio-LiSICON/Polymer Composite Electrolyte and Sulfurized Polyacrylonitrile Cathode. *Adv Funct Mater* 30:1910123. https://doi.org/10.1002/adfm.201910123
[184] Xie H, Tang Z, Li Z, et al (2008) PVDF-HFP Composite Polymer Electrolyte with Excellent Electrochemical Properties for Li-ion Batteries. *J Solid State Electrochem* 12:1497–1502. https://doi.org/10.1007/s10008-008-0511-9
[185] Li X, Wang D, Wang H, et al (2019) Poly(ethylene oxide)-Li10SnP2S12 Composite Polymer Electrolyte Enables High-Performance All-Solid-State Lithium Sulfur Battery. *ACS Appl Mater Interfaces* 11:22745–22753. https://doi.org/10.1021/acsami.9b05212
[186] Han F, Zhu Y, He X, et al (2016) Electrochemical Stability of Li10GeP2S12 and Li7La3Zr2O12 Solid Electrolytes. *Adv Energy Mater* 6:1–9. https://doi.org/10.1002/aenm.201501590
[187] Chen L, Venkatram S, Kim C, et al (2019) Electrochemical Stability Window of Polymeric Electrolytes. *Chem Mater* 31:4598–4604. https://doi.org/10.1021/acs.chemmater.9b01553
[188] Wang C, Yang Y, Liu X, et al (2017) Suppression of Lithium Dendrite Formation by Using LAGP-PEO (LiTFSI) Composite Solid Electrolyte and Lithium Metal Anode Modified by PEO (LiTFSI) in All-Solid-State Lithium Batteries. *ACS Appl Mater Interfaces* 9:13694–13702. https://doi.org/10.1021/acsami.7b00336
[189] Seidl L, Grissa R, Zhang L, et al (2022) Unraveling the Voltage-Dependent Oxidation Mechanisms of Poly(Ethylene Oxide)-Based Solid Electrolytes for Solid-State Batteries. *Adv Mater Interfaces* 9:1–10. https://doi.org/10.1002/admi.202100704
[190] Bae J, Li Y, Zhao F, et al (2018) Designing 3D Nanostructured Garnet Frameworks for Enhancing Ionic Conductivity and Flexibility in Composite Polymer Electrolytes for Lithium Batteries. *Energy Storage Mater* 15:46–52. https://doi.org/10.1016/j.ensm.2018.03.016
[191] Yang T, Zheng J, Cheng Q, et al (2017) Composite Polymer Electrolytes with Li7La3Zr2O12 Garnet-Type Nanowires as Ceramic Fillers: Mechanism of Conductivity Enhancement and Role of Doping and Morphology. *ACS Appl Mater Interfaces* 9:21773–21780. https://doi.org/10.1021/acsami.7b03806
[192] Méry A, Rousselot S, Lepage D, Dollé M (2021) A Critical Review for an Accurate Electrochemical Stability Window Measurement of Solid Polymer and Composite Electrolytes. *Materials* 14:3840. https://doi.org/10.3390/ma14143840
[193] Peng X, Huang K, Song S, et al (2020) Garnet-Polymer Composite Electrolytes with High Li+ Conductivity and Transference Number via Well-Fused Grain Boundaries in Microporous Frameworks. *ChemElectroChem* 7:2389–2394. https://doi.org/10.1002/celc.202000202
[194] Zhao E, Guo Y, Xin Y, et al (2021) Enhanced Electrochemical Properties and Interfacial Stability of Poly(ethylene oxide) Solid Electrolyte Incorporating Nanostructured Li1.3Al0.3Ti1.7(PO4)3 Fillers for All Solid State Lithium Ion Batteries. *Int J Energy Res* 45:6876–6887. https://doi.org/10.1002/er.6278
195. Chen F, Jing M xiang, Yang H, et al (2021) Improved Ionic Conductivity and Li Dendrite Suppression of PVDF-Based Solid Electrolyte Membrane by LLZO Incorporation

and Mechanical Reinforcement. *Ionics* 27:1101–1111. https://doi.org/10.1007/s11581-020-03891-0
[196] Rusling JF, Suib SL (1994) Characterizing Materials with Cyclic Voltammetry. *Adv Mater* 6:922–930. https://doi.org/10.1002/adma.19940061204
[197] Evans DH, O'Connell KM, Petersen RA, Kelly MJ (1983) Cyclic Voltammetry. *J Chem Educ* 60:290–293. https://doi.org/10.1021/ed060p290
[198] Elgrishi N, Rountree KJ, McCarthy BD, et al (2018) A Practical Beginner's Guide to Cyclic Voltammetry. *J Chem Educ* 95:197–206. https://doi.org/10.1021/acs.jchemed.7b00361
[199] Li Y, Zhang W, Dou Q, et al (2019) Li7La3Zr2O12 Ceramic Nanofiber-Incorporated Composite Polymer Electrolytes for Lithium Metal Batteries. *J Mater Chem A Mater* 7:3391–3398. https://doi.org/10.1039/c8ta11449h
[200] Liu W, Lin D, Sun J, et al (2016) Improved Lithium Ionic Conductivity in Composite Polymer Electrolytes with Oxide-Ion Conducting Nanowires. *ACS Nano* 10:11407–11413. https://doi.org/10.1021/acsnano.6b06797
[201] Tian L, Liu Y, Su Z, et al (2021) A Lithiated Organic Nanofiber-Reinforced Composite Polymer Electrolyte Enabling Li-ion Conduction Highways for Solid-state Lithium Metal Batteries. *J Mater Chem A Mater* 9:23882–23890. https://doi.org/10.1039/d1ta06269g
[202] Liu W, Lee SW, Lin D, et al (2017) Enhancing Ionic Conductivity in Composite Polymer Electrolytes with Well-aligned Ceramic Nanowires. *Nat Energy* 2:1–7. https://doi.org/10.1038/nenergy.2017.35
[203] Liu W, Liu N, Sun J, et al (2015) Ionic Conductivity Enhancement of Polymer Electrolytes with Ceramic Nanowire Fillers. *Nano Lett* 15:2740–2745. https://doi.org/10.1021/acs.nanolett.5b00600
[204] Zhai H, Xu P, Ning M, et al (2017) A Flexible Solid Composite Electrolyte with Vertically Aligned and Connected Ion-Conducting Nanoparticles for Lithium Batteries. *Nano Lett* 17:3182–3187. https://doi.org/10.1021/acs.nanolett.7b00715
[205] Al-Salih H, Huang A, Yim C-H, et al (2020) A Polymer-Rich Quaternary Composite Solid Electrolyte for Lithium Batteries. *J Electrochem Soc* 167:070557. https://doi.org/10.1149/1945-7111/ab7fb8
[206] Zheng J, Dang H, Feng X, et al (2017) Li-ion transport in a representative ceramic-polymer-plasticizer composite electrolyte: Li7La3Zr2O12-polyethylene oxide-tetraethylene glycol dimethyl ether. *J Mater Chem A Mater* 5:18457–18463. https://doi.org/10.1039/c7ta05832b
[207] Alarco PJ, Abu-Lebdeh Y, Abouimrane A, Armand M (2004) The Plastic-Crystalline Phase of Succinonitrile as a Universal Matrix for Solid-state Ionic Conductors. *Nat Mater* 3:476–481. https://doi.org/10.1038/nmat1158
[208] Guo Q, Xu F, Shen L, et al (2021) Poly(ethylene glycol) Brush on Li6.4La3Zr1.4Ta0.6O12 Towards Intimate Interfacial Compatibility in Composite Polymer Electrolyte for Flexible All-Solid-State Lithium Metal Batteries. *J Power Sources* 498:229934. https://doi.org/10.1016/j.jpowsour.2021.229934
[209] Hu J, Wang W, Zhou B, et al (2019) Poly(ethylene oxide)-based Composite Polymer Electrolytes Embedding with Ionic Bond Modified Nanoparticles for All-Solid-State Lithium-ion Battery. *J Memb Sci* 575:200–208. https://doi.org/10.1016/j.memsci.2019.01.025
[210] Yang YN, Li YX, Li YQ, Zhang T (2020) On-surface Lithium Donor Reaction Enables Decarbonated Lithium Garnets and Compatible Interfaces within Cathodes. *Nat Commun* 11:1–10. https://doi.org/10.1038/s41467-020-19417-1
[211] Duan H, Chen WP, Fan M, et al (2020) Building an Air Stable and Lithium Deposition Regulable Garnet Interface from Moderate-Temperature Conversion Chemistry. *Angewandte Chemie* 59:12069–12075. https://doi.org/10.1002/anie.202003177

[212] He K, Cheng SH, Hu J, et al (2021) In-Situ Intermolecular Interaction in Composite Polymer Electrolyte for Ultralong Life Quasi-Solid-State Lithium Metal Batteries. *Angewandte Chemie* 133:12223–12230. https://doi.org/10.1002/ange.202103403
[213] Wu B, Wang S, Lochala J, et al (2018) The Role of the Solid Electrolyte Interphase Layer in Preventing Li Dendrite Growth in Solid-state Batteries. *Energy Environ Sci* 11:1803–1810. https://doi.org/10.1039/c8ee00540k
[214] Homann G, Stolz L, Nair J, et al (2020) Poly(Ethylene Oxide)-based Electrolyte for Solid-State-Lithium-Batteries with High Voltage Positive Electrodes: Evaluating the Role of Electrolyte Oxidation in Rapid Cell Failure. *Sci Rep* 10:2–10. https://doi.org/10.1038/s41598-020-61373-9
[215] Zhou W, Wang Z, Pu Y, et al (2019) Double-Layer Polymer Electrolyte for High-Voltage All-Solid-State Rechargeable Batteries. *Adv Mater* 31:1–7. https://doi.org/10.1002/adma.201805574
[216] Zou L, Shi K, Xu Z, et al (2022) Double-Layer Solid Composite Electrolytes Enabling Improved Room-Temperature Cycling Performance for High-Voltage Lithium Metal Batteries. *ACS Omega* 7:994–1002. https://doi.org/10.1021/acsomega.1c05576
[217] Yu X, Li J, Manthiram A (2020) Rational Design of a Laminated Dual-Polymer/Polymer–Ceramic Composite Electrolyte for High-Voltage All-Solid-State Lithium Batteries. *ACS Mater Lett* 2:317–324. https://doi.org/10.1021/acsmaterialslett.9b00535
[218] Liang JY, Zeng XX, Zhang XD, et al (2019) Engineering Janus Interfaces of Ceramic Electrolyte via Distinct Functional Polymers for Stable High-Voltage Li-Metal Batteries. *J Am Chem Soc* 141:9165–9169. https://doi.org/10.1021/jacs.9b03517
[219] Jiang T, He P, Wang G, et al (2020) Solvent-Free Synthesis of Thin, Flexible, Nonflammable Garnet-Based Composite Solid Electrolyte for All-Solid-State Lithium Batteries. *Adv Energy Mater* 10:1–10. https://doi.org/10.1002/aenm.201903376
[220] Chen A, Qu C, Shi Y, Shi F (2020) Manufacturing Strategies for Solid Electrolyte in Batteries. *Front Energy Res* 8:1–18. https://doi.org/10.3389/fenrg.2020.571440
[221] Bubulinca C, Kazantseva NE, Pechancova V, et al (2023) Development of All-Solid-State Li-Ion Batteries: From Key Technical Areas to Commercial Use. *Batteries* 9:157–181. https://doi.org/10.3390/batteries9030157
[222] Chen J, Wu J, Wang X, et al (2021) Research Progress and Application Prospect of Solid-State Electrolytes in Commercial Lithium-Ion Power Batteries. *Energy Storage Mater* 35:70–87. https://doi.org/10.1016/j.ensm.2020.11.017

8 Industrial Applications of Ceramic Membranes

Energy and Environmental Prospects

Shalini Singh, Pubali Mandal, Abhradeep Majumder, Partha Sarathi Ghosal, and Manoj Kumar Yadav

8.1 INTRODUCTION

Rapid industrialization and population growth over the past century have necessitated the use of membrane-based technologies for various applications. Membrane technology is a versatile separation process that can be applied in various industries and processes. Membrane technologies have been widely used in food processing and dairy industries, semiconductor manufacturing, microbial fuel cells, and others. Also, water and wastewater treatment using membranes have gained immense popularity recently due to their high efficiency and less treatment time. In comparison to conventional water treatment technologies, membrane technologies have the benefits of reduced thermal inputs, a smaller environmental impact, less chemical sludge effluent, and greater removal efficiencies (Li et al., 2020). There are different kinds of membrane processes, such as microfiltration (MF), ultrafiltration (UF), nanofiltration (NF), and reverse osmosis (RO). These processes are used in a variety of applications, including seawater desalination, wastewater treatment, biotechnology and food industry separations, and chemical production. Other hybrid systems like membrane emulsification and membrane extraction are also expected to become more prevalent in industries such as food and beverage, cosmetics, pharmaceuticals, and biofuels (Charcosset, 2016). The food and beverage industry commercially uses ceramic membranes for microfiltration (Asif & Zhang, 2021). In addition, membrane technology is also used in gas separation, where membranes selectively permeate certain gases such as oxygen or nitrogen (Strathmann, 2001). Based on the material of fabrication, ceramic and polymeric membranes are the two most used types of membranes. Polymeric membranes may come at a lower cost but are subjected to biofouling and have low thermal and chemical stability.

On the other hand, high thermal and chemical stability make ceramic membranes better suited to withstand harsh operating conditions, like high temperatures and

DOI: 10.1201/9781003731672-8

acidic or alkaline environments. Ceramic membranes are widely used in numerous separation processes due to their superior characteristics over polymeric membranes. As a result, ceramic membranes are often preferred over polymeric membranes (Hubadillah et al., 2018).

A ceramic membrane is a type of filter membrane made from inorganic materials such as alumina, zirconia, titania, and silica. It is used for separation processes in various industries, including chemical, petrochemical, pharmaceutical, food and beverage, and water treatment. They have high mechanical strength, making them durable and long-lasting (Lokhandwala et al., 2010; Marchese et al., 2000). Ceramic membranes allow only specific molecules or particles to pass through while blocking others based on their size and charge. The separation of large molecules, such as proteins, peptides, and polymers, is more effectively achieved by them than smaller molecules like ions and solvents.

Ceramic membranes are highly favored over organic membranes due to their many advantages. One such advantage is that they are easy to clean and regenerate, which results in a longer membrane lifespan and reduced process downtime. Also, they have excellent mechanical strength, making them highly durable and resistant to physical damage. Another advantage is their thermal stability, which enables them to withstand high temperatures and thermal shock, making them ideal for use in high-temperature separation processes. Ceramic membranes are highly resistant to chemical corrosion, which is essential in applications where the membrane may come into contact with corrosive chemicals (Dong et al., 2022). Additionally, their highly porous structure allows for reduced fouling during the separation process, leading to a high flux rate and a more efficient separation process overall. Therefore, ceramic membranes are preferred in industrial separation processes because of their convenient cleaning and regeneration, robust mechanical strength, thermal stability, resistance to corrosion, and reduced fouling properties (Hofs et al., 2011; Kaur et al., 2016).

Ceramic membranes are a new type of high-efficiency, energy-saving, environmentally friendly separation material that has emerged as a crucial element in the process industry's efforts to save energy and reduce carbon emissions (Gu et al., 2020; Zhang et al., 2021). Despite significant laboratory and full-scale testing of ceramic MF and UF ceramic membranes for industrial wastewater treatment, their usage on a broad scale in wastewater treatment is constrained by their high capital costs (Zoubeik et al., 2018). Furthermore, often the solvents used for the preparation of the membranes are not eco-friendly. Hence, bio-based and non-toxic solvents (green solvents) may be used for the preparation of the membranes. These ceramic membranes made from green solvents are known as green ceramic membranes (Naziri Mehrabani et al., 2022). Additionally, several efforts have been made to lower the cost of manufacturing to promote the large-scale application of ceramic membranes, such as using clay as the membrane material and reducing the sintering temperature (Hubadillah et al., 2018).

8.2 PREPARATION METHODS OF CERAMIC MEMBRANES

Typically, there are three layers in ceramic membranes. The porous support layer in the inner layer gives the manufactured membrane a high level of mechanical strength.

The second one is the intermediate layer, which is coated over the supporting layer and has smaller pores. The intermediate layer serves as a connection between the support layer and the top layer, as there is a difference in pore size between these two layers. The top layer, which is the final, is where the separation happens (Peng, 2008).

There are four key processes in the fabrication of ceramic membranes:

(i) Selection of appropriate raw materials
(ii) Mechanical processing of raw materials involves actions such as crushing, sieving, mixing., etc., to create a homogenous mixture and to knead when making the paste.
(iii) Using the proper manufacturing procedure to shape the membrane to achieve the necessary geometry; and
(iv) Thermally treating the green membrane to create a very stable ceramic membrane.

The production of ceramic membranes can vary depending on the specific materials, desired membrane structure, and application requirements. Different manufacturing methods, such as slip casting, tape casting, pressing, extrusion, and freeze casting, are commonly used (Ning, 2011).

8.2.1 Slip Casting Method

Slip casting can be considered as a fundamental and cost-effective method for producing advanced ceramic membranes. With this technique, it is possible to shape complicated geometries and asymmetrical structures while simultaneously obtaining acceptable material uniformity. After thorough mixing, a suspension of particles is poured into a permeable mold. The solvents are then able to permeate through the pores of the mold, facilitated by capillary action. This results in the formation of a layer of particles on the inner surface of the mold, which consolidates quickly to prevent the particles from penetrating through the mold (Wardell, 2007). This technique has been utilized to produce a ceramic membrane using inexpensive materials like kaolin and fly ash (Jedidi et al., 2009; Mohammadi & Pak, 2003).

8.2.2 Tape Casting

Tape casting is the technique used to produce ceramic membranes in the form of planar sheets. To create a tape-cast suspension, the ceramic material, a liquid-dispersing medium, and organic additives are blended together. This provides the suspension with the ability to behave similarly to plastic. The suspension is then transferred into a reservoir, which is controlled by a blade whose height can be altered while it is positioned over a mobile carrier film. The suspension film is then exposed to a controlled atmosphere to cure it. Tape casting is a technology that, in addition to coating layers onto membrane supports, can manufacture flat-sheet membranes. This makes the method more versatile (Lee et al., 2015). Several researchers have chosen to use this method to create ceramic membranes (Das & Maiti, 1998). The properties of ceramic membranes are influenced by different fabrication techniques as given in Table 8.1.

TABLE 8.1
Membrane Types, Fabrication Techniques, and Properties of Commercial Ceramic Membranes

Membrane	Raw Material Used	Fabrication Technique	Sintering Temperature (°C)	Pore Size (μm)	Porosity (%)	Mechanical Strength (MPa)	References
Mullite-TiO_2	Bauxite, Fly Ash	Wet/dry spinning	1400	0.11	51.0	85.8	Zhu et al., (2016)
Mullite	Bauxite, Kaolin	Dry pressing	1450	0.62	31.6	100.2	Zhu et al., (2016)
Mullite	Bauxite, fly ash	Wet/dry spinning	1400	0.48	44.6	68.7	Chen et al., (2016)
MF	Natural Moroccan bentonite	Uniaxial pressing method	950	1.70	32.12	22	Bouazizi et al., (2016)
MF	Clay and banana peel powder	Spin coating	1000	0.45	40.5	19.2 ± 2.6/ 8.5 ± 1.8	Mouiya et al., (2019)
MF	Natural clay and pyrrhotite ash solid waste	Dry pressing method	1000	2.5	34	27.42	Hatimi et al., (2020)
Aluminium titanate	Attapulgite clay and aluminium titanate	_	1300	4.5–8	32.41	45.48	Wang et al., (2018)
Hollow fiber	Palm oil fuel ash	Wet/Dry spinning	1050	1.51	33.2	60	Tai et al., (2019)
Mullite-zeolite	Kaolin, zeolite	Paste extrusion	1150	0.32	_	70	Jafari et al., (2020)
Hollow fiber	Bauxite	Wet/dry spinning	1300	0.36	50.9	35.6	Chen et al., (2020)
Hollow fiber	Palm Oil Fuel Ash	Wet/dry spinning	1150	0.03–0.04	_	_	Yusof et al., (2018)
MF	Moroccan natural magnesite	Dry uniaxial pressing	1100	1.12	48	6.1	Manni et al., (2020)
MF	Natural clay paste	Extrusion technique	_	0.309	53	_	Kumar et al., (2016)
MF	Clay material such as kaolin and quartz	_	950	1.49	_	_	Deka et al., (2023)
UF	Tunisian kaolin clay	Extrusion technique	1000	0.035	26	13	Bousbih et al., (2021)
MF	Natural Pozzolan, micronized phosphate	Uniaxial pressing	1050	1.3	32.07	16	Beqqour et al., (2019)
MF	α-alumina	Uniaxial pressing	1450	0.4	30	35.5	Silva et al., (2016)
MF	Moroccan natural perlite	Dry pressing technique	950	1.70	52.11	21	Saja et al., (2018)

MF: MicrofiltrationUF: Ultrafiltration

8.3 APPLICATION OF CERAMIC MEMBRANES

Ceramic membrane technology is becoming increasingly popular in industries due to its numerous benefits. Ceramic membranes are made of inorganic materials, such as aluminum oxide, silicon carbide, or zirconium oxide, which make them highly resistant to extreme temperatures, chemical corrosion, and fouling. These properties make ceramic membranes suitable for a variety of industrial applications, including oil removal, toxic heavy metal removal, and wastewater treatment of various industries, as shown in Figure 8.1.

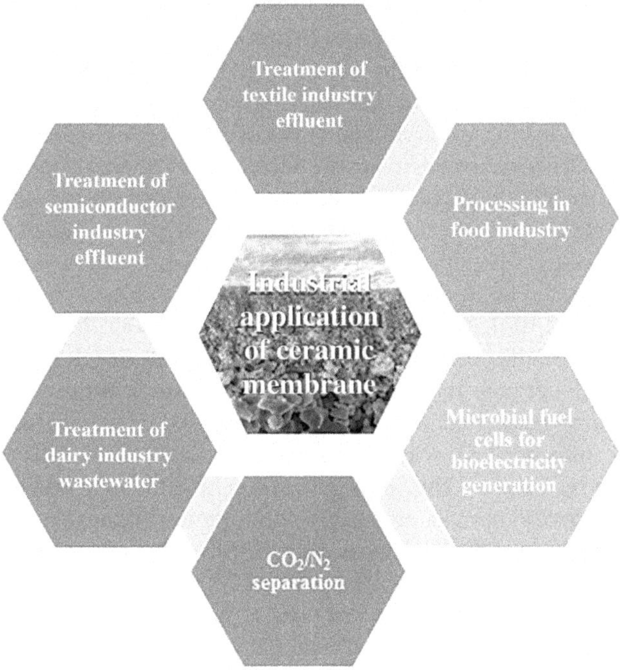

FIGURE 8.1 Industrial applications of ceramic membrane.

8.3.1 Water and Wastewater Treatment

8.3.1.1 Water Treatment

To treat surface water or secondary-treated effluent, which may contain a wide variety of contaminants such as dissolved organic matter, suspended particles, heavy metals, microorganisms, and algae, both MF and UF ceramic membranes can be utilized on a small scale as well as on a large scale for the treatment process. The smaller-scale implementation could be in the form of portable membrane filtration systems that utilize MF or UF membrane technology, which are also referred to as package treatment plants. These plants are able to meet the requirements of a smaller population. It is possible to employ MF/UF ceramic membranes for large-scale applications as a component of a water treatment plant (WTP) (Asif & Zhang, 2021). The use of ceramic membranes in the field of water purification is growing, and this trend is anticipated to increase in the coming years. According to the quality of the water used as a source and the quality of drinking water that is desired, the production of drinking water can be pretty costly and necessitates several different treatment processes (Asif & Zhang, 2021). Some of these processes include filtration, coagulation, disinfection, and oxidation. Ceramic membranes can be an appealing alternative and have the ability to lessen the impact that traditional wastewater treatment plants have on the surrounding environment by taking the place of the sedimentation and disinfection processes or both. This could be because of the fact that anti-fouling qualities of ceramic membranes are superior and that they have a longer life duration in comparison to polymeric membranes (Alresheedi et al., 2019; Park et al., 2015). Also, ceramic membranes have the potential to reject a vast range of microorganisms, such as viruses and harmful bacteria, which can result in a significant reduction in the amount of disinfectant used and in the overall cost of operation. It has been demonstrated that ceramic membranes are adequate for the pretreatment of saltwater, which has the capacity to increase the effectiveness of the reverse osmosis process, which is now utilized all over the world (and particularly in the Middle East) to produce drinking water. The efficient removal (greater than 90%) of suspended solids, viruses, and bacteria can be obtained by membranes with a pore size greater than 10 nm, 10–30 nm, and 10–1,100 nm, respectively (Asif & Zhang, 2021). The size of the suspended particles influences the effectiveness of the rejection. It has been observed that a single MF ceramic membrane with hole sizes of 150 nm and 200 nm can altogether remove germs. Independent MF and UF ceramic membranes may significantly remove suspended particles (over 99%) and produce filtrate that complies with drinking water regulations. This is a significant achievement. In addition, the standalone ceramic membranes can effectively remove microorganisms, natural organic matter (NOM), and disinfection by-products (DBPs). The effectiveness of MF and UF ceramic membranes in removing pollutants and increasing permeate flux may depend on the membrane properties and effluent water characteristics, including pH levels and inorganic content (Asif & Zhang, 2021).

To reduce the expense of replacing membranes, flat sheet ceramic membranes were used in place of polymeric membranes in a sizable (42,000 m^3/d) WTP in Qassim, Saudi Arabia. Japan has the ceramic-based WTP that has been running the longest (up to 100,000 m^3/d) (Asif & Zhang, 2021). Ceramic MF membranes made by Metawater in Japan are used in this factory to coagulate, filter, and disinfect water. Ceramic membrane filtration devices made by Metawater clean the water in 150 places in Japan. Metawater also has ceramic membrane filtration systems in Vanvikan WTP in Norway, Base Creek WTP, and Reuter-Hess in the United States. The largest ceramic membrane WTP in the world is in Singapore and has a capacity of 180,000 m^3/d. The CeraMac® ceramic membrane (PWNT water technologies, Netherlands) and biologically activated carbon are all used at the Choa Chu Kang WTP (Galjaard et al., 2012). CeraMac® ceramic membranes are utilized in WTPs in the Netherlands, Switzerland, and the United Kingdom (Mayflower) (Asif & Zhang, 2021).

8.3.1.2 Textile Industry

The textile industry is a heavy user of water and generates a significant amount of wastewater as a result. According to Berradi et al. (2019), wastewater frequently contains a variety of pollutants, including suspended particles, chemical oxygen demand (COD), total nitrogen, total phosphorus, non-biodegradable organic compounds, heavy metals, residual bleaching agents, total suspended solids (TSS), and fibers from the textile manufacturing process (Berradi et al., 2019). Each fabric product tone requires approximately 150 m^3 of water throughout the finishing and dyeing operations in a traditional textile mill. Eight thousand types of distinct chemicals and 3,600 types of dyes are used in addition to the water in the numerous bleaching, finishing, dyeing, and printing operations carried out in textile mills. Because of the contamination of soil and water, these chemicals may have an impact on human health and aquatic ecosystems (Hussain & Wahab, 2018). The textile industry needs a lot of energy in addition to water since baths must be heated during wet processing, which uses a tremendous quantity of water (Kocabas et al., 2009). Hot wastewater reclamation is crucial for energy efficiency and wastewater quantity reduction, which further lowers energy, water, and wastewater expenses (Voigt et al., 2001). The development of membrane technology is an effective and affordable means of treating textile effluents and a replacement for traditional separation techniques (Fersi et al., 2005; Van der Bruggen et al., 2001). Over polymeric membranes, ceramic membranes have various benefits, like reduced fouling, high mechanical strength, thermal stability, corrosion resistance, and ease of cleaning and regeneration.

By using the uniaxial pressing technique, Bouazizi et al. (2016) created a ceramic membrane from natural Moroccan bentonite, which had a removal rate of 94% to 99% for suspended particles but was ineffective at removing soluble salts from effluents. According to Manni et al. (2020), the sintering process had a substantial effect on the microstructure of the membrane because the ceramic membrane made from Moroccan natural magnesite had a high removal efficiency for turbidity (99.9%) and COD (69.0%) when subjected to dry uniaxial pressing. da Silva et al. (2016) reported that ceramic membrane from α-alumina, when subjected to uniaxial pressing, exhibited excellent separation characteristics with an extremely high level of chemical and physical stability. The study also revealed that the material had an impressive COD removal efficiency of 73% and a turbidity removal efficiency of 95%, as given in Table 8.2 (da Silva et al., 2016). Bouazizi et al. (2017) created a ceramic support using bentonite, which was subsequently covered with a layer of ultrafiltration. This layer was prepared using a dip-coating process with nano-TiO_2 powder. The resulting membrane had hydrophilic properties, which contributed to a high permeate flux and less fouling (Bouazizi et al., 2017). The combination of raw clay and banana peel was utilized to create an environmentally friendly ceramic membrane support that demonstrated remarkable efficacy in purifying textile effluents. According to Mouiya et al. (2019), the membrane was successful in removing pollutants, reducing turbidity, residual dye content, suspended particles, and biological and chemical oxygen demand.

TABLE 8.2
Application and Performance of Ceramic Membrane in Different Industries

Membrane	Geometry	Application	Flux ($L/m^2.h$)	Membrane Properties/ Operating Conditions [Pressure, Temperature, pH, etc.]	Performance (%)	Remark/Advantages/ Disadvantage	Reference
MF	Flat ceramic membrane	Treatment of textile wastewater	520	Pressure: 0.12 bar pH: 6.93	Removal efficiency of suspended matter: 94–99%	Not efficient in eliminating soluble salts from wastewater	Bouazizi et al., (2016)
MF	-	Treatment of tannery wastewater	550	Pressures: 0.5–2 bar pH: 10.2	Turbidity rejection: 99% Residual dye removal: 96.7% Reduction of COD < 85.72%	Chemical cleaning is needed for the effective elimination of the accumulated layers	Mouiya et al., (2019)
UF	Tubular ceramic membrane	Portable water	–	–	Removal efficiency Turbidity– 85% Ferrous–75%	–	Özdemir, (2016)
MF	Flat ceramic membranes	Treatment of dairy wastewater	2288	Pressure: 0.6 Bar pH: 9.81	Removal efficiency Organic oil and fat: 94%, Turbidity: 100%,	The membrane demonstrated nearly complete turbidity, oil, and fat reduction at high flow	Hatimi et al., (2020)
MF	Flat ceramic membranes	Agro-food effluents	1433.46	Pressure: 1.2 Bar pH: 5.09	Removal efficiency Turbidity: 97% Conductivity: 19%	Prepared membrane can be used as a substitute in the treatment of industrial effluents	Saja et al., (2018)

MF	_	Paper and pulp industry wastewater	_	_	Pore size: 1.01 mm Porosity: 44% Removal of COD: 87.6% Reduction of toxicity: 94.5%	_	Goswami et al., (2019)
MF	Symmetric membrane	Industrial oily wastewaters	75	_	Removal efficiency TOC: 94% COD: 89%	Result shows high TOC and COD removal	Shokrkar et al., (2011)
MF	_	Oily wastewaters	5360	Pressure: 68.95 Kpa	Oil rejection efficiency: 98.8%	Relevant for treating oil–water emulsions to produce infiltrate streams that can comply with stringent environmental regulations	Nandi et al., (2010)
MF	Tubular ceramic membrane	Dairy wastewater	54.3	Pressure: 13.78 Kpa	Conductivity: 2mS/cm COD removal: 52.08% Turbidity: 45 NTU	Fouling mechanism determined by conventional pore-blocking models	Deka et al., (2023)
MF	Flat ceramic membrane	Textile wastewater	_	_	Removal efficiency Turbidity: 99.9% Cod: 69%	Sintering process has a significant impact on membrane microstructure	Manni et al., (2020)
UF	Tubular ceramic membrane	Textile effluent	28–50	Pressure: 3–7 bar	Turbidity rejection: 98% COD Rejection: 80%	Reversible fouling is dominant	Bousbih et al., (2021)
MF	Flat ceramic membrane	Tannery industry wastewater	1732.5	pH (Feed) =7.6	Turbidity rejection: 97% pH of permeate: 8.93	Excellent removal efficiency and low cost	Beqqour et al., (2019)
MF	_	Textile industry wastewater	_	_	COD rejection: 73% Turbidity rejection: 95%	Strong chemical and physical stability with excellent separation properties	Silva et al., (2016)

8.3.1.3 Municipal Wastewater Treatment

Over the last 20 years, significant research has been conducted on ceramic membrane bioreactors (CMBRs) for wastewater treatment from municipal sources in both the laboratory and on a pilot scale (Asif & Zhang, 2021). Ceramic membranes are a desirable substitute for polymeric membranes because they last longer than polymeric membranes. Additionally, CMBRs have a longer operating life than polymeric membranes because of their resistance to fouling brought on by chemical agents during cleaning and their lower fouling rate during the treatment of municipal wastewater. As a result, the operation is more dependable, and the amount of chemicals used to clean the membrane is decreased (Kashif et al., 2020). It is possible to employ either the flat sheet or the tabular arrangement. However, the flat-sheet configuration has the flexibility to be either immersed or used as a standalone device (i.e., side stream). It is interesting that even at high mixed liquor suspended solid (MLSS) concentrations and large fluxes, CMBR may function effectively due to the location of the ceramic membrane. It has been observed that the flux in side stream CMBRs ranges from 10 LMH to 200 LMH, which is much higher than the flux in submerged CMBRs. Under the same circumstances of operation, it was discovered that the permeate flux of the ceramic membrane was much larger (up to 10 times) than that of the polymeric membrane (Ghyoot & Verstraete, 1997).

Most CMBR studies have centered on organic removal, and they have demonstrated that nutrients like nitrogen and phosphorus can be successfully removed (80–95%) as well (Asif & Zhang, 2021). The studies on CMBRs have also focused primarily on the rejection of nutrients. In research by Asif et al. (2020), the effectiveness of CMBR in cleaning up micropollutants was analyzed. They reported clearance rates of specific micropollutants that ranged from 60% to 99%. In a separate piece of research, a lab- and pilot-scale CMBR was used to evaluate the degree to which nutrient removal could be achieved through step-feeding and internal recirculation. By distributing organic matter evenly and uniformly across the CMBR, this technique improved the denitrification rate and enabled total nitrogen removal to reach 74% and 81% at the pilot- and lab scale, respectively (Chung et al., 2014).

8.3.1.4 Industrial Applications of Ceramic Nanofiltration Membranes for Water Treatment in Oil Sand Mines

Ceramic nanofiltration membranes have a range of industrial applications, including water treatment in oil sand mines. Oil sand mines generate large volumes of wastewater, which contains a variety of contaminants, including suspended solids, organic matter, and heavy metals. Before the water may be reused in industrial operations or properly released into the environment, these impurities must be eliminated. Loganathan (2017) investigated the treatment of oil sands recycling process water using ceramic ultrafiltration membranes. They discovered that adding a coagulant was required for solids removal and that including a softening stage decreased membrane fouling. According to (Loganathan, 2017), nanofiltration technology can also reduce organic components and electrolytes by up to 95%. The current water treatment effluent standards can potentially be met without the need for chemical pretreatment because of the small pore size of these membranes (Duraisamy et al., 2013; Mueller & Witte, 2008). The precise planned application and the potential long-term cost reductions would determine the practicality of utilizing ceramic nanofiltration membranes in the oil sands industry. Factors to consider include whether they will be used further to refine the feed to existing traditional reverse osmosis systems or to treat warm thickener overflow water. In an oil sand mine in Canada, a commercial pilot unit using titania ceramic nanofiltration membranes (TNU) was conducted for nearly two years under real-world conditions without any pretreatment.

Additionally, it appeared that the TNU preferred to reject ions with high charge densities. As a result, less process water would need to be stored in tailings ponds, and the resulting permeate stream may be used as a feed for existing traditional reverse osmosis systems (Cabrera et al., 2021). Overall, ceramic nanofiltration membranes could be successfully implemented in the oil sand mining process to improve water quality.

8.3.1.5 Application of Ceramic Membrane in the Semiconductor Industry

Water back grinding, sawing, die attachment, encapsulating, electroplating, wire bonding, trimming, shaping, and labeling are only a few of the intricate steps involved in producing semiconductor integrated circuits (Lin & Yang, 2004). These procedures generate roughly equal amounts of wastewater and necessitate large amounts of ultrapure water (UPW) (Tsai et al., 2007). UPW usage can account for up to 40% of the total water consumed in semiconductor factories for rinsing and cleaning (Drouiche et al., 2007). In a pilot-scale study, Huang et al. (2011) utilized a spiral-wound UF membrane made of polyether sulfone (PES) to treat wastewater from the semiconductor industry. According to their findings, 96–99% of the turbidity was successfully removed by the membrane, allowing the treated wastewater to be reused (Huang et al., 2011). Despite the membrane technology's demonstrated sustainability, membrane fouling continues to be a major obstacle to its widespread application. Membrane fouling causes a reduction in permeate flux, which shortens the membrane's lifespan and increases energy usage. As a result, operating costs are elevated (Teow et al., 2020). The study developed a sustainable approach using ceramic membranes to recover silica and treat wastewater from the semiconductor industry. Two types of membranes were tested, and both showed a rejection of more than 60% of water quality parameters. However, Membrane B showed better performance in rejecting all parameters. Alkaline cleaning with NaOCl effectively cleaned the membrane with a high flux recovery rate. In the production of semiconductor wafers, recycled and repurposed silica from wastewater can be used as an insulator. This approach provides a breakthrough in semiconductor industry wastewater treatment by recovering silica simultaneously for recycling and reuse (Vijayan et al., 2022). According to studies, ceramic membranes have the potential to be commercialized and work in conjunction with ceramic enterprises to give a long-term performance with less maintenance (Jadhav et al., 2022).

8.3.2 Food Industry

Ceramic membranes simplify process flow sheets by eliminating the need for more complex steps that could otherwise subject the products to chemical stress (Tomasula et al., 2011). The use of ceramic membranes in the food industry can help produce high-quality foods and develop new fractions that may be used in novel dietary supplements, enabling the industry to better meet the changing consumer demands in the market (Daufin et al., 2001). It can be used to create fruit juices, resulting in premium, preservative-free juices with a fresh, natural flavor to stand storage well. Given that most fruit juices have an acidic pH of 3–5, this is especially advantageous. Because of their exceptional chemical and thermal stability, low-cost ceramic membranes are a good choice because they can withstand the high temperatures needed to increase permeate flow in some juice varieties. To microfilter centrifuged kiwi juice, Qin et al. (2015) created ceramic membranes employing fly ash particles (Qin et al., 2015).

8.3.3 Dairy Industry

The use of ceramic membrane filtration is primarily utilized in dairy processing within the food industry (Mohammad et al., 2012). The dairy industry utilizes membrane processes that operate through the application of pressure, including MF, UF, NF, and RO. The most frequently used technique in milk processing is MF with ceramic membranes. This technique is helpful in concentrating milk and eliminating bacteria (Charcosset, 2016). Almandoz et al. (2015) created cylindrical composite MF membranes using natural aluminosilicates, which they then used to pasteurize goat milk. The results of their tests concluded that a substantial portion of the bacteria could be successfully eliminated by the membranes, ranging from 87% to 99% (Almandoz et al., 2015). To handle dairy wastewater, Kumar et al. (2016) used a brand-new, reasonably priced tubular ceramic membrane, which successfully reduced the COD level from 1,462 mg/L to an allowable limit of 135 mg/L (Kumar et al., 2016). Deka et al. (2023) treated dairy effluent with a COD removal rate of 52.08%, turbidity of 45 NTU, a pressure of 13.78 kPa, and conductivity of 2 mS/cm using a tubular ceramic membrane with MF capabilities. Using traditional pore-blocking models, they were able to determine the fouling mechanism (Deka et al., 2023).

8.3.4 Use of Ceramic Membrane in Microbial Fuel Cells for Bioelectricity Generation

A microbial fuel cell (MFC) is a bioelectrochemical system that utilizes bacteria to oxidize the organic waste present in water. The process generates electricity while simultaneously treating wastewater. MFCs can potentially provide a solution for producing clean, renewable energy at a low cost (Slate et al., 2019). Therefore, it is crucial to prioritize the substitution of sustainable energy sources for fossil fuels, emphasizing the need for dedicated efforts toward achieving this objective (Gao et al., 2018). MFCs comprise two compartments, an anodic and a cathodic compartment, which are separated by cation exchange membranes. These cells provide a versatile platform for various electrochemical reactions and utilize bacterial metabolism to decompose wastewater while simultaneously generating electricity (Jadhav et al., 2015). The electromotive force produced by the voltage difference, the concentration gradient, and the pH gradient profile are the main driving forces behind the electrochemical reactions in MFCs. All of these processes are made possible by the effective operation of the membranes. When compared to ceramic membranes made with Kalporgan's soil (KS) and commercial ceramic powder (CCP), those made with leached Kalporgan's soil (LKS) have the lowest internal resistance and offer optimum performance (Cheraghipoor et al., 2019). LKS is an excellent choice of membrane material for MFCs.

8.3.5 Ceramic Membrane Uses for Separation of CO_2/N_2

Ceramic membranes have also undergone thorough investigation because of their exceptional mechanical properties, high thermal and chemical resilience, and successful CO_2/N_2 separation. Ceramic membranes can be used for CO_2/N_2 separation in various processes, such as post-combustion carbon capture, natural gas purification, and biogas upgrading. In these processes, the ceramic membrane functions as a specific barrier that separates CO_2 from N_2 according to its size and affinity for the membrane surface (Kárászová et al., 2020). An amine-silica-based membrane with a greater CO_2 permeance of 100 GPU and a selectivity of 44 was created by Xomeritakis et al. in 2005. Due to its demand for lower temperatures and the usage of expensive commercial-scale gamma-alumina-coated tubular membranes in some situations, the membrane has limited applicability (Xomeritakis et al., 2005). Sharma et al. (2022) developed cost-effective ceramic membrane support using two industrial-grade alpha-alumina powders with different particle sizes. The optimized processing parameters and sintering facets resulted in suitable mechanical and physical properties for creating the ceramic support system. On the graded alpha-alumina substrate, a silica membrane impregnated with amine demonstrated reliable gas separation performance (Sharma et al., 2022). To lower the pore size for gas penetration, Cerneaux and Germain (2015) modified commercial gamma-alumina membranes by covering them with tetragonal zirconia. Chemical grafting of (3,3,4,4,5,5,6,6,7,7,8,8,8)-tridecafluorooctyl perfluorophosphonic acid makes the membrane superhydrophobic, resulting in significant CO_2 over N_2 selectivity. These modified ceramic membranes are suitable for the first stage of CO_2 enrichment and gas cleansing steps in power plant's raw exhaust gas treatment (Cerneaux & Germain, 2015).

8.4 ENVIRONMENTAL BENEFITS OF CERAMIC MEMBRANES

Ceramic membranes have emerged as a preferred and more ecologically friendly solution in industrial applications as compared to polymer counterparts owing to their different properties. Ceramic membranes are notable for their great thermal and chemical resilience, which makes them durable in harsh industrial conditions. Unlike polymer membranes, which degrade or lose efficiency when exposed to high temperatures or harsh chemicals, ceramic membranes retain their integrity across a variety of situations (Artun & Askin, 2022; Guechi et al., 2016). This durability not only enhances the lifespan of the membranes but also minimizes the need for frequent replacements, thereby reducing material waste and the associated environmental impact (Freeman & Shorney-Darby, 2011; Li et al., 2014). Ceramic membranes demonstrate enhanced durability in comparison to polymer membranes in industries where filtration processes entail the use of aggressive substances or difficult conditions. By virtue of their resistance to severe environments, the probability of membrane failure, leakage, or contamination is significantly diminished. This practice not only guarantees the preservation of the filtration process' integrity but also mitigates the environmental hazards that may arise from inadvertent discharges of contaminants (Freeman & Shorney-Darby, 2011; Lee et al., 2015; Li et al., 2014). In addition, ceramic membranes are very sustainable because they last a long time and don't need much upkeep. These membranes usually last longer than polymer alternatives, and because they don't get clogged up easily, they can work well for longer periods of time with little upkeep. This makes the membrane last longer, which not only uses fewer resources but also makes less waste, which is in line with the ideas of a more environmentally friendly industry process (Freeman & Shorney-Darby, 2011; Guechi et al., 2016). One unique benefit of ceramic membranes is that they don't require chemicals for filtering. Ceramic membranes enable a more environmentally friendly and pure separation procedure than some polymer membranes, which could depend on chemical additives for enhanced performance. This feature promotes a safer and cleaner industrial operation by removing the environmental impact of chemical usage and disposal (Mueller et al., 2010; Shurygin et al., 2021).

Another notable factor that contributes to the environmental advantage of ceramic membranes is their capacity to be recycled. The membranes possess inherent recyclability, and their composition facilitates effective retrieval of components at the conclusion of their life cycle (Lee et al., 2015). This is in contrast to certain polymer membranes that may present difficulties in terms of recyclability. The recyclability of ceramic membranes is in accordance with the concepts of a circular economy, which advocates for a more sustainable approach to resource utilization and waste reduction (Li et al., 2015; Zhu et al., 2021). Furthermore, ceramic membranes play a role in conserving energy in industrial processes. These membranes provide excellent thermal stability, allowing filtration to be conducted at elevated temperatures. As a result, the energy needed for cooling is significantly reduced (Artun & Askin, 2022; Guechi et al., 2016). Additionally, the high permeability of ceramic membranes can lead to more efficient filtration processes, potentially reducing energy consumption (Li et al., 2020). The resulting energy savings translate into reduced greenhouse gas emissions, supporting global efforts to mitigate climate change and promoting sustainable industrial practices.

8.5 CERAMIC MEMBRANES: ENERGY AND ENVIRONMENTAL PROSPECTS

Ceramic membranes are efficient, energy-saving, environmentally friendly, and stable separation materials with high-temperature resistance. They have become a commonly used support for reducing energy consumption and emissions in the process industry (Dong et al., 2022). Membrane processes are suitable for both reducing environmental pollution caused by wastewater and recovering valuable substances from sewage. The extensive use of ceramic membranes in numerous applications is hampered by their high cost and low efficiency. Therefore, it is essential to consider both factors to ensure the efficient and cost-effective use of ceramic membranes (Chen et al., 2020; Wu et al., 2022). Three costs can be categorized under the cost of ceramic membranes: raw material costs, labor costs, and method operating costs. Energy use is one of these that most significantly raises the entire cost. In their article from 2021, Zou and Fan explore ways to reduce energy usage for low-cost ceramic membrane sintering. Future goals include achieving a balance between performance and cost, enhancing the performance of inexpensive membranes, speeding up the industrial process, and creating novel techniques to reduce sintering energy consumption (Zou & Fan, 2021). Hubadillah et al. conducted a thorough study using low-cost ceramic membranes made from the widely available mineral kaolin, which is commonly used in traditional ceramics for domestic and building purposes. The review covers the mineral characteristics of kaolin, various techniques for fabricating ceramic membranes, and the applications of such membranes in treating oily wastewater and supplementing existing membrane processes. Kaolin-based ceramic membrane has an advantage in its thermal properties, which enable it to undergo sintering at lower temperatures than alumina (Hubadillah et al., 2018). Issaoui and Limousy's (2019) thorough assessment of low-cost ceramic membrane preparation methods and important industrial and environmental applications placed special emphasis on their economic and environmental viewpoints. Their literature suggested ceramic membranes are effective filters in food and beverage processing, especially in protein separation, juice clarification, alcoholic beverages, and wastewater treatment. They have strong mechanical properties, high permeability, and can use inexpensive materials. Although clay- and ash-based ceramic membranes are still under development, it is anticipated that new generations will be created for environmental and food applications. The overall cost of the treatment may be further reduced by recovering nutrients and heavy metals from wastewater (Qiu et al., 2015). The review covered various preparation techniques of ceramic membranes and their potential applications in industries such as water and wastewater treatment, as well as in environmental remediation (Issaoui & Limousy, 2019). Potential uses for dense ceramic membrane reactors include the creation of hydrogen, the conversion of methane, the capture and use of CO_2, and the treatment of hazardous gases. Before commercialization can take place, however, a number of concerns must be resolved, including membrane cost, high-temperature sealing, large-scale production, and reliability. The manufacturing of high-performance membrane materials from naturally abundant, inexpensive elements, the creation of innovative reactor topologies and catalysts, and the coupling of processes to boost reaction productivity and energy efficiency are the main areas of research. Future research should focus on membrane materials, membrane configuration, high-temperature sealing, and membrane reaction processes to further promote commercialization (Dong et al., 2011).

8.6 FUTURE TRENDS AND INNOVATIONS

Within the domain of ceramic membranes for industrial applications, numerous encouraging trends and innovations are influencing the future landscape. Researchers are investigating methods to incorporate nanomaterials into ceramic membranes to improve their selectivity, permeability, and resistance to fouling, thereby highlighting the significant contribution of nanotechnology (Khulbe & Matsuura, 2021; Li et al., 2020; Ying et al., 2017). Moreover, the primary focus is on advancing material design to create ceramics that exhibit exceptional durability and resilience when subjected to challenging operating circumstances. The utilization of additive manufacturing, often known as 3D printing, is becoming increasingly prominent in the production of complex membrane structures, allowing for exact manipulation of pore size and distribution (Chen et al., 2019; Nunes et al., 2020). Hybrid membrane systems, which involve the combination of ceramic membranes with other technologies, are now being investigated to improve efficiency and cost-effectiveness in certain industrial applications. Efforts are also focused on enhancing resistance to fouling, prolonging the lifespan of the membrane, and lowering maintenance requirements (Li et al., 2020; Mustafa et al., 2018). As the ceramic membrane sector advances, there is an increasing focus on expanding production and tackling issues with cost, scalability, and integration with current processes. In line with the principles of the circular economy, the ceramic membrane industry is progressively taking into account the complete life cycle of membranes, which encompasses recycling and disposal to reduce environmental harm. In addition, advancements in energy-efficient procedures strive to minimize the overall environmental impact of industrial applications by concentrating on optimizing designs for reduced energy usage and investigating the integration of renewable energy sources. Therefore, the future of ceramic membranes in industrial applications is characterized by the merging of advanced technologies, the consideration of sustainability, and a dedication to meeting the changing requirements of various industries.

8.7 CONCLUSIONS

In this chapter, the multifaceted applications of ceramic membranes have been discussed. The use of ceramic membranes has not been simply limited to water and wastewater treatment but has been used in innovative ways in the food industry, dairy industry, gas separation, and generation of electricity using microbial fuel cells. For water and wastewater treatment, it was observed that these membranes had shown a high rejection of organic matter, turbidity, microorganisms, and other specific pollutants. Although more expensive than polymeric membranes, ceramic membranes offer various advantages, like easy cleaning and regeneration, thermal stability, high mechanical strength, chemical resistance, and reduced fouling properties. They are widely used in separation processes, especially in the food and beverage industry, where they are preferred due to their superior characteristics over organic membranes. However, their high capital cost limits their broader-scale application in the treatment of water. It is essential to create membrane materials that are both affordable and of high quality for use in the future to reduce the price of the ceramic membrane and make it more widely available. To achieve this, it would be necessary to find ways to lower the cost of sourcing materials while also enhancing the performance of the ceramic membranes. Ceramic membranes are a kind of separation material that is both energy-efficient and environmentally friendly. As a result, they have become a crucial part of the process industry's efforts to conserve energy and cut carbon emissions. Both the separation effectiveness and permeability of ceramic membranes can be enhanced by improving the technology used in the processing of films. The goal is to decrease the amount of pollution produced by ceramic membranes and enable their regeneration, thereby extending their useful life. Clay- and ash-based low-cost ceramic membranes have the potential to fulfill the requirements of both the industrial and environmental sectors. However, further research is required to commercialize such membranes. Ceramic membranes have massive potential in the field of water and wastewater treatment and other industries. However, for ceramic membranes to have a sustainable future, it is essential to develop low-cost ceramic membranes that can compete with polymeric membranes in terms of pollutant rejection and ease of manufacture.

REFERENCES

Almandoz, M. C., Pagliero, C. L., Ochoa, N. A., & Marchese, J. (2015). Composite ceramic membranes from natural aluminosilicates for microfiltration applications. *Ceramics International*, *41*(4), 5621–5633. https://doi.org/10.1016/j.ceramint.2014.12.144

Alresheedi, M. T., Barbeau, B., & Basu, O. D. (2019). Comparisons of NOM fouling and cleaning of ceramic and polymeric membranes during water treatment. *Separation and Purification Technology*, *209*(March 2018), 452–460. https://doi.org/10.1016/j.seppur.2018.07.070

Artun, G., & Askin, A. (2022). Studies on production of low-cost ceramic membranes and their uses in wastewater treatment processes. *The European Journal of Research and Development*, *2*(2), 126–140.

Asif, M. B., Ren, B., Li, C., Maqbool, T., Zhang, X., & Zhang, Z. (2020). Powdered activated carbon – Membrane bioreactor (PAC-MBR): Impacts of high PAC concentration on micropollutant removal and microbial communities. *Science of the Total Environment*, *745*, 141090. https://doi.org/10.1016/j.scitotenv.2020.141090

Asif, M. B., & Zhang, Z. (2021). Ceramic membrane technology for water and wastewater treatment: A critical review of performance, full-scale applications, membrane fouling and prospects. *Chemical Engineering Journal*, *418*, 129481. https://doi.org/10.1016/j.cej.2021.129481

Beqqour, D., Achiou, B., Bouazizi, A., Ouaddari, H., Elomari, H., Ouammou, M., Bennazha, J., & Younssi, S. A. (2019). Enhancement of microfiltration performances of pozzolan membrane by incorporation of micronized phosphate and its application for industrial wastewater treatment. *Journal of Environmental Chemical Engineering*, *7*(2), 102981.

Berradi, M., Hsissou, R., Khudhair, M., Assouag, M., Cherkaoui, O., El Bachiri, A., & El Harfi, A. (2019). Textile finishing dyes and their impact on aquatic environs. *Heliyon*, *5*(11), e02711.

Bouazizi, A., Breida, M., Karim, A., Achiou, B., Ouammou, M., Calvo, J. I., Aaddane, A., Khiat, K., & Younssi, S. A. (2017). Development of a new TiO2 ultrafiltration membrane on flat ceramic support made from natural bentonite and micronized phosphate and applied for dye removal. *Ceramics International*, *43*(1), 1479–1487.

Bouazizi, A., Saja, S., Achiou, B., Ouammou, M., Calvo, J. I., Aaddane, A., & Younssi, S. A. (2016). Elaboration and characterization of a new flat ceramic MF membrane made from natural Moroccan bentonite. Application to treatment of industrial wastewater. *Applied Clay Science*, *132*, 33–40.

Bousbih, S., Errais, E., Darragi, F., Duplay, J., Trabelsi-Ayadi, M., Daramola, M. O., & Ben Amar, R. (2021). Treatment of textile wastewater using monolayered ultrafiltation ceramic membrane fabricated from natural kaolin clay. *Environmental Technology*, *42*(21), 3348–3359.

Cabrera, S. M., Winnubst, L., Richter, H., Voigt, I., & Nijmeijer, A. (2021). Industrial application of ceramic nanofiltration membranes for water treatment in oil sands mines. *Separation and Purification Technology*, *256*, 117821. https://doi.org/10.1016/J.SEPPUR.2020.117821

Cerneaux, S., & Germain, V. (2015). Turning commercial ceramic membranes into a first stage of membranes for post-combustion CO2 separation. *Journal of Membrane Science & Technology*, *05*(02). https://doi.org/10.4172/2155-9589.1000139

Charcosset, C. (2016). Ultrafiltration, microfiltration, nanofiltration and reverse osmosis in integrated membrane processes. *Integrated Membrane Systems and Processes*, 1–22.

Chen, M., Zhu, L., Chen, J., Yang, F., Tang, C. Y., Guiver, M. D., & Dong, Y. (2020). Spinel-based ceramic membranes coupling solid sludge recycling with oily wastewater treatment. *Water Research*, *169*, 115180. https://doi.org/10.1016/J.WATRES.2019.115180

Chen, M., Zhu, L., Dong, Y., Li, L., & Liu, J. (2016). Waste-to-resource strategy to fabricate highly porous whisker-structured mullite ceramic membrane for simulated oil-in-water emulsion wastewater treatment. *ACS Sustainable Chemistry and Engineering*, *4*(4), 2098–2106. https://doi.org/10.1021/acssuschemeng.5b01519

Chen, Z., Li, Z., Li, J., Liu, C., Lao, C., Fu, Y., Liu, C., Li, Y., Wang, P., & He, Y. (2019). 3D printing of ceramics: A review. *Journal of the European Ceramic Society*, *39*(4), 661–687. https://doi.org/10.1016/j.jeurceramsoc.2018.11.013

Cheraghipoor, M., Mohebbi-Kalhori, D., Noroozifar, M., & Maghsoodlou, M. T. (2019). Comparative study of bioelectricity generation in a microbial fuel cell using ceramic membranes made of ceramic powder, Kalporgan's soil, and acid leached Kalporgan's soil. *Energy*, *178*, 368–377. https://doi.org/10.1016/j.energy.2019.04.124

Chung, J., Kim, G., Seo, K. W., Jin, J., & Choi, Y. S. (2014). Effects of step-feeding and internal recycling on nitrogen removal in ceramic membrane bioreactors, and their hydraulic backwashing characteristics. *Separation and Purification Technology*, *138*, 219–226. https://doi.org/10.1016/j.seppur.2014.10.005

da Silva, V. J., da Silva, M. F., Gonçalves, W. P., de Menezes, R. R., de Araújo Neves, G., de Lucena Lira, H., & de Lima Santana, L. N. (2016). Porous mullite blocks with compositions containing kaolin and alumina waste. *Ceramics International*, *42*(14), 15471–15478.

Das, N., & Maiti, H. S. (1998). Formatation of pore structure in tape-cast alumina membranes – Effects of binder content and firing temperature. *Journal of Membrane Science*, *140*(2), 205–212. https://doi.org/10.1016/S0376-7388(97)00282–2

Daufin, G., Escudier, J.-P., Carrère, H., Bérot, S., Fillaudeau, L., & Decloux, M. (2001). Recent and emerging applications of membrane processes in the food and dairy industry. *Food and Bioproducts Processing*, *79*(2), 89–102.

Deka, A., Rasul, A., Baruah, A., Malakar, H., & Basumatary, A. K. (2023). Treatment of dairy wastewater with tubular ceramic membrane. *Materials Today: Proceedings*, *72*, 2773–2779.

Dong, X., Jin, W., Xu, N., & Li, K. (2011). Dense ceramic catalytic membranes and membrane reactors for energy and environmental applications. *Chemical Communications*, *47*(39), 10886–10902.

Dong, Y., Wu, H., Yang, F., & Gray, S. (2022). Cost and efficiency perspectives of ceramic membranes for water treatment. *Water Research*, *220*, 118629. https://doi.org/10.1016/J.WATRES.2022.118629

Drouiche, N., Ghaffour, N., Lounici, H., & Mameri, M. (2007). Electrocoagulation of chemical mechanical polishing wastewater. *Desalination*, *214*(1–3), 31–37. https://doi.org/10.1016/J.DESAL.2006.11.009

Duraisamy, R. T., Beni, A. H., & Henni, A. (2013). State of the art treatment of produced water. *Water Treatment, 199*.

Fersi, C., Gzara, L., & Dhahbi, M. (2005). Treatment of textile effluents by membrane technologies. *Desalination, 185*(1–3), 399–409.

Freeman, S., & Shorney-Darby, H. (2011). What's the buzz about ceramic membranes? *Journal-American Water Works Association, 103*(12), 12–13.

Galjaard, G., Clement, J., Ang, W. S., & Lim, M. H. (2012). Ceramac®-19 demonstration plant ceramic microfiltration at Choa Chu Kang Waterworks. *Water Practice and Technology, 7*(4).

Gao, N., Qu, B., Xing, Z., Ji, X., Zhang, E., & Liu, H. (2018). Development of novel polyethylene air-cathode material for microbial fuel cells. *Energy, 155*, 763–771. https://doi.org/10.1016/J.ENERGY.2018.05.055

Ghyoot, W. R., & Verstraete, W. H. (1997). Coupling membrane filtration to anaerobic primary sludge digestion. *Environmental Technology, 18*(6), 569–580. https://doi.org/10.1080/09593331808616575

Goswami, L., Kumar, R. V., Pakshirajan, K., & Pugazhenthi, G. (2019). A novel integrated biodegradation – Microfiltration system for sustainable wastewater treatment and energy recovery. *Journal of Hazardous Materials, 365*, 707–715. https://doi.org/10.1016/J.JHAZMAT.2018.11.029

Gu, Q., Ng, T. C. A., Zang, W., Zhang, L., Lyu, Z., Zhang, Z., Ng, H. Y., & Wang, J. (2020). Surface engineered alumina microfiltration membranes based on rationally constructed core-shell particles. *Journal of the European Ceramic Society, 40*(15), 5951–5958. https://doi.org/10.1016/J.JEURCERAMSOC.2020.07.007

Guechi, A., Harabi, A., Condoum, S., Zenikheri, F., Boudaira, B., Bouzerara, F., & Foughali, L. (2016). Elaboration and characterization of tubular supports for membranes filtration. *Desalination and Water Treatment, 57*(12), 5246–5252.

Hatimi, B., Mouldar, J., Loudiki, A., Hafdi, H., Joudi, M., Daoudi, E. M., Nasrellah, H., Lançar, I. T., Mhammedi, M. A. El, & Bakasse, M. (2020). Low cost pyrrhotite ash/clay-based inorganic membrane for industrial wastewaters treatment. *Journal of Environmental Chemical Engineering, 8*(1), 103646. https://doi.org/10.1016/J.JECE.2019.103646

Hofs, B., Ogier, J., Vries, D., Beerendonk, E. F., & Cornelissen, E. R. (2011). Comparison of ceramic and polymeric membrane permeability and fouling using surface water. *Separation and Purification Technology, 79*(3), 365–374. https://doi.org/10.1016/J.SEPPUR.2011.03.025

Huang, C. J., Yang, B. M., Chen, K. S., Chang, C. C., & Kao, C. M. (2011). Application of membrane technology on semiconductor wastewater reclamation: A pilot-scale study. *Desalination, 278*(1), 203–210. https://doi.org/10.1016/j.desal.2011.05.032

Hubadillah, S. K., Othman, M. H. D., Matsuura, T., Ismail, A. F., Rahman, M. A., Harun, Z., Jaafar, J., & Nomura, M. (2018). Fabrications and applications of low cost ceramic membrane from kaolin: A comprehensive review. *Ceramics International, 44*(5), 4538–4560. https://doi.org/10.1016/J.CERAMINT.2017.12.215

Hussain, T., & Wahab, A. (2018). A critical review of the current water conservation practices in textile wet processing. *Journal of Cleaner Production, 198*, 806–819.

Issaoui, M., & Limousy, L. (2019). Low-cost ceramic membranes: Synthesis, classifications, and applications. *Comptes Rendus Chimie, 22*(2–3), 175–187. https://doi.org/10.1016/J.CRCI.2018.09.014

Jadhav, D. A., Ghadge, A. N., & Ghangrekar, M. M. (2015). Enhancing the power generation in microbial fuel cells with effective utilization of goethite recovered from mining mud as anodic catalyst. *Bioresource Technology, 191*, 110–116. https://doi.org/10.1016/J.BIORTECH.2015.04.109

Jadhav, D. A., Park, S. G., Eisa, T., Mungray, A. K., Madenli, E. C., Olabi, A. G., Abdelkareem, M. A., & Chae, K. J. (2022). Current outlook towards feasibility and sustainability of ceramic membranes for practical scalable applications of microbial fuel cells.

Renewable and Sustainable Energy Reviews, *167*, 112769. https://doi.org/10.1016/J.RSER.2022.112769

Jafari, B., Abbasi, M., & Hashemifard, S. A. (2020). Development of new tubular ceramic microfiltration membranes by employing activated carbon in the structure of membranes for treatment of oily wastewater. *Journal of Cleaner Production*, *244*, 118720. https://doi.org/10.1016/J.JCLEPRO.2019.118720

Jedidi, I., Saïdi, S., Khemakhem, S., Larbot, A., Elloumi-Ammar, N., Fourati, A., Charfi, A., Salah, A. Ben, & Amar, R. Ben. (2009). Elaboration of new ceramic microfiltration membranes from mineral coal fly ash applied to waste water treatment. *Journal of Hazardous Materials*, *172*(1), 152–158. https://doi.org/10.1016/j.jhazmat.2009.06.151

Kárászová, M., Zach, B., Petrusová, Z., Červenka, V., Bobák, M., Šyc, M., & Izák, P. (2020). Post-combustion carbon capture by membrane separation, Review. *Separation and Purification Technology*, *238*, 116448. https://doi.org/10.1016/J.SEPPUR.2019.116448

Shahid, M. K., Kashif, A., Rout, P. R., Aslam, M., Fuwad, A., Choi, Y., Banu J. R., Park, J. H., & Kumar, G. (2020). A brief review of anaerobic membrane bioreactors emphasizing recent advancements, fouling issues and future perspectives. *Journal of Environmental Management*, *270*(August 2019), 110909. https://doi.org/10.1016/j.jenvman.2020.110909

Kaur, H., Bulasara, V. K., & Gupta, R. K. (2016). Effect of carbonates composition on the permeation characteristics of low-cost ceramic membrane supports. *Journal of Industrial and Engineering Chemistry*, *44*, 185–194. https://doi.org/10.1016/J.JIEC.2016.08.026

Khulbe, K. C., & Matsuura, T. (2021). *Nanotechnology in Membrane Processes* (pp. 199–343). Springer International Publishing.

Kocabas, A. M., Yukseler, H., Dilek, F. B., & Yetis, U. (2009). Adoption of European Union's IPPC Directive to a textile mill: Analysis of water and energy consumption. *Journal of Environmental Management*, *91*(1), 102–113.

Kumar, R. V., Goswami, L., Pakshirajan, K., & Pugazhenthi, G. (2016). Dairy wastewater treatment using a novel low cost tubular ceramic membrane and membrane fouling mechanism using pore blocking models. *Journal of Water Process Engineering*, *13*, 168–175. https://doi.org/10.1016/j.jwpe.2016.08.012

Lee, M., Wu, Z., & Li, K. (2015). 2 – Advances in ceramic membranes for water treatment. In A. Basile, A. Cassano, & N. K. B. T.-A. in M. T. for W. T. Rastogi (Eds.), *Woodhead Publishing Series in Energy* (pp. 43–82). Woodhead Publishing. https://doi.org/10.1016/B978-1-78242-121-4.00002-2

Li, C., Sun, W., Lu, Z., Ao, X., & Li, S. (2020). Ceramic nanocomposite membranes and membrane fouling: A review. *Water Research*, *175*, 115674.

Li, G. Y., Wang, X. H., & Wei, Y. J. (2014). Application of inorganic ceramic membrane in wastewater treatment I: Process research and parameter optimization. *Applied Mechanics and Materials*, *454*, 158–162.

Li, Y., Ding, W., Jin, X., Yu, J., Hu, X., & Huang, Y. (2015). Toward extensive application of Pd/ceramic membranes for hydrogen separation: A case study on membrane recycling and reuse in the fabrication of new membranes. *International Journal of Hydrogen Energy*, *40*(8), 3528–3537. https://doi.org/10.1016/j.ijhydene.2014.09.017

Lin, S. H., & Yang, C. R. (2004). Chemical and physical treatments of chemical mechanical polishing wastewater from semiconductor fabrication. *Journal of Hazardous Materials*, *108*(1–2), 103–109. https://doi.org/10.1016/J.JHAZMAT.2004.01.014

Loganathan, K. (2017). Ozone-based advanced oxidation processes for the removal of harmful algal bloom (HAB) toxins: A review. *Desalination and Water Treatment*, *59*, 65–71. https://doi.org/10.5004/dwt.2016.0346

Lokhandwala, K. A., Pinnau, I., He, Z., Amo, K. D., DaCosta, A. R., Wijmans, J. G., & Baker, R. W. (2010). Membrane separation of nitrogen from natural gas: A case study from membrane synthesis to commercial deployment. *Journal of Membrane Science*, *346*(2), 270–279. https://doi.org/10.1016/J.MEMSCI.2009.09.046

Manni, A., Achiou, B., Karim, A., Harrati, A., Sadik, C., Ouammou, M., Alami Younssi, S., & El Bouari, A. (2020). New low-cost ceramic microfiltration membrane made from natural magnesite for industrial wastewater treatment. *Journal of Environmental Chemical Engineering*, *8*(4), 103906. https://doi.org/10.1016/j.jece.2020.103906

Marchese, J., Almandoz, C., Amaral, M., Palacio, L., Calvo, J. I., Pradanos, P., & Hernández, A. (2000). Fabricación y caracterización de membranas cerámicas tubulares para microfiltración. *Boletín de La Sociedad Española de Cerámica y Vidrio*, *39*(2), 215–219. https://doi.org/10.3989/cyv.2000.v39.i2.869

Mohammad, A. W., Ng, C. Y., Lim, Y. P., & Ng, G. H. (2012). Ultrafiltration in food processing industry: Review on application, membrane fouling, and fouling control. *Food and Bioprocess Technology*, *5*, 1143–1156.

Mohammadi, T., & Pak, A. (2003). Effect of calcination temperature of kaolin as a support for zeolite membranes. *Separation and Purification Technology*, *30*(3), 241–249. https://doi.org/10.1016/S1383-5866(02)00146-6

Mouiya, M., Bouazizi, A., Abourriche, A., Benhammou, A., El Hafiane, Y., Ouammou, M., Abouliatim, Y., Younssi, S. A., Smith, A., & Hannache, H. (2019). Fabrication and characterization of a ceramic membrane from clay and banana peel powder: Application to industrial wastewater treatment. *Materials Chemistry and Physics*, *227*, 291–301.

Mueller, U., Biwer, G., & Baldauf, G. (2010). Ceramic membranes for water treatment. *Water Science and Technology: Water Supply*, *10*(6), 987–994.

Mueller, U., & Witte, M. (2008). *Ceramic Membrane Applications for Spent Filter Backwash Water Treatment*. Techneau.

Mustafa, G., Wyns, K., Janssens, S., Buekenhoudt, A., & Meynen, V. (2018). Evaluation of the fouling resistance of methyl grafted ceramic membranes for inorganic foulants and co-effects of organic foulants. *Separation and Purification Technology*, *193*, 29–37.

Nandi, B. K., Moparthi, A., Uppaluri, R., & Purkait, M. K. (2010). Treatment of oily wastewater using low cost ceramic membrane: Comparative assessment of pore blocking and artificial neural network models. *Chemical Engineering Research and Design*, *88*(7), 881–892.

Naziri Mehrabani, S. A., Vatanpour, V., & Koyuncu, I. (2022). Green solvents in polymeric membrane fabrication: A review. *Separation and Purification Technology*, *298*, 121691. https://doi.org/10.1016/j.seppur.2022.121691

Ning, R. Y. (2011). *Expanding Issues in Desalination*. BoD–Books on Demand.

Nunes, S. P., Culfaz-Emecen, P. Z., Ramon, G. Z., Visser, T., Koops, G. H., Jin, W., & Ulbricht, M. (2020). Thinking the future of membranes: Perspectives for advanced and new membrane materials and manufacturing processes. *Journal of Membrane Science*, *598*, 117761.

Özdemir, K. (2016). A ceramic ultrafiltration membrane system for producing high quality drinking water. *Karaelmas Science & Engineering Journal*, *6*(1).

Park, S. H., Park, Y. G., Lim, J.-L., & Kim, S. (2015). Evaluation of ceramic membrane applications for water treatment plants with a life cycle cost analysis. *Desalination and Water Treatment*, *54*(4–5), 973–979. https://doi.org/10.1080/19443994.2014.912162

Peng, L. C. (2008). Bimodal porous ceramic membrane via nano-sized polystyrene templating: Synthesis, characterization and performance evaluation. *Proceedings of the Volume Doctor of Philosophy, USM*.

Qin, G., Lü, X., Wei, W., Li, J., Cui, R., & Hu, S. (2015). Microfiltration of kiwifruit juice and fouling mechanism using fly-ash-based ceramic membranes. *Food and Bioproducts Processing*, *96*, 278–284.

Qiu, G., Law, Y. M., Das, S., & Ting, Y. P. (2015). Direct and complete phosphorus recovery from municipal wastewater using a hybrid microfiltration-forward osmosis membrane bioreactor process with seawater brine as draw solution. *Environmental Science and Technology*, *49*(10), 6156–6163. https://doi.org/10.1021/es504554f

Saja, S., Bouazizi, A., Achiou, B., Ouammou, M., Albizane, A., Bennazha, J., & Younssi, S. A. (2018). Elaboration and characterization of low-cost ceramic membrane made from natural Moroccan perlite for treatment of industrial wastewater. *Journal of Environmental Chemical Engineering*, *6*(1), 451–458.

Sharma, S. K., Das, P., Mandal, B., & Sanfui, B. K. (2022). Fabrication, characterization and optimization of industrial alpha alumina powders based ceramic membrane supports and its applicative potential for CO2/N2 separation. *Journal of CO_2 Utilization*, *63*, 102121. https://doi.org/10.1016/j.jcou.2022.102121

Shokrkar, H., Salahi, A., Kasiri, N., & Mohammadi, T. (2011). Mullite ceramic membranes for industrial oily wastewater treatment: Experimental and neural network modeling. *Water Science and Technology*, *64*(3), 670–676.

Shurygin, M., Guenther, C., Fuchs, S., & Prehn, V. (2021). Effective treatment of the wastewater from ceramic industry using ceramic membranes. *Water Science and Technology*, *83*(5), 1055–1071.

Slate, A. J., Whitehead, K. A., Brownson, D. A. C., & Banks, C. E. (2019). Microbial fuel cells: An overview of current technology. *Renewable and Sustainable Energy Reviews*, *101*, 60–81. https://doi.org/10.1016/J.RSER.2018.09.044

Strathmann, H. (2001). Membrane separation processes: Current relevance and future opportunities. *AIChE Journal*, *47*(5), 1077–1087.

Tai, Z. S., Hubadillah, S. K., Othman, M. H. D., Dzahir, M. I. H. M., Koo, K. N., Tendot, N. I. S. T. I., Ismail, A. F., Rahman, M. A., Jaafar, J., & Aziz, M. H. A. (2019). Influence of pre-treatment temperature of palm oil fuel ash on the properties and performance of green ceramic hollow fiber membranes towards oil/water separation application. *Separation and Purification Technology*, *222*, 264–277. https://doi.org/10.1016/J.SEPPUR.2019.04.046

Teow, Y. H., Ooi, B. S., Ahmad, A. L., & Lim, J. K. (2020). Investigation of anti-fouling and UV-cleaning properties of PVDF/TiO2 mixed-matrix membrane for humic acid removal. *Membranes*, *11*(1), 16.

Tomasula, P. M., Mukhopadhyay, S., Datta, N., Porto-Fett, A., Call, J. E., Luchansky, J. B., Renye, J., & Tunick, M. (2011). Pilot-scale crossflow-microfiltration and pasteurization to remove spores of Bacillus anthracis (Sterne) from milk. *Journal of Dairy Science*, *94*(9), 4277–4291.

Tsai, J. C., Kumar, M., Chen, S. Y., & Lin, J. G. (2007). Nano-bubble flotation technology with coagulation process for the cost-effective treatment of chemical mechanical polishing wastewater. *Separation and Purification Technology*, *58*(1), 61–67. https://doi.org/10.1016/J.SEPPUR.2007.07.022

Van der Bruggen, B., De Vreese, I., & Vandecasteele, C. (2001). Water reclamation in the textile industry: Nanofiltration of dye baths for wool dyeing. *Industrial & Engineering Chemistry Research*, *40*(18), 3973–3978.

Vijayan, Y., Teow, Y. H., Mohammad Al-Rajabi, M., & Wikramasinghe, S. R. (2022). A sustainable development approach of silica recovery and treatment of semiconductor-industry wastewater using ceramic membranes. *Separation Science and Technology*, 1–12.

Voigt, I., Stahn, M., Junghans, A., Rost, J., & Voigt, W. (2001). Integrated cleaning of coloured waste water by ceramic NF membranes. *Separation and Purification Technology*, *25*(1–3), 509–512.

Wang, Y. H., Chen, G., Wang, Z. S., Liu, J. W., & Luo, P. F. (2018). Improvement of microcracks resistance of porous aluminium titanate ceramic membrane support using attapulgite clay as additive. *Ceramics International*, *44*(2), 2077–2084.

Wardell, S. (2007). *Slipcasting*. University of Pennsylvania Press.

Wu, H., Sun, C., Huang, Y., Zheng, X., Zhao, M., Gray, S., & Dong, Y. (2022). Treatment of oily wastewaters by highly porous whisker-constructed ceramic membranes:

Separation performance and fouling models. *Water Research*, *211*, 118042. https://doi.org/10.1016/J.WATRES.2022.118042

Xomeritakis, G., Tsai, C. Y., & Brinker, C. J. (2005). Microporous sol–gel derived aminosilicate membrane for enhanced carbon dioxide separation. *Separation and Purification Technology*, *42*(3), 249–257. https://doi.org/10.1016/J.SEPPUR.2004.08.003

Ying, Y., Ying, W., Li, Q., Meng, D., Ren, G., Yan, R., & Peng, X. (2017). Recent advances of nanomaterial-based membrane for water purification. *Applied Materials Today*, *7*, 144–158.

Yusof, M. S. M., Othman, M. H. D., Mustafa, A., Rahman, M. A., Jaafar, J., & Ismail, A. F. (2018). Feasibility study of cadmium adsorption by palm oil fuel ash (POFA)-based low-cost hollow fibre zeolitic membrane. *Environmental Science and Pollution Research*, *25*(22), 21644–21655. https://doi.org/10.1007/s11356-018-2286-6

Zhang, Z., Ng, T. C. A., Gu, Q., Zhang, L., Lyu, Z., Zhang, X., Ng, H. Y., & Wang, J. (2021). Ultrathin TiO2 microfiltration membranes supported on a holey intermediate layer to raise filtration performance. *Journal of the European Ceramic Society*, *41*(2), 1622–1628. https://doi.org/10.1016/J.JEURCERAMSOC.2020.09.053

Zhu, H., Pan, Y., Sun, X., Liu, G., Qiu, M., Ding, X., Fan, Y., & Jin, W. (2021). Recycle of ceramic substrate of PDMS/ceramic composite membranes towards alcohol-permselective pervaporation. *Journal of Membrane Science*, *640*, 119835.

Zhu, Z., Wei, Z., Sun, W., Hou, J., He, B., & Dong, Y. (2016). Cost-effective utilization of mineral-based raw materials for preparation of porous mullite ceramic membranes via in-situ reaction method. *Applied Clay Science*, *120*, 135–141. https://doi.org/10.1016/J.CLAY.2015.11.020

Zou, D., & Fan, Y. (2021). State-of-the-art developments in fabricating ceramic membranes with low energy consumption. *Ceramics International*, *47*(11), 14966–14987. https://doi.org/10.1016/J.CERAMINT.2021.02.195

Zoubeik, M., Ismail, M., Salama, A., & Henni, A. (2018). New developments in membrane technologies used in the treatment of produced water: A review. In *Arabian Journal for Science and Engineering* (Vol. 43, Issue 5, pp. 2093–2118). Springer Verlag. https://doi.org/10.1007/s13369-017-2690-0

9 Application of Ceramics in Sensors and Batteries

Bhavna Rohilla, Aryan Boora, Priya Malik, Supriya Sehrawat, and Surender Duhan

9.1 INTRODUCTION

Ceramics have been an essential part of human civilization for thousands of years. From pottery and ceramics art to modern technological applications, ceramics have found many uses due to their unique properties. Ceramics are inorganic, nonmetallic materials that are typically made from clay, minerals, and other natural materials [1]. The manufacturing process for ceramics involves high-temperature treatment and often results in a material that is hard, brittle, and can withstand high temperatures. One area where ceramics have made a significant impact is in the fields of sensors and batteries [2–5]. The properties of ceramics like high-level temperature resistance, chemical stability, and low electrical conductivity make them ideal for use in these applications. Ceramics have found many applications in sensors due to their ability to detect changes in physical or chemical properties and convert them into electrical signals [6, 7]. Sensors are used in various industries, such as healthcare, automotive, aerospace, and manufacturing, among others, to monitor and control various processes.

In recent years, the use of ceramics in sensors has been growing steadily because of the development of new materials and the availability of advanced mass production techniques. Ceramic materials such as alumina, zirconia, and silicon nitride are commonly used in pressure sensors due to their high strength, wear resistance, and stability toward higher temperatures. These types of materials have found applications in the aerospace and automotive industries, where they are used to monitor pressure changes in engines and other components [8].

Ceramic materials are also used in gas sensors due to their high selectivity and sensitivity. Zirconia and tin oxide are generally used as detector materials in gas sensors due to their ability to detect specific gases such as oxygen, carbon monoxide, and nitrogen oxides [9, 10]. These materials have found applications in industries such as to access environmental conditions and chemical activities, where they are used to detect and monitor toxic gases. Temperature sensors are another area where ceramics have found applications due to their high-temperature stability. Materials such as alumina and silicon carbide are commonly used in temperature sensors as they can withstand temperatures up to 1700 °C. These materials have found applications in industries such as aerospace and metallurgy, where they are used to monitor temperature changes in engines and other high-temperature environments [11–13].

DOI: 10.1201/9781003731672-9

Ceramics are widely used in batteries due to their unique properties, leading to the evolution of high-performance batteries with improved specific energy and stability. Batteries convert chemical energy into electricity and are utilized in diverse applications such as electronic devices, electric automobiles, and renewable energy storage. A new form of battery called a solid-state battery utilizes a solid electrolyte rather than a liquid one, and ceramic materials such as lithium garnets and NASICON (sodium superionic conductor) are commonly used as solid electrolytes [14]. These materials have good electrical conductivity and chemical resistance stability, making them a potential replacement for traditional lithium-ion batteries due to their improved safety and stability. In lithium-ion batteries, ceramic materials like lithium cobalt oxide, lithium manganese oxide, and lithium iron phosphate are commonly used as cathode materials due to their high energy density and stability [15]. They have numerous applications in industries such as electronics and electric vehicles, where they are used to power various devices. Fuel cell devices are equipment that transforms a fuel's chemical energy into electrical current; ceramic materials such as yttria-stabilized zirconia and doped ceria are commonly used as electrolyte materials in fuel cells because of their excellent ionic conductivity and chemical resistance stability. These materials have found applications in industries such as transportation and energy generation, where they are used to generate electricity from renewable sources.

9.2 CLASSIFICATION OF THE CERAMIC MATERIALS FOR SENSORS

Ceramic materials for sensors can be classified based on their composition and functionality. Here are some common classifications.

- **Oxides**: These ceramics are composed of metal cations and oxygen anions. Examples include alumina, zirconia, and titania. Oxide ceramics are commonly used in gas sensors and pressure sensors.
- **Nitrides**: These ceramics are composed of metal cations and nitrogen anions. Examples include silicon nitride and aluminum nitride. Nitride ceramics are commonly used in temperature sensors and pressure sensors.
- **Carbides**: These ceramics are composed of metal cations and carbon anions. Examples include silicon carbide and tungsten carbide. Carbide ceramics are commonly used in pressure sensors and temperature sensors.
- **Silicates**: These ceramics are composed of silicon cations and oxygen anions, along with other metal cations. Examples include mullite and cordierite. Silicate ceramics are commonly used in temperature sensors and gas sensors.
- **Ferrites**: These ceramics are composed of iron cations and oxygen anions, along with other metal cations. Examples include spinel ferrites and garnet ferrites. Ferrite ceramics are commonly used in magnetic sensors and temperature sensors.
- **Glass-ceramics**: These ceramics are produced by the controlled crystallization of glasses. Examples include lithium disilicate and leucite. Glass-ceramic materials are commonly used in temperature sensors, pressure sensors, and chemical sensors.

The choice of ceramic material depends on the specific application and required functionality. Each ceramic material has its distinctive properties, like electrical conductivity, thermal insulation, and resistance to chemicals, which makes it suitable for particular sensor applications.

9.3 PROPERTIES OF CERAMIC MATERIALS

Ceramic materials are a class of inorganic, nonmetallic solids that are typically crystalline in structure and composed of a combination of metallic and nonmetallic elements. The following are some common features of ceramic materials.

- **High Hardness**: Ceramic materials are known for their hardness and abrasion resistance, which make them ideal for cutting tools, ball bearings, and other applications where wear resistance is critical.
- **High Melting Point**: Ceramic materials typically have very high melting points, which make them suitable for use in high-temperature applications such as furnace linings, gas turbine engines, and space shuttle heat shields.
- **Low Thermal Expansion**: Because ceramics have a relatively small coefficient of thermal expansion, temperature changes do not cause them to significantly expand or contract. They are the best choice for usage in precise applications where stability in dimension is important due to this feature.
- **High Stiffness**: Ceramic materials are very stiff and rigid, meaning they do not bend or deform easily. This property makes them ideal for use in structural applications such as aerospace components and body armor.
- **Excellent Electrical and Thermal Insulation**: Ceramic materials are excellent insulators of both electricity and heat, which make them useful in applications such as electronic components, thermal barriers, and insulators for high-voltage power lines.
- **Brittle**: Ceramic materials are generally brittle, which means they are prone to cracking and fracturing under stress. This property can limit their use in applications where toughness and resilience are required.
- **Chemically Stable**: Ceramic materials are generally chemically stable and resistant to corrosion, which make them ideal for use in harsh environments such as chemical processing plants and nuclear reactors.

Figure 9.1 shows some common features of ceramic materials.

9.4 APPLICATIONS OF CERAMICS IN SENSORS

Ceramics have found numerous applications in sensors due to their unique properties. The ability of ceramics to detect changes in physical or chemical properties and convert them into electrical signals makes them ideal for use in sensors. Ceramic materials possess properties such as high-temperature resistance, chemical stability, and low electrical conductivity, which make them suitable for use in various sensing applications. This section will explore the various ways in which ceramics are used in sensors.

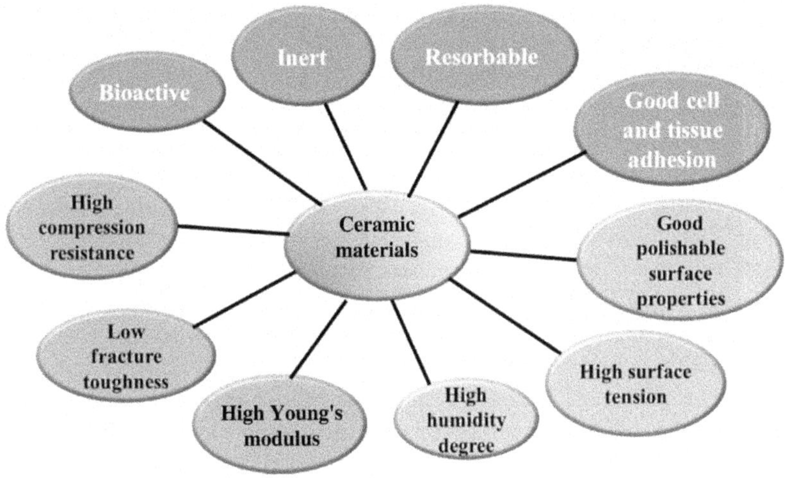

FIGURE 9.1 Properties of ceramic materials.

9.4.1 Pressure Sensors

Pressure sensors are used to measure changes in pressure and convert them into electrical signals. Ceramic materials such as alumina, zirconia, and silicon nitride are commonly used in pressure sensors due to their high strength, wear resistance, and high-temperature stability. Alumina is a ceramic material that is commonly used in pressure sensors due to its high strength and wear resistance. Alumina-based pressure sensors are commonly used in applications such as hydraulic systems, refrigeration systems, and medical equipment. These sensors can withstand high pressure and temperature changes without being damaged, making them ideal for use in harsh environments. Zirconia is another ceramic material that is commonly used in pressure sensors due to its high-temperature stability and chemical resistance. Zirconia-based pressure sensors are commonly used in applications such as combustion engines, gas turbines, and chemical processing, where they are used to monitor pressure changes in harsh environments. Zirconia sensors are also used in oxygen sensors due to their ability to detect oxygen levels. Silicon nitride is a ceramic material that is commonly used in pressure sensors due to its high strength and low density. Silicon nitride-based pressure sensors are commonly used in applications such as automotive engines and aerospace, where they are used to monitor pressure changes in harsh environments. Silicon nitride sensors can withstand high temperatures and pressure changes without being damaged, making them ideal for use in high-performance applications.

The basic principle behind pressure sensors is the use of a sensing element that changes its electrical properties in response to a change in pressure. There are several types of pressure sensors available, each with their own unique features and capabilities. One of the most common types of pressure sensors is the piezoresistive pressure sensor. This type of sensor uses a thin membrane that deforms in response to changes in pressure. The membrane is coated with a thin layer of piezoresistive material, which changes its electrical resistance in response to the deformation of the membrane. This change in resistance is measured and used to calculate the pressure. Piezoresistive pressure sensors are widely used in automotive and industrial applications, where they are used to measure the pressure of fluids and gases.

Another type of pressure sensor is the capacitive pressure sensor. This type of sensor uses a diaphragm that changes its capacitance in response to changes in pressure. The diaphragm is made of a material that is electrically conductive, and is placed between two electrodes. As pressure is applied to the diaphragm, it moves closer to one of the electrodes, changing the capacitance of the sensor. Capacitive pressure sensors are widely used in medical applications, where they are used to measure blood pressure and other physiological parameters. Another type of pressure sensor is the strain gauge pressure sensor [16]. This type of sensor uses a metal diaphragm that deforms in response to changes in pressure. The deformation of the diaphragm causes a change in the resistance of a strain gauge that is bonded to the diaphragm. This change in resistance is measured and used to calculate the pressure. Strain gauge pressure sensors are commonly used in industrial applications, where they are used to measure the pressure of fluids and gases.

In addition to these types of pressure sensors, there are several other types of sensors available, including optical pressure sensors, acoustic pressure sensors, and thermal pressure sensors. Each of these sensors uses a different principle to measure pressure, and each has its own unique advantages and disadvantages [17]. Pressure sensors are used in a wide variety of applications, including process control, automotive, aerospace, medical, and environmental monitoring. They are used to measure the pressure of fluids and gases and are often integrated into larger systems to provide real-time monitoring and control. For example, pressure sensors are used in automotive engines to measure the pressure of engine oil and coolant and are used in medical equipment to measure blood pressure and other physiological parameters.

One of the key advantages of pressure sensors is their accuracy and precision. Pressure sensors can be calibrated to provide highly accurate and precise measurements, which is critical in many applications. They are also highly reliable, with a long service life and minimal maintenance requirements.

9.4.2 Porous Ceramic Sensors

Porous ceramics are widely used as sensing materials for a variety of applications due to their unique structural and chemical properties. Porous ceramic sensors are particularly useful for gas and humidity sensing, as their high surface area and porosity allow for efficient interaction with the target molecules.

In gas sensing applications, porous ceramics can be used as sensitive materials for detecting a wide range of gases, including toxic and flammable gases in industrial and environmental settings. The gas molecules diffuse into the porous structure of the ceramic material, where they interact with the sensing sites on the ceramic surface [18]. This interaction leads to changes in the electrical or optical properties of the ceramic material, which can be measured and used to detect the presence and concentration of the target gas.

In humidity sensing applications, porous ceramics can be used as sensitive materials for monitoring the relative humidity of the surrounding environment. The porous structure of the ceramic material can absorb water vapor molecules from the environment, leading to changes in the electrical or optical properties of the ceramic material that can be correlated with the relative humidity. Various types of porous ceramic materials have been developed and used for sensing applications, including alumina, zirconia, silica, and titania. The properties of the porous ceramic materials can be tailored through control of the processing parameters, such as the particle size, porosity, and surface area.

Overall, porous ceramic sensors have shown great potential for a wide range of sensing applications, including gas and humidity sensing, and ongoing research is expected to lead to the development of new and improved sensing materials and devices for various industries.

9.4.2.1 Humidity Sensors

Ceramic materials are commonly used in humidity sensors due to their unique properties such as stability, durability, and sensitivity to humidity changes. These materials are typically classified as either porous or non-porous ceramics, with each type having distinct advantages and limitations in humidity sensing applications. Non-porous ceramics, such as aluminum oxide and titanium dioxide, have high resistance to humidity and are commonly used as humidity-sensitive materials in capacitive humidity sensors. These materials adsorb or absorb water molecules on their surface, changing the dielectric properties of the material and producing an electrical signal that is proportional to the humidity level.

Porous ceramics, such as zeolites and porous alumina, have high surface area and are commonly used in resistive humidity sensors. These materials adsorb or desorb water vapor molecules into their pores, causing changes in electrical resistance that can be measured and correlated to the humidity level [18, 19]. Porous ceramics also have the advantage of being highly selective to specific water vapor molecules, allowing for the detection of trace amounts of certain gases or pollutants.

Other ceramic materials, such as barium titanate and lead zirconate titanate, are also used in humidity sensors due to their piezoelectric properties. These materials generate an electrical signal in response to changes in pressure or mechanical stress, allowing them to be used as humidity-sensitive materials in piezoelectric humidity sensors.

Figure 9.2 shows the schematic of the humidity sensor. The sensor measures the moisture content in the surrounding environment using a ceramic substrate as a base and silver-palladium (Ag-Pd) electrodes as the sensing material [20]. The ceramic substrate provides a stable base for the sensor, while the Ag-Pd electrodes are highly sensitive to changes in humidity levels. When the humidity level changes, the electrical resistance of the Ag-Pd electrodes changes as well, allowing the sensor to detect and measure the change in humidity. The Ag-Pd electrodes are typically coated with a thin layer of a humidity-sensitive polymer, which helps to enhance the sensor's sensitivity to humidity changes. The sensor is designed to be compact, lightweight, and easy to use, making it ideal for a wide range of applications, including environmental monitoring, food storage, and industrial process control.

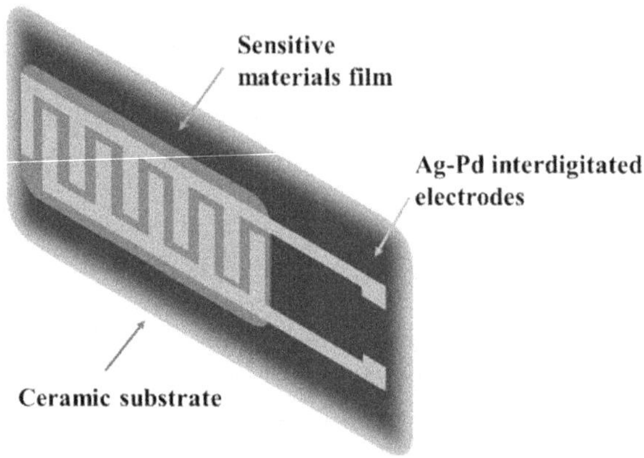

FIGURE 9.2 Schematic diagram of the humidity sensor with Ag-Pd interdigitated electrodes.

9.4.2.2 Gas Sensors

Gas sensors are used to detect the presence of gases in the environment and convert them into electrical signals. Ceramic materials such as zirconia and tin oxide are commonly used as sensing materials in gas sensors due to their high selectivity and sensitivity. Zirconia-based gas sensors are commonly used in applications such as automotive emissions monitoring, industrial gas monitoring, and environmental monitoring [21]. Zirconia sensors can detect the presence of specific gases such as oxygen, carbon monoxide, and nitrogen oxides with high accuracy, making them ideal for use in environmental monitoring and chemical processing. Tin oxide is another ceramic material that is commonly used in gas sensors due to its ability to detect a wide range of gases. Tin oxide-based gas sensors are commonly used in applications such as industrial gas monitoring, environmental monitoring, and medical diagnostics [22]. Tin oxide sensors can detect the presence of gases such as hydrogen, carbon monoxide, and methane with high sensitivity and accuracy.

A ceramic-tube gas sensor is shown schematically in Figure 9.3. Ceramic tube gas sensor is commonly used for detecting combustible gases such as methane, propane, and butane, as well as toxic gases such as carbon monoxide and nitrogen dioxide [23]. The sensitivity of the sensor depends on the specific material used for the ceramic tube and the operating temperature. In operation, the ceramic tube gas sensor is typically connected to a circuit that measures the change in electrical resistance of the tube and converts it into a signal that can be used to determine the concentration of the gas being detected. The sensor can be designed to operate in a wide range of temperatures and humidities, making it suitable for use in a variety of industrial and environmental monitoring applications.

9.5 PROGRESS IN CERAMICS HUMIDITY AND GAS SENSORS

In the recent decades, significant advancement in the development of ceramic humidity and gas sensors has been observed. This progress has been driven by advances

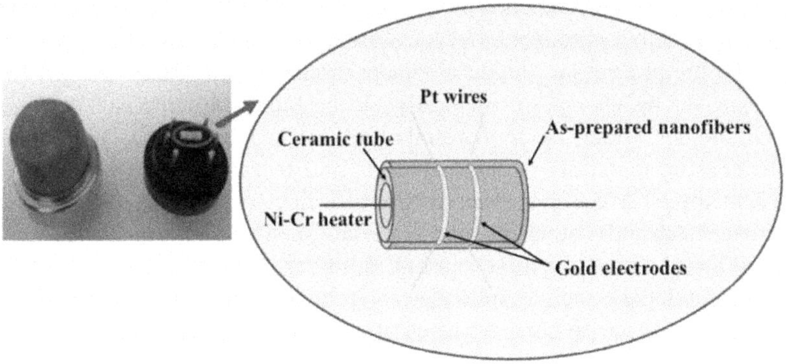

FIGURE 9.3 Schematic diagram for the ceramic-tube gas sensor.

in materials science and the development of new sensing technologies [24]. One of the key advances in the evolution of ceramic humidity sensors has been the utilization of oxides of metal such as tin dioxide $\left(SnO_2\right)$ and titanium dioxide (TiO_2) as sensing materials. These materials have high sensitivity to changes in humidity and can be used to develop highly accurate and reliable sensors. Recent research has the attention on upgrading the performance of these metal oxide sensors by modifying their surface properties. For example, researchers have developed techniques to coat the surface of metal oxide sensors with thin layers of other materials, such as gold or graphene, to enhance their sensitivity and selectivity. Another area of progress in the development of ceramic humidity sensors has been the use of ceramic materials such as alumina (Al_2O_3) and zirconia (ZrO_2) as substrates for the sensors. These materials have high mechanical strength and thermal stability, which makes them suitable for use in harsh environments.

In addition to humidity sensors, there has also been progress in the growth of ceramic gas sensors. Gas sensors perform by finding out alterations in the electrical or optical properties of the detecting material in response to the presence of a gas. Metal oxide materials such as SnO_2 and TiO_2 are also commonly used as the materials for sensing in ceramic gas sensors. Recent research has focused on developing new methods for synthesizing these materials and modifying their properties to improve the gas sensor's performance. For example, researchers have developed methods to modify the surface properties of gas sensors based on metal oxides using techniques such as doping with other materials or applying thin coatings of other materials. These modifications can improve the sensitiveness and selectiveness of gas sensors, making them more effective at detecting specific gases.

In addition to metal oxide materials, researchers have also investigated other ceramic materials such as tungsten oxide $\left(WO_3\right)$ and molybdenum oxide $\left(MoO_3\right)$ for use in gas sensors. These materials have distinct sensing properties, making them ideal for detecting specific gases, such as nitrogen dioxide or hydrogen [25]. There has also been progress in the evolution of ceramic gas sensors that use solid electrolytes, such as yttria-stabilized zirconia (YSZ) or ceria $\left(CeO_2\right)$, as the sensing material. These sensors function by identifying changes in the solid electrolyte's electrical conductivity brought on by the presence of a gas.

Overall, the progress in the development of ceramic humidity and gas sensors has been driven by advances in materials science and the development of new sensing technologies. Metal oxide materials such as SnO_2 and TiO_2 are commonly used as sensing materials in these sensors, and recent research has focused on modifying their surface properties to improve their performance. In addition to metal oxide materials, other ceramic materials such as alumina and zirconia are also used as substrates for these sensors. Researchers have also investigated new sensing technologies, such as solid electrolyte gas sensors, to further improve the performance of these devices. As research in this area continues, it is likely that we will see further progress in the development of ceramic humidity and gas sensors, leading to more accurate, reliable, and versatile sensors for a wide range of applications.

9.5.1 Temperature Sensors

Temperature sensors are used to measure changes in temperature and convert them into electrical signals [26]. Ceramic materials such as alumina and silicon carbide are commonly used in temperature sensors due to their high-temperature stability and chemical resistance.

Alumina-based temperature sensors are commonly used in applications such as aerospace, metallurgy, and chemical processing, where they are used to monitor temperature changes in harsh environments. Alumina sensors can withstand temperatures up to 1700 °C, making them ideal for use in high-temperature applications.

Silicon carbide is another ceramic material that is commonly used in temperature sensors due to its high-temperature stability and chemical resistance. Silicon carbide-based temperature sensors are commonly used in applications such as automotive engines and aerospace, where they are used to monitor temperature changes in harsh environments [27]. Silicon carbide sensors can withstand temperatures up to 2500 °C, making them ideal for use in extremely high temperature.

Temperature measurement scenarios typically involve three key components: a temperature sensor, the object being measured (the thermal target), and external factors that may affect the accuracy and speed of the measurement. To achieve optimal design, it is essential to establish perfect thermal coupling between the temperature sensor (such as the STTS22 sensor in Figure 9.4) and the object being measured. This requires minimizing the thermal resistance between the two while simultaneously maximizing the thermal resistance between these objects and other sources of heat that can be considered external noise factors (as shown in Figure 9.4).

9.6 APPLICATIONS OF CERAMICS IN BATTERIES

Ceramics have been widely used in various battery applications due to their unique properties such as high thermal stability, chemical resistance, and high-energy storage capacity. Ceramic materials such as lithium-ion ceramics, solid-state electrolytes,

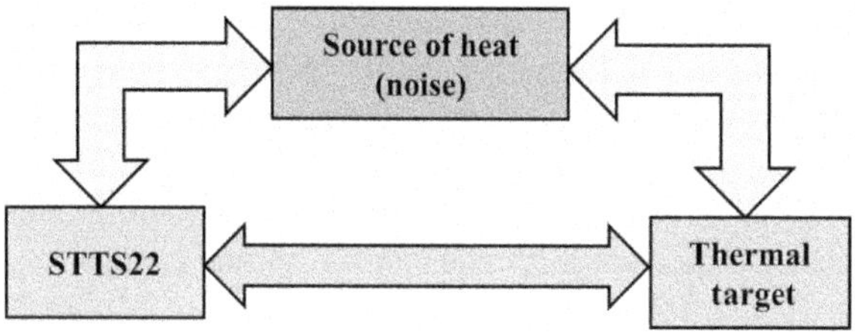

FIGURE 9.4 Applicative scenario of temperature measurement.

and ceramic coatings have found numerous applications in batteries. Some of the key ways ceramics are used in batteries include the following.

- **Electrolytes**: Ceramics are used as solid-state electrolytes in batteries, which can provide a range of benefits over liquid electrolytes. For example, solid-state electrolytes can offer higher energy density, better stability, and improved safety compared to liquid electrolytes. Additionally, ceramics can be used to create flexible, thin-film electrolytes that can be integrated into a range of battery designs.
- **Separator Materials**: Ceramic materials can be used as separators in batteries to prevent the electrodes from coming into direct contact with each other. This helps to prevent short circuits and improve battery safety.
- **Anode Materials**: Some types of ceramics, such as tin-based ceramics, have been investigated as potential anode materials for lithium-ion batteries. These materials can offer improved performance compared to traditional graphite anodes, such as higher energy density and longer cycle life.
- **Cathode Materials**: Ceramics can also be used as cathode materials in batteries. A common cathode material used in lithium-ion batteries is lithium cobalt oxide (LCO), which has a ceramic structure.

A family of ceramics known as polymer-derived ceramics (PDCs) is formed from polymer precursors. These materials are advantageous for a variety of applications, including energy storage, due to their distinctive features. Due to their high specific capacity (the amount of energy they can store per unit of mass), silicon-based PDCs (Si-PDCs) are particularly promising for energy storage applications. Si-PDCs are also very stable and have excellent cycling performance, which means they can be charged and discharged many times without degrading [28]. One application for Si-PDCs in energy storage is its use as anode components in lithium-ion batteries. The high specific capacity of Si-PDCs means that they can store more energy per unit of volume than traditional graphite anodes, which could lead to higher energy density batteries. Si-PDCs also have the advantage of being able to operate at higher voltages than graphite anodes, which could lead to higher power output. Si-PDCs can also be used in supercapacitors, which are devices that can store and release large amounts of energy very quickly. The high specific capacity and stability of Si-PDCs make them well-suited for use in supercapacitors [29]. In addition to their potential use in lithium-ion batteries and supercapacitors, Si-PDCs could also be used in other energy storage applications, such as hydrogen storage and fuel cells. Si-PDCs have a high hydrogen storage capacity, which could make them a useful material for hydrogen storage. Additionally, Si-PDCs have a high thermal stability and good chemical resistance, which are important properties for use in fuel cells.

However, there are still some challenges that need to be addressed before Si-PDCs can be widely used in energy storage applications. One challenge is the development of scalable synthesis methods for Si-PDCs. Current synthesis methods are typically limited to small-scale production, which makes it difficult to produce Si-PDCs in large quantities for commercial use. Another challenge is the improvement of the electrochemical performance of Si-PDCs. While Si-PDCs have shown promise in laboratory studies, their performance in practical applications needs to be improved for widespread use.

9.6.1 Lithium-Ion Ceramics

Portable electronic devices frequently employ lithium-ion batteries owing to their significant amount of energy and long lifespan. They utilize lithium-ion ceramics as cathodes, solid-state electrolytes, and anode electrodes as they have a high energy storage capacity and thermal stability.

Two electrodes (a cathode and an anode), a separator, and an electrolyte make up the fundamental components of a lithium-ion battery [30]. The cathode is frequently made of lithium manganese oxide, lithium cobalt oxide, or lithium iron phosphate, while the anode is frequently made of graphite. The passage of lithium ions among the electrodes in the course of charging and discharging is made possible by the electrolyte, which is a lithium salt solution in an organic solvent. Lithium ions are transported from the cathode to the anode during charging via the electrolyte, where they are then stored in the graphite anode. Lithium ions migrate back to the cathode during discharge, generating an electron flow that can be utilized to power electrical devices. This process is reversible, allowing the battery to be recharged for multiple cycles.

Significant energy density, or the quantity of energy that a battery can store in a given volume or weight, is one of the main benefits of lithium-ion batteries [31]. They are perfect for portable applications since they can offer sustained power for electronic gadgets. Additionally, because of its low self-discharge rate, lithium-ion batteries may hold their charge for a long time when not in use. Long cycle life, which refers to how many charges and discharges a battery may withstand before losing power, is another benefit of lithium-ion batteries. This makes them a cost-effective option for electronic devices that require frequent recharging. Additionally, lithium-ion batteries have a high-power density, which allows them to deliver a high amount of power quickly, making them suitable for use in electric vehicles.

Lithium-ion batteries do, however, also have significant drawbacks. One of the main concerns is the health risks of batteries powered by lithium-ions. A thermal runaway reaction that results from a defective or overheated battery may result in an explosion or fire. To mitigate this risk, manufacturers have implemented safety features such as thermal management systems and protective circuits. Another concern with lithium-ion batteries is their environmental impact [32]. The production and disposal of lithium-ion batteries can have negative environmental effects due to the mining and processing of the raw materials and the disposal of the batteries that are nearest to the last stage of their life cycle. To address this, researchers are exploring new materials and recycling methods to make lithium-ion batteries more sustainable.

A typical ceramic material used as a cathode in lithium-ion batteries is lithium cobalt oxide $\left(LiCoO_2\right)$. $LiCoO_2$-based batteries are the best option for usage in portable electronic devices due to their significant energy density and prolonged lifespan. Additional ceramic materials that are frequently utilized as cathodes in lithium-ion batteries are lithium manganese oxide $\left(LiMn_2O_4\right)$ and lithium iron phosphate $\left(LiFePO_4\right)$. A popular ceramic material used as an anode in lithium-ion batteries is lithium titanate $\left(Li_4Ti_5O_{12}\right)$. $Li_4Ti_5O_{12}$-based batteries offer high-energy storage capacity and long lifespan, making them ideal for use in electric vehicles and other high-power applications [33].

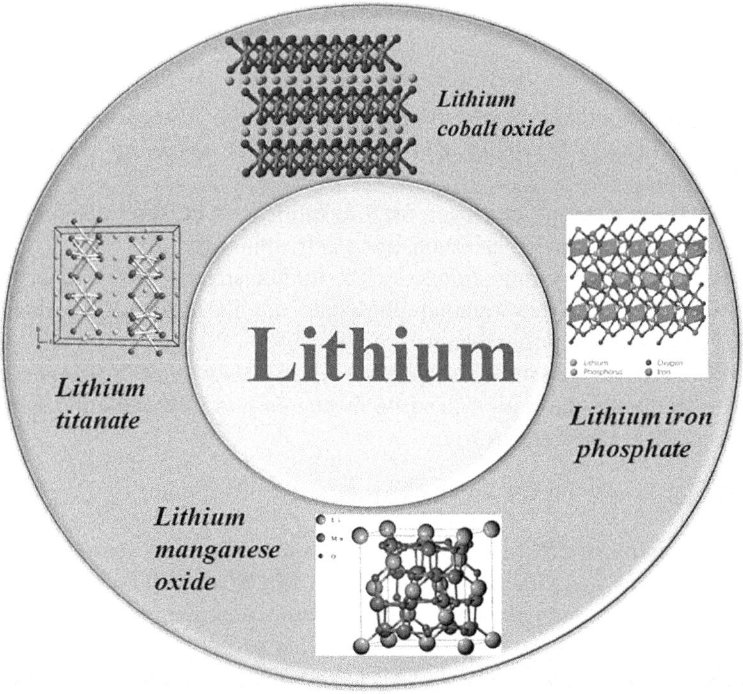

FIGURE 9.5 Lithium-ion ceramics.

9.6.2 Solid-State Electrolytes

Solid-state electrolytes are used in batteries rather than liquid electrolytes, which can pose safety risks and have limited lifespans. Ceramic materials such as lithium-ion ceramics and sulfide ceramics are commonly used as solid-state electrolytes in batteries due to their strong ionic conductivity and chemical resistance stability [34].

As solid-state electrolytes in lithium-ion batteries, lithium-ion ceramics like lithium lanthanum zirconate $\left(Li_6La_3Zr_2O\right)$ and lithium aluminium titanium phosphate $\left(Li_{1.5}Al_{0.5}Ti_{1.5}\left(PO_4\right)_3\right)$ are frequently employed. These materials are excellent for use in high-power and high-temperature applications for the reason that they have a strong ionic conductivity and chemical resistance stability. As solid-state electrolytes in batteries, sulfide ceramics like lithium sulfide $\left(Li_2S\right)$ and sodium sulfide $\left(Na_2S\right)$ are also employed. Sulfide-based electrolytes are excellent for use in significant energy density batteries like lithium-sulfur batteries because they have high ionic conductivity and good stability. [35].

9.6.2.1 Development of Lithium Battery Solid Electrolytes Made of Ceramic/Polymer Composites and Their Future Prospects

Solid electrolytes made of ceramic/polymer composites (CPCSEs) have garnered a lot of interest as a potential means of enhancing the performance, stability, and safety of lithium batteries. Here is some information on the progress and perspectives of CPCSEs for lithium batteries.

Progress:

- CPCSEs have been demonstrated to exhibit high lithium-ion conductivity, mechanical flexibility, and good interfacial suitability with lithium electrodes.
- Several approaches have been used to fabricate CPCSEs, including melt blending, in-situ polymerization, and electrospinning.
- Various types of ceramic fillers, such as lithium silicate, lithium phosphate, and lithium aluminum titanium phosphate, are added to polymer matrices to enhance the ionic conductivity of CPCSEs.
- CPCSEs have been shown to upgrade the harmlessness of lithium batteries by eliminating the risk of dendrite formation and improving the stability toward heat of the electrolyte.

Perspectives:

- The evolution of CPCSEs with excessive ionic conductivity at room temperature remains a significant challenge. Most CPCSEs exhibit high conductivity only at elevated temperatures, which limits their practical applications.
- The fabrication process of CPCSEs needs to be optimized to achieve uniform spreading of ceramic fillers in the polymer matrix and prevent the formation of agglomerates, which can negatively affect the performance of the electrolyte.

- To reduce interfacial resistance and increase the durability of the electrode/electrolyte contact, the interfacial solidarity within CPCSEs and electrode materials needs to be further upgraded.
- The long-term stability of CPCSEs needs to be evaluated under various conditions, including high temperatures, humidity, and cycling, to ensure their durability in practical applications.

9.6.2.2 Impact of Morphology of Ceramic on Ionic Conductivity of Solid Composite Electrolytes

The morphology of ceramic particles can have a remarkable effect on the ionic conductivity of solid composite electrolytes. Some of the effects are discussed here.

- **Particle Size**: The size of the ceramic particles can affect the ionic conductivity of the composite electrolyte. Smaller particles can improve the interfacial contact area between the particles, which can enhance the ionic conductivity. However, excessively small particles can lead to particle agglomeration, which can limit the interfacial area and decrease ionic conductivity.
- **Particle Shape**: The shape of ceramic particles can also affect the ionic conductivity of composite electrolytes. Spherical particles have a high packing density, which can increase the ionic conductivity and interfacial contact area. On the other hand, irregularly shaped particles can lead to particle agglomeration and decrease the ionic conductivity.
- **Porosity**: The porosity of the ceramic particles can affect the conductivity of ions in the composite electrolyte. Higher porosity can improve the electrolyte's ability to absorb and transport ions, which can enhance the ionic conductivity. However, excessive porosity can lower the mechanical toughness of the composite electrolyte.
- **Distribution**: The ionic conductivity of the composite electrolyte can also be impacted by the dispersion of ceramic particles in between the polymer matrix. A uniform distribution of ceramic particles can provide an optimal interfacial contact area, enhancing the ionic conductivity. In contrast, an uneven distribution can lead to particle agglomeration and decrease the ionic conductivity.

9.6.3 Ceramic Coatings

Ceramic coatings are increasingly being used in battery applications due to their ability to improve battery performance and longevity. Ceramic coatings provide a range of benefits to batteries, including increased stability, enhanced conductivity, and improved safety. The use of ceramic coatings in batteries will be covered in this subsection.

An important advantage of ceramic coatings is their capability to improve the stability and durability of batteries. Ceramic coatings can prevent the degradation of battery materials and protect them from harmful reactions, leading to improved battery longevity. For example, ceramic coatings can prevent the formation of lithium dendrites, which are needle-like structures that may develop on the outer layer of anode and cause short circuits in lithium-ion batteries [36]. Ceramic coatings can also protect battery materials from corrosion, leading to improved battery stability and longer cycle life. Another benefit of ceramic coatings is their ability to enhance the conductivity of battery materials. Ceramic coatings can improve the electrodes' surface area, allowing for more efficient ion transport and leading to improved battery performance. Ceramic coatings can also improve the junction of the electrode and the electrolyte, leading to lower resistance and improved conductivity.

Ceramic coatings can also improve the safety of batteries by preventing thermal runaway and other dangerous reactions. They can act as a barrier between the battery materials and the environment, preventing unwanted reactions and decreasing the possibility of fires and explosions. For instance, ceramic coatings can prevent the electrolyte's reaction with oxygen, reducing the risk of thermal runaway in lithium-ion batteries [37]. Ceramic coatings can be applied to batteries using a range of methods, including vapor deposition, sol-gel method, and electrophoretic deposition [38]. The method of application is determined by the particular application and the desired characteristics of the ceramic coating.

There are several types of ceramic coatings that are commonly used in battery applications, including lithium aluminate, lithium phosphate, and lithium titanate. These materials have unique properties that make them well-suited for use in batteries, including high thermal stability, good chemical compatibility, and high ionic conductivity.

9.7 ENVIRONMENTAL AND SUSTAINABILITY CONSIDERATIONS IN CERAMIC SENSORS

Ceramic sensors play a pivotal role in various industries, from automotive to healthcare, owing to their exceptional durability, stability, and high-temperature resistance. As the demand for sensor technologies continues to rise, it becomes imperative to scrutinize the environmental and sustainability aspects associated with the production, use, and disposal of ceramic sensors.

9.8 RAW MATERIAL SOURCING

The environmental footprint of ceramic sensors begins with the extraction and processing of raw materials. Most ceramic sensors incorporate materials such as alumina, zirconia, and silicon carbide. The extraction of these materials, particularly alumina, involves mining bauxite, which can result in habitat disruption and soil erosion. Sustainable sourcing practices, such as responsible mining and recycling programs, are essential to mitigate the environmental impact of raw material extraction.

9.9 MANUFACTURING PROCESSES

The manufacturing of ceramic sensors involves energy-intensive processes, including sintering and firing. Sintering, in particular, requires high temperatures, leading to substantial energy consumption. Transitioning to cleaner energy sources and adopting advanced manufacturing technologies, such as microwave sintering or spark plasma sintering, can significantly reduce the environmental impact of ceramic sensor production. Moreover, optimizing production processes to minimize waste generation and emissions is crucial for sustainable manufacturing.

9.10 LIFE CYCLE ASSESSMENT

A comprehensive life cycle assessment (LCA) is essential to evaluate the overall environmental impact of ceramic sensors. This assessment considers factors such as raw material extraction, manufacturing, transportation, and end-of-life disposal. By conducting LCAs, manufacturers can identify hotspots in the life cycle and implement targeted measures to reduce environmental impact. Strategies may include using recycled materials, improving energy efficiency, and optimizing transportation logistics.

9.11 ENERGY EFFICIENCY

Ceramic sensors are frequently employed in applications where energy efficiency is paramount, such as in automotive engines and industrial processes. However, the energy required for sensor operation should be minimized to enhance overall system sustainability. Engineers and researchers are exploring ways to design ceramic sensors with lower power consumption without compromising performance. This may involve developing sensors with enhanced sensitivity, enabling them to provide accurate measurements with lower energy inputs.

9.12 END-OF-LIFE CONSIDERATIONS

The disposal of electronic components, including ceramic sensors, poses a challenge due to the potential environmental impact of electronic waste (e-waste). Manufacturers should design sensors with recyclability in mind, promoting the use of materials that are easily recoverable or biodegradable. Additionally, initiatives for proper e-waste management and recycling programs should be encouraged to prevent the accumulation of ceramic sensors in landfills.

9.13 APPLICATION IN ENVIRONMENTAL MONITORING

Interestingly, ceramic sensors themselves find application in environmental monitoring systems. They can be employed to measure air and water quality, contributing to sustainable practices by providing crucial data for pollution control and resource conservation. By utilizing ceramic sensors for environmental monitoring, industries can actively contribute to sustainability efforts.

9.14 CONCLUSION

A variety of technical applications have found ceramics to be a versatile material with special features. Their use in sensors and batteries has revolutionized the development of advanced devices with enhanced properties, making them a valuable material in these fields. In sensors, ceramics have enabled the development of high-precision devices used in industrial applications. Ceramic sensors have high sensitivity, low drift, and high accuracy, making them ideal for use in harsh environments, where other materials would fail. For example, ceramic sensors can be used to measure temperature, pressure, and gas concentration in industrial processes, providing accurate and reliable data that can be used to optimize operations and improve safety.

The development of high-performance batteries with significant energy density and stability has been greatly aided by the use of ceramics in batteries. Lithium-ion ceramics, solid-state electrolytes, and ceramic coatings have found numerous applications in batteries, making them more efficient, reliable, and safer. Due to their significant energy density and great lifespan, lithium-ion batteries in particular have taken over as the standard technology in portable electronic gadgets. High-performance batteries, used in electric vehicles and other high-power applications, have been made possible by the use of ceramic materials such as lithium cobalt oxide $\left(LiCoO_2\right)$ and lithium titanate $\left(Li_4Ti_5O_{12}\right)$. Solid-state electrolytes made of ceramic materials such as lithium lanthanum zirconate $\left(Li_6La_3Zr_2O_{12}\right)$ and lithium aluminum titanium phosphate $\left(Li_{1.5}Al_{0.5}Ti_{1.5}\left(PO_4\right)_3\right)$ have also enabled the development of batteries that are safer, more efficient, and environmentally friendly. These electrolytes offer high ionic conductivity and good stability, making them suitable for utility in high-power and high-temperature applications. Ceramic coatings have also played an important role in the evolution of batteries. These coatings improve the stability and durability of electrodes, prevent electrode degradation, and enhance the safety of batteries. Ceramic coatings such as aluminum oxide $\left(Al_2O_3\right)$ and titanium dioxide (TiO_2) improve the stability of electrodes and prevent electrode degradation, improving the performance and lifespan of batteries. Dendrite growth, which can result in short circuits and battery failure, can be halted by ceramic coatings.

ACKNOWLEDGMENTS

Bhavna Rohilla and Aryan Boora express their gratitude to the University Grants Commission (UGC) for their support in funding the research through the Junior Research Fellowship (JRF) program, with reference numbers 221610112839 and 221610122664, respectively.

REFERENCES

[1] Wu H, Pan W, Lin D, Li H. Electrospinning of ceramic nanofibers: Fabrication, assembly and applications. *Journal of Advanced Ceramics*. 2012 Mar;1(1):2–3.

[2] Vivekananthan V, Alluri NR, Purusothaman Y, Chandrasekhar A, Selvarajan S, Kim SJ. Biocompatible collagen nanofibrils: An approach for sustainable energy harvesting and battery-free humidity sensor applications. *ACS Applied Materials & Interfaces*. 2018 May 10;10(22):18650–6.

[3] Paulowicz I, Hrkac V, Kaps S, Cretu V, Lupan O, Braniste T, Duppel V, Tiginyanu I, Kienle L, Adelung R, Mishra YK. Three-dimensional SnO2 nanowire networks for multifunctional applications: From high-temperature stretchable ceramics to ultra-responsive sensors. *Advanced Electronic Materials*. 2015 Aug;1(8):1500081.

[4] Zhu Y, Gonzalez-Rosillo JC, Balaish M, Hood ZD, Kim KJ, Rupp JL. Lithium-film ceramics for solid-state lithionic devices. *Nature Reviews Materials.* 2021 Apr;6(4):313–31.
[5] Esfahani H, Jose R, Ramakrishna S. Electrospun ceramic nanofiber mats today: Synthesis, properties, and applications. *Materials.* 2017 Oct 27;10(11):1238.
[6] Sudevalayam S, Kulkarni P. Energy harvesting sensor nodes: Survey and implications. *IEEE Communications Surveys & Tutorials*. 2010 Jul 26;13(3):443–61.
[7] Tuller H. Ionic conduction and applications. In *Handbook of electronic and photonic materials*, vol. 1. Springer; 2017 Oct 4.
[8] Kim JH, Kim JH, Kim JM, Lee YG, Lee SY. Superlattice crystals–mimic, flexible/functional ceramic membranes: Beyond polymeric battery separators. *Advanced Energy Materials.* 2015 Dec;5(24):1500954.
[9] Su D, Xie X, Dou S, Wang G. CuO single crystal with exposed {001} facets-A highly efficient material for gas sensing and Li-ion battery applications. *Scientific Reports.* 2014 Aug 29;4(1):5753.
[10] Pascariu P, Homocianu M. ZnO-based ceramic nanofibers: Preparation, properties and applications. *Ceramics International.* 2019 Jun 15;45(9):11158–73.
[11] Baranov A, Spirjakin D, Akbari S, Somov A. Optimization of power consumption for gas sensor nodes: A survey. *Sensors and Actuators A: Physical.* 2015 Sep 1;233:279–89.
[12] Malik R, Tomer VK, Chaudhary V, Dahiya MS, Nehra SP, Rana PS, Duhan S. Ordered mesoporous In-(TiO2/WO3) nanohybrid: An ultrasensitive n-butanol sensor. *Sensors and Actuators B: Chemical.* 2017 Feb 1;239:364–73.
[13] Tufekcioglu E, Dogan A. A flextensional piezo-composite structure for energy harvesting applications. *Sensors and Actuators A: Physical.* 2014 Sep 1;216:355–63.
[14] Hueso KB, Armand M, Rojo T. High temperature sodium batteries: Status, challenges and future trends. *Energy & Environmental Science.* 2013;6(3):734–49.
[15] Akhtar F, Rehmani MH. Energy replenishment using renewable and traditional energy resources for sustainable wireless sensor networks: A review. *Renewable and Sustainable Energy Reviews.* 2015 May 1;45:769–84.
[16] Zang Y, Zhang F, Di CA, Zhu D. Advances of flexible pressure sensors toward artificial intelligence and health care applications. *Materials Horizons.* 2015;2(2):140–56.
[17] Mannsfeld SC, Tee BC, Stoltenberg RM, Chen CV, Barman S, Muir BV, Sokolov AN, Reese C, Bao Z. Highly sensitive flexible pressure sensors with microstructured rubber dielectric layers. *Nature Materials.* 2010 Oct;9(10):859–64.
[18] Blank TA, Eksperiandova LP, Belikov KN. Recent trends of ceramic humidity sensors development: A review. *Sensors and Actuators B: Chemical.* 2016 Jun 2;228:416–42.
[19] Tomer VK, Duhan S. Ordered mesoporous Ag-doped TiO2/SnO2 nanocomposite based highly sensitive and selective VOC sensors. *Journal of Materials Chemistry A.* 2016;4(3):1033–43.
[20] Tomer VK, Devi S, Malik R, Nehra SP, Duhan S. Highly sensitive and selective volatile organic amine (VOA) sensors using mesoporous WO3–SnO2 nanohybrids. *Sensors and Actuators B: Chemical.* 2016 Jun 28;229:321–30.
[21] Wang C, Yin L, Zhang L, Xiang D, Gao R. Metal oxide gas sensors: Sensitivity and influencing factors. *Sensors.* 2010 Mar 15;10(3):2088–106.
[22] Hossein-Babaei F, Amini A. Recognition of complex odors with a single generic tin oxide gas sensor. *Sensors and Actuators B: Chemical.* 2014 Apr 1;194:156–63.
[23] Ju D, Xu H, Qiu Z, Guo J, Zhang J, Cao B. Highly sensitive and selective triethylamine-sensing properties of nanosheets directly grown on ceramic tube by forming NiO/ZnO PN heterojunction. *Sensors and Actuators B: Chemical.* 2014 Sep 1;200:288–96.
[24] Nenov T, Yordanov SP. *Ceramic sensors: Technology and applications.* CRC Press; 2020 Dec 17.
[25] Mirzaei A, Leonardi SG, Neri G. detection of hazardous volatile organic compounds (VOCs) by metal oxide nanostructures-based gas sensors: A review. *Ceramics International.* 2016 Nov 1;42(14):15119–41.

[26] Dong B, Cao B, He Y, Liu Z, Li Z, Feng Z. Temperature sensing and in vivo imaging by molybdenum sensitized visible upconversion luminescence of rare-earth oxides. *Advanced Materials*. 2012 Apr 17;24(15):1987–93.
[27] Zhang N, Lin CM, Senesky DG, Pisano AP. Temperature sensor based on 4H-silicon carbide pn diode operational from 20 C to 600 C. *Applied Physics Letters*. 2014 Feb 17;104(7):073504.
[28] Hasegawa G, Kanamori K, Nakanishi K. Porous polymer-derived ceramics: Flexible morphological and compositional controls through sol–gel chemistry. *Journal of the American Ceramic Society*. 2022 Jan;105(1):5–34.
[29] Litasov KD, Bekker TB, Kagi H. "Kamchatite" diamond aggregate from northern Kamchatka, Russia: New find of diamond formed by gas phase condensation or chemical vapor deposition – Discussion. *American Mineralogist*. 2020 Jan 1;105(1):141–3.
[30] Li S, Zhang SQ, Shen L, Liu Q, Ma JB, Lv W, He YB, Yang QH. Progress and perspective of ceramic/polymer composite solid electrolytes for lithium batteries. *Advanced Science*. 2020 Mar;7(5):1903088.
[31] Choi JA, Kim SH, Kim DW. Enhancement of thermal stability and cycling performance in lithium-ion cells through the use of ceramic-coated separators. *Journal of Power Sources*. 2010 Sep 15;195(18):6192–6.
[32] Xu RC, Xia XH, Yao ZJ, Wang XL, Gu CD, Tu JP. Preparation of Li7P3S11 glass-ceramic electrolyte by dissolution-evaporation method for all-solid-state lithium ion batteries. *Electrochimica Acta*. 2016 Nov 20;219:235–40.
[33] Li Y, Wang Z, Cao Y, Du F, Chen C, Cui Z, Guo X. W-doped Li7La3Zr2O12 ceramic electrolytes for solid state Li-ion batteries. *Electrochimica Acta*. 2015 Oct 20;180:37–42.
[34] Zhao Q, Stalin S, Zhao CZ, Archer LA. Designing solid-state electrolytes for safe, energy-dense batteries. *Nature Reviews Materials*. 2020 Mar;5(3):229–52.
[35] Liu H, Cheng XB, Huang JQ, Yuan H, Lu Y, Yan C, Zhu GL, Xu R, Zhao CZ, Hou LP, He C. Controlling dendrite growth in solid-state electrolytes. *ACS Energy Letters*. 2020 Feb 7;5(3):833–43.
[36] Shin WK, Kim DW. High performance ceramic-coated separators prepared with lithium ion-containing SiO2 particles for lithium-ion batteries. *Journal of Power Sources*. 2013 Mar 15;226:54–60.
[37] Shi C, Dai J, Shen X, Peng L, Li C, Wang X, Zhang P, Zhao J. A high-temperature stable ceramic-coated separator prepared with polyimide binder/Al2O3 particles for lithium-ion batteries. *Journal of Membrane Science*. 2016 Nov 1;517:91–9.
[38] Narita F, Fox M. A review on piezoelectric, magnetostrictive, and magnetoelectric materials and device technologies for energy harvesting applications. *Advanced Engineering Materials*. 2018 May;20(5):1700743.

10 Advances in Photocatalytic Ceramic Materials for Energy and Environmental Applications

Sampa Chakrabarti, Pallavi Mitra, and Anirban Roy

10.1 INTRODUCTION

Ceramic materials are conventionally defined as inorganic non-metallic substances, hardened by using high temperature. Such materials generally exhibit low thermal and electrical conductivity, chemical and thermal resistance, brittleness, as well as hardness. Glass, bricks, and cement are referred to as ceramic materials. In human civilization, 'ceramics' were first produced from clay about 20,000 years ago. Traditional ceramics are generally made from natural materials, like quartz, sand, or clay minerals. They are used for producing clay tile and brick, china tableware, refractory linings, and industrial abrasives. With the advancement of civilization, many uses of ceramic materials were introduced other than for the containers. Depending on the targeted usage, special properties were imbibed within the ceramic materials. In 1993, the Versailles Project on Advanced Materials and Standards (VAMAS) defined advanced ceramic materials as "an inorganic, nonmetallic (ceramic), basically crystalline material of rigorously controlled composition and manufactured with detailed regulation from highly refined and/or characterized raw materials giving precisely specified attributes". Advanced ceramics are alternatively referred to as high-tech ceramic, technical ceramic, or high-performance ceramics. They are used in electronics, telecommunication, cutting tools, electrodes, chemical, medical, energy, automotive, and other industries. After the discovery of nanotechnology, such advanced ceramic materials became especially significant. They found their usage for harvesting renewable energy, abatement of environmental pollution, manufacture of surgical implants, manufacture of novel composites, and surface coating.

Four major differences are generally noted between traditional and advanced ceramics. The first one is based on raw materials; while traditional ceramics are made from natural materials, bases of advanced ceramics are synthetic powders like silicon carbide, silicon nitride, or aluminium oxide. The composition of natural

DOI: 10.1201/9781003731672-10

clay decides the structure and texture of traditional ceramics, and the raw materials generally do not need any pretreatment. Hence traditional ceramics generally have imperfections. However, the structure of advanced ceramics can be engineered by modifying the raw material. Additional post-sintering processing or secondary machining may also be required. Hence, the microstructure of advanced ceramics is uniform and mostly perfect. Another point of difference is based on the use. Products made from traditional ceramic are mainly used for building materials and daily use household items, whereas advanced ceramic materials, having exclusive mechanical and physical properties, are utilized for manufacturing various specialized products for different industry requirements [1].

The two major critical issues faced by human civilization are energy and the environment. In 2015, 17 sustainable development goals were adopted by the UN, of which goal numbers 6 and 7 are Clean Water and Sanitation and Affordable and Clean Energy. However, due to the recent pandemic and the subsequent war, energy is no longer affordable for most of the world's population. People in poorer countries do not have access to energy; a large proportion of people in countries with a GDP per capita of less than $25,000 do not have access to electricity and clean cooking fuels [2]. Hence, they mostly rely on fossil fuels and biomass. Again, due to the announced pledges scenario, countries do not want to use fossil fuels to meet their national net zero emission pledges [3]. More efficient equipment as well as a transition to renewable energy, and hydrogen fuel, at affordable prices, are the demands of the day.

It is universally accepted that air, water, and soil – all are polluted by anthropogenic activities. The generation of energy is eventually associated with air pollution. According to the World Health Organization (WHO), 99% of the world's population is compelled to breathe in polluted air. Every year 0.7 million people die of air pollution, whereas 90% of them are from developing countries; many of them are children below 5 years [4]. In 2020, 74% of the world's population has access to safe drinking water. Two million people do not have access to safely managed sources of water. The most important chemical hazards that can be present in drinking water are arsenic, fluoride, or nitrate; however, emerging pollutants such as pharmaceuticals, pesticides, per- and polyfluoroalkyl substances (PFASs), and microplastics generate public concern [5]. Treatment of contaminated water and seawater is needed for an abundant supply of water to the growing population. Soil is polluted by chemical fertilizers, construction wastes, and waste plastics. As many as 3,200 million people are suffering from soil or land pollution [6].

Against this backdrop, humankind must prevent and abate environmental pollution on the one hand and generate the energy necessary for the advancement of civilization without polluting the environment on the other, using suitable technology. For the generation of sustainable energy, renewable sources, such as sunlight, air, and biogas can be explored. The generation of hydrogen sustainably is also required. Efficiency enhancement of conventional energy generation may also be considered. Environmental pollution may be controlled sustainably using solar photocatalysis. All these advanced and sustainable technologies need efficient and suitable materials – advanced ceramic materials can play significant roles in sustainable energy and a cleaner environment. Among the various technologies involving advanced ceramic materials for energy and environment, photocatalysis is the most important one.

A few compounds like TiO_2, ZnO, CdS, and WO_3 have a unique property to get excited with light having an energy larger than the band-gap energy of the compound. Upon excitation, hole–electron pair(s) are generated from the filled valance band and empty conduction band. The photogenerated hole, with the hydroxide ion of water, generates hydroxyl radical, which is a very powerful oxidant. With this oxidant, organic pollutants in wastewater, air, or soil can be degraded. The photogenerated electron, on the other hand, may remove the reducible pollutant by reduction [7]. It may be noted that most of the photocatalysts are ceramic materials; moreover, doping with various metals and non-metals is often done in the matrix of the base compounds. Nanotechnology, additionally, inspires nano-photocatalysts to get engineered for specific purposes.

10.2 TYPES OF ADVANCED CERAMIC MATERIALS

Classification of advanced ceramic materials may be done based on various criteria. According to the composition, ceramics can be classified as silicate ceramics, oxide ceramics, non-oxide ceramics, and glass ceramics. Advanced ceramic materials can also be classified [8] based on application as shown in Figure 10.1.

10.2.1 Metal Oxides

From the mechanism of photocatalysis described in the previous section, it is clear that a photocatalyst must have mobile electrons in the valance band as they are in semiconductors. To date, metal oxides and carbon-based materials are mostly being used as photocatalysts. Undoped, doped, and composite compounds (both micro and nanosized) of TiO_2 and ZnO are widely used as solar or UV-light-enhanced photocatalysts for energy and environmental applications. Metal oxide photocatalysts, both in micro and nano sizes, can be used as a photocatalyst to decompose toxic organic compounds, in photovoltaics, to prevent fogging of glass and even to split water into hydrogen and oxygen. Hence, they are highly important in both the fields of energy and the environment [7]. Heterogeneous photocatalysis is also applicable

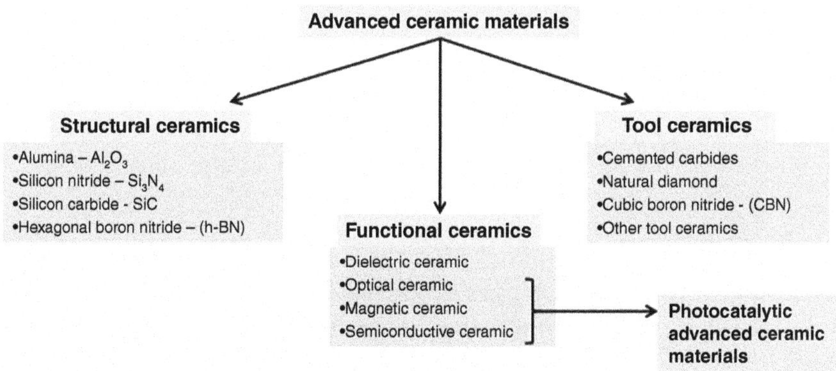

FIGURE 10.1 Classification of advanced ceramic materials.

to self-cleaning, antibacterial, and deodorization systems and especially for purification of wastewater. The physicochemical properties of the metal oxides, like shape, size, morphology, and composition, are critical for the photocatalytic performance. In the case of synthetic and engineered metal oxide particles, such properties may be controlled. Several applications of metal oxide photocatalysts are available in the literature. Some of those applications will be described in the following sections. However, there are a few limitations of the metal oxide photocatalysts like photocorrosion, self-oxidation, narrow catalytic zone, and thermal instability. Photocatalysts other than the metal-oxides are classified as follows.

10.2.2 Metal sulfides

Metal sulfides have empty valance bands and small effective mass; they therefore have a high potential for visible light harvesting. Quantum dots of CdS, $CuInS_2$, and PdS are well-explored. Many metal sulfides, like sulfides of cadmium and cobalt, are used for environmental applications and hydrogen generation. However, such sulfides are often toxic, vulnerable to recombination, and have low photo response. To overcome the limitations, many techniques like doping, sensitization, formation of heterojunction, and composite catalysts are being explored. Zinc and silver sulfides are free from toxicity, though.

10.2.3 Metal and non-metal nitrides, oxynitrides

Metal nitrides and oxynitrides are nowadays getting popular for photocatalytic applications including hydrogen generation by water splitting. N-doping enhances photocatalytic activity in visible regions. Among nitrides and oxynitrides, BN, Si_3N_4, GaON, AlON, Ga–Zn oxynitride, Ta_3N_5, TaNO, Ge_3N_4, Ga–In–Zn oxy-nitride, and $LaTiO_2N$ are proven photocatalysts.

10.2.4 Metal phosphides and black phosphorous

Phosphorus has three common allotropes – red, white, and black. Among them, black phosphorus was originally used in field transistors. Black phosphorus has a very low and flexible bandgap (~0.3–2.1 eV) and is thereby active in visible and near-infrared regions as a photocatalyst [9]. However, due to low bandgap, it has the drawback of rapid recombination of the photogenerated hole and electron(s). Therefore, in general, black phosphorus is utilized as co-catalysts or composites. Metal phosphides are used for various catalytic applications – of them, phosphides of iron, boron, cobalt, copper, and tungsten are suitable for photocatalytic applications.

10.2.5 Metal oxyhalides

Oxyhalide-based photocatalysts generally include bismuth oxyhalides (BiOX , X= F, Cl, Br, and I). Bismuth oxyhalides (BiOX), having a layered structure, represent a category of materials that are suitable for energy harvest/conversion and environment remediation. Due to the similarity of Cl, Br, and I, BiOCl, BiOBr, and BiOI possess similar properties. However, BiOX possesses variable bandgaps; such as 1.8 eV for BiOI, 2.8 eV for BiOBr, and nearly 3.5 eV for BiOCl.

10.2.6 Magnetically Recoverable Ferrites

During various applications of the photocatalysts, especially for wastewater treatment, post-treatment separation of the catalyst is a critical issue. For nano-sized photocatalysts, this issue tells upon the cost of the whole process. To address this issue, engineered photocatalysts are often made magnetic, so that they can be separated easily and economically after the treatment. Transition metal ferrites are the most popular members of magnetic photocatalysts; they can be categorized as MFe_2O_4 (spinel), MFe_2O_4 (orthoferrite), $MFe_{12}O_{19}$ (hexaferrites like $SrFe_{12}O_{19}$ and $BaFe_{12}O_{19}$), and $M_3Fe_5O_{12}$ (garnet) [10]. Among these, spinel ferrites have high adsorption capacity and good photocatalytic activities, especially in wastewater treatment.

10.2.7 Carbon-Based

Metal-free photocatalysts have recently been introduced as an alternative class of metal-based materials owing to their properties such as adjustable bandgap, high surface area, easy availability, and low toxicity. This class of photocatalysts often include photocatalysts based on C, S, B, N, Se, and P. Among the carbon-based materials, there are graphene oxide (GO), reduced graphene oxide (rGO), carbon quantum dots, organic conducting polymers, and graphitic carbon nitride (g-C_3N_4) (Table 10.1). Graphene oxide (GO) is structurally similar to graphene and is used for enhancing photocatalytic reactions as promoters. Impregnation of GO/RGO in the catalyst not only reduces the recombination of photogenerated hole–electron pairs but also increases the number of active sites to increase absorption on the catalyst surface.

TABLE 10.1
A Few C-Based and Perovskite-Based Photocatalytic Ceramic Materials

S. No.	Type	Photocatalyst	Application
1.	Carbon-based	Amorphous g-C_3N_4	H_2 production
2.		g-C_3N_4-CNTs	H_2 production
3.		$C_3N_3S_3$/RGO	Degradation of benzylic alcohols
4.		g-C_3N_4/GO aerogel	Degradation of dye pollutants and bromate
5.	Perovskite-based	$SrTiO_3$	Water splitting
6.		$LaFeO_3$	Hydrogen evolution
7.		$NiTiO_3$	Degradation of nitrobenzene
8.		$BiFeO_3$	Degradation of organic contaminants under visible light

10.2.8 Perovskite-based

Perovskites are a group of compounds with a general formula ABO_3, where A denotes a larger cation and B denotes a comparatively smaller cation. Perovskites are stable crystals but have flexible compositions; they have applications in both the energy and environment fields, mainly using photocatalysis. Perovskites have a few advantages over the other photocatalysts – they have suitable bandgap for hydrogen generation; they are flexible in the structure where there are two sites, A and B, for doping; and they have ferroelectric and piezoelectric properties. Some of the perovskites are $PbTiO_3$, $PbZrO_3$, $BaTiO_3$, and $SrTiO_3$ (Table 10.1). A few of them are used for the degradation of pollutants by photocatalysis, and a few are used for photocatalytic water splitting to generate hydrogen that can address the fuel crisis [11].

10.3 Role of Advanced Ceramic Materials for Both Renewable and Non-Renewable Energy

Advanced ceramic materials play an important role in technologies related to energy production and harvesting of sunlight. Nuclear power plants generally use ceramic electrolytes and insulators. Thermal barrier coatings, made of ceramic materials, are often applied to gas turbines or diesel engines to increase efficiency and reduce maintenance requirements, thereby increasing operating lifespan. Zirconia composites are often selected for such services due to their low sintering, stability, chemical resistance, and low heat conductivity. However, the current challenge for using zirconium composites is their low operating temperature range. Piezoelectric ceramics are another type of materials that are essential for energy harvesting systems due to their ability to convert and store mechanical forces into electrical energy and to run devices with the same. Ceramic materials possess outstanding thermo-mechanical properties and chemical stability; for this, they are indispensable in high-temperature applications. Advanced ceramic matrix composites (CMCs) are reinforced with fibers and used in gas turbines, as they are suitable for static and moving parts as lightweight structural materials [12]. Metal and ceramic substrates can be protected from extreme environmental, corrosive, or thermal conditions using barrier coatings made of advanced ceramic materials. In nuclear power plants, ceramics are used as neutron-control materials, for sintering nuclear fuels, and as moderators and barriers.

10.3.1 Energy harvesting using advanced ceramics

Heat energy wasted from many industrial sources can be converted into electricity by thermoelectric modules made of ceramic materials. Ceramics are used as transparent conductive coatings (TCOs) in the solar panels. Indium-tin oxide (ITO) constitutes the most popular TCO; next comes aluminium-doped oxide of zinc (AZO) followed by fluorine-doped oxide of tin (FTO). Solar cells based on perovskite materials are considered to be highly efficient with prospects. In the latest technology, many layers are deposited on ceramic substrates to form solar cells – advanced ceramic photocatalytic materials are extensively used in such cells. Advanced ceramic oxides are generally stable at high temperatures and seldom contain toxic or critical elements. In the case of both solar photovoltaic cells and photoelectron catalytic water splitting, functional ceramic metal-oxide layers act as the electron transferring medium.

A few usages of advanced ceramic materials in energy harvesting other than photocatalytic are thermoelectricity, ion conduction, and supercapacitors. The n-type perovskite materials have high electrical conductivity but low thermal conductivity, and they can convert heat to electricity; hence they are good candidates for fabrication of thermoelectric generators. Electroceramics are fast ion conductors, and they have been explored because of their roles in fuel cells, batteries, and sensor technologies. Ceramic capacitors and supercapacitors are used for energy storage. Supercapacitors have high capacitance ranging from microfarads to kilofarads. Typically, high-temperature supercapacitors contain a ceramic separator that prevents the recombination of positive and negative charges. Supercapacitors are used as power sources in electronic devices like mobile phones, solar cells, and wind turbines to store energy, pure or hybrid electronic vehicles, as well as in power grids [13].

10.3.2 Ceramic Fuel Cells

Photofuel cells (PFCs) or photocatalytic fuel cells are generally used for the combustion or oxidation of a so-called fuel that produces either electric power or storable chemical energy mainly in the form of hydrogen. A photoelectrochemical cell has a photo-anode that embeds a photocatalyst. Perovskite-based photocatalysts are generally used for water splitting. During the splitting of water, water molecules are adsorbed on the surface of the anode, coated or impregnated with the photocatalyst. Perovskite-titanates like $SrTiO_3$ can be used for photocatalytic splitting of pure water under UV light. One of such perovskites, lanthanum titanate $La_2Ti_2O_7$, achieved 120% photo-efficiency for splitting water. Many nitrides, mixed chalcogenides, and perovskites, with or without co-catalysts, have been used for H_2 production by water splitting [11].

PFCs can achieve degradation of organic pollutants with simultaneous generation of electricity by redox reactions. Electron–hole pairs are generated by absorbing photons on the electrodes; the external bias over the photo-electrodes drives the photo-generated electrons from the anode to the cathode, and the protons arrest the electrons to form hydrogen [14].

10.4 ROLE OF ADVANCED CERAMIC MATERIALS FOR CO_2 REDUCTION, CONTROL OF WATER, AND SOIL POLLUTION

With anthropogenic activities, the CO_2 level is increasing in the environment; eventually, there is a need to decrease the quantity of CO_2 in the atmosphere by reverse combustion, whereby CO_2 can be reduced to an organic fuel. Again air, water, and soil are getting contaminated by organic pollutants discharged from various industries. Though many techniques are being explored for the remediation of such pollution, here the photocatalytic route for the remediation will be discussed using advanced ceramic materials [15].

10.4.1 Photocatalytic CO_2 Reduction with Layered Perovskites

In recent years, new classes of materials, other than the most popular TiO_2, are being explored as photocatalysts. One of them is the layered perovskite, $ALa_4Ti_5O_{15}$ (where A = Ca, Sr, and Ba) These photocatalysts have bandgaps in the range 3.79–3.85 eV and were proved to be effective for water splitting [16]. These photocatalysts have also been utilized for CO_2 reduction by Iizuka et al. (2011) [17]. In this process, HO acted as a reducing agent. Also, they examined the role of the co-catalysts. They found that $ALa_4Ti_5O_{15}$ (A = Ca, Sr, and Ba), loaded with Ag-co-catalysts, reduced CO_2 into CO , where H_2O acted as an electron donor. For this purpose, $BaLa_4Ti_4O_{15}$, loaded with Ag-co-catalyst was the most effective photocatalyst.

10.4.2 Hybrid and Supported Ceramic Photocatalysts

One of the limitations of the photocatalysts, especially nano-photocatalysts used for environmental applications, is their separation from the substrate after the use. To overcome this drawback, photocatalysts are often immobilized on different ceramic supports. The other advantages of ceramic supports are their easy availability and low cost, controllability of physical structure and properties, inertness that does not affect the catalyst, high specific surface area, and adsorption capacity [18–20]. A combination of two or more photocatalysts, in the form of hybrid catalysts, is also found to enhance the life and effectivity of the catalyst.

Conventional catalysts are often supported or impregnated on the ceramic matrix like cement or clay or the ceramic membranes [21], a few of them are presented in Table 10.2. Ceramic membranes based on Al_2O_3, TiO_2, SiO_2, and ZrO_2 show good performance in water and wastewater treatments [22, 23]. Many reports indicate the effectivity of ceramic-supported TiO_2 for photocatalytic degradation of organic pollutants [24]. Photocatalytic activity of TiO_2 was found enhanced by dispersing TiO_2 nanoparticles on clay mineral surfaces for large specific surface areas, high

TABLE 10.2
A Few Ceramics-Supported [21] and Hybrid Photocatalysts [11]

S. No.	Type	Photocatalyst	Application
1.	Supported	Bacterial cellulose (BC)-supported MoS_2	Removal of methylene blue dye
2.		SiO_2-supported TiO_2	Removal of crystal violet dye
3.		Clay-nanocomposite-supported TiO_2	Treatment of wastewater containing dye and heavy metals
4.		Al_2O_3-supported CdS/Ni_9S_8	CO_2 photoreduction
5.	Hybrid	$Ag-AgCl-CaTiO_3$	H_2 evolution
6.		CdS/MoS_2	H_2 evolution rate
7.		$Pd/GdCrO_3$	Photocatalytic reduction of nitrate into N_2
8.		$BiPO_4/Bi_2S_4$	Complete removal of methylene blue dye.

adsorption capacity, large pore volumes, good chemical stability, and high mechanical strength [25]. Photo-induced hydrophilicity on TiO_2 -coated porous substrates, like clay tiles, can effectively prevent bacterial adhesion [26, 27].

Hybrid photocatalysts obtained from combining more than one semiconductor oxide, perovskite, or quantum dots with other supports improve the photoactivity as well as adsorption and reduce charge recombination. In this category, heterojunctions of p- and n-type semiconductors have significant importance since they improve the movement of electrons and facilitate charge separation [11].

Layered composites of TiO_2 prepared by the deposition of TiO_2 nanoparticles into layers can be used both for reducing air pollution and to decompose dyes. The porosity and roughness of the surface increased efficiency [28]. Single-flow ceramic membranes, with TiO_2 deposited onto them, were found to be effective for the removal of organic dye from water. Many techniques are adopted for the synthesis and surface-deposition of TiO_2 -nanocomposites, including dip-coating, sol-gel, impregnation, chemical vapor deposition (CVD), electrophoretic deposition (EPD), and electrochemical treatment.

10.5 ROLE OF CERAMIC PHOTOCATALYSTS IN SOIL PURIFICATION

Soil is generally polluted by toxic chemicals like pesticides and also by plastic waste. Landfill leachates are another source of soil pollution that contains mixed types of pollutants. Advanced ceramic photocatalysts may be utilized for the remediation of soil pollution. A major challenge for the removal of soil contaminants by in situ solar photocatalysis is the penetration of light beyond the surface. However, a few reports like the one by Higarashi and Jardim (2000) [29] reported their attempts on the laboratory scale. They exposed soil samples contaminated with pesticides, mixed with TiO_2 , to sunlight and artificial blacklight. Roy et al. (2016) [30] attempted the degradation of PVC with Fe-doped ZnO under sunlight to address the problem of plastic pollution. The same researcher also reported simultaneous degradation of PVC film and dye pollutants under sunlight with Fe-doped ZnO nanoparticles immobilized on PVC matrix [31, 32].

10.6 SYNTHESIS OF ADVANCED CERAMIC MATERIALS

10.6.1 Metal oxides

The properties of metal oxides pertinent for photocatalytic applications depend on their size and shape, which are mainly controlled by the synthesis routes adopted. These methods of synthesis can be broadly divided into three categories: solution-based techniques, vapor-phase synthesis, and biological methods.

- Solution-based synthesis

Under solution phase synthesis, the following techniques are used.

10.6.2 Sonochemical method

In sonochemical methods, ultrasound with high intensity can be used for the production of nano-sized metal oxides. The most significant steps for this kind of synthesis are consequences of 'acoustic cavitation' which propagates through the stages of formation, growth, and implosive collapse of bubbles. Initially, the precursor materials, which are generally metallic salts, are exposed to ultrasonic vibrations. This intensive sonication of a liquid generates bubbles by cavitation, which collapse and generate an enormous amount of energy locally in the solution. Transient temperatures of 5000 K and pressure of ~1800 atm are produced momentarily inside the collapsing cavity [33]. This 'hot spot' is accountable for the homogeneous sonochemical reactions. Many advantages achieved with the sonochemical methods include a higher specific surface area, uniform size distribution, faster synthesis time, and improved crystal purity. Meshram et al. (2017) [34] studied a simpler sonochemical method using sodium succinate and polyethylene glycol (PEG) for the synthesis of Ce-doped ZnO-nanorod photocatalysts. Different characterization results showed successful synthesis of nanoparticles with a few exceptions, where cubic CeO_2 impurity phases appeared only in the 6 and 8 (atomic) % Ce-doped samples. Roy et al. (2012) [30] also successfully synthesized iron-doped zinc oxide nanoparticles by using the sonochemical method from zinc acetyl acetonate solution in an aqueous ethanol medium. The effective synthesis of nanoparticles was confirmed by several characterization data (Figure 10.2).

10.6.3 Co-precipitation method

In the co-precipitation method, nucleation of metallo-organic precursor takes place which is followed by aggregation or Ostwald ripening to form metal oxides. In addition to low cost, this process has other advantages such as mild reaction conditions

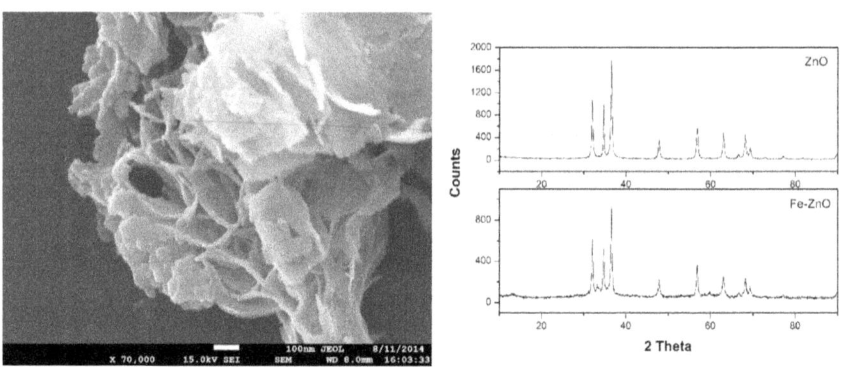

FIGURE 10.2 FESEM and XRD images of ZnO/Fe-doped ZnO nanoparticles.

Source: Authors' unpublished work.

like low temperature, simplicity of processing, direct synthesis in water, ease of scaling up, and versatile modulation of core and surface properties. Using the co-precipitation method, LaGrow et al. (2019) [35] has successfully synthesized iron oxide nanoparticles. The synthesis has been done with an abrupt increase in pH via the addition of sodium carbonate.

10.6.4 Solvothermal method

In this method, the precursor material is dispersed in an appropriate solvent and then exposing it into elevated temperature and pressure, the final material is formed. Reactions carried out with water as a solvent or medium are called hydrothermal synthesis [36]. The major benefit of this synthesis is that it has high vapor pressures and minimum loss of materials; hence the process is well-controlled. High temperatures and expensive equipment are some of its disadvantages. Metal oxides and metal powders were successfully synthesized by Komarneni et al. (2010) [37] using both conventional-hydrothermal and microwave-hydrothermal methods. It has been discovered that the presence of a microwave field during hydrothermal operations accelerates the crystallization rate.

10.6.5 Sol-gel method

During sol-gel processing, liquid solutions of molecular precursors undergo a sequence of chemical reactions and physical processes (phase separation, dissolution, evaporation, phase transition, etc.) that finally form desired solids. To be brief, the basic idea of sol-gel synthesis is to dissolve a compound in an appropriate liquid and then return it to its original form as a solid by a controlled process [38]. By using sol-gel synthesis, materials can be prepared with various shapes, including porous structures, thin fibers, dense powders, and thin films. In comparison to conventional methods, sol-gel synthesis produces high-quality materials with homogeneity and purity at lower temperatures [39]. Synthesis of $TiO_2 - MgO$ nanoparticles with different magnesium contents has been carried out by Bayal and Jeevanandam (2014) using a simple sol-gel method [40].

10.6.6 Microwave-assisted method

This technique has been popular due to its minimal energy and time requirements. The reaction periods were effectively reduced from a few hours to several minutes without affecting the size or purity of the nanoparticles [41]. The selection of precursor materials, such as acetates, chlorides, and isopropyls, as well as solvents, such as ethylene glycol and benzene, can influence the outcome of the reaction in terms of the size and crystal structure of the particles [42]. The microwave-assisted method is a greener and environmentally friendly way to synthesize. Roy and Bhattacharya (2011) [43] successfully synthesized nanostructured CaO with cubic morphology from $Ca(NO_3)_2$ and NaOH by employing microwave radiation in an ambient atmosphere.

10.6.7 Microemulsion method

This method involves the formation of emulsion; when a small amount of a suitable surfactant is mechanically agitated with two immiscible phases, that is, oil and water, it results in a two-phase dispersion – one phase as droplets coated by surfactant that is dispersed in the other continuous phase [44]. Micro-emulsion technologies are thermodynamically stable and reversible. Additionally, the method also enables manipulation of the size and morphology of nanoparticles by adjusting the experimental conditions such as type and dosing of surfactant, nature of the continuous phase, concentration of precursors, and the molar ratio of surfactant to water. Numerous metal oxide nanoparticles including iron oxide [45], nickel oxide (NiO) [46], cerium oxide (CeO_2) [47], titanium oxide (TiO_2) [48], zinc oxide (ZnO) [49], and copper oxide (CuO) [50] have been synthesized using this technique.

- Vapor state synthesis

This type of synthesis method includes the following methods.

10.6.8 Laser ablation method

In this method, bulk materials are suspended into aqueous or non-aqueous solvents and then irradiated by laser to produce colloidal solutions of nanoparticles. The intensity of laser irradiation and the nature of the solvent are the two parameters that can control the size of the nanoparticles [51]. In addition to high production rates, laser ablation provides good particle size control and mono-dispersity. By applying the laser ablation process and altering the power of the laser, iron oxide nanoparticles were produced by Rivera-Chaverra et al. (2020) [52].

10.6.9 Chemical vapor-based methods

This technique brings precursors in the vapor phase into a hot-wall reactor where conditions are adjusted so that it favors the nucleation of particles rather than the deposition of films. In ambient conditions, precursors can exist as solids, liquids, or gases, but they are brought into the reactor as vapors. Energy sources such as resistance heating, lasers, and plasma induce chemical reactions among gas molecules in a reaction chamber when a mixture of gas reactants is introduced [53, 54]. An important feature of chemical vapor synthesis is that it allows the formation of doped or multi-component nanoparticles with multiple precursors.

10.6.10 Combustion method

The combustion method is an effective, low-cost method for the production of various oxide materials. This method involves heating a metallic precursor to evaporate it into a gas where an oxidizing agent is mixed. When the metal is partially heated, the synthesis starts via oxidation. Subsequently, the heat necessary for the evaporation of the metal is produced in situ by the combustion/oxidation reactions itself [55]. Major advantages of such a technique include relatively simple equipment, cheap precursors and oxidants, fast and exothermic self-sustaining reactions, and highly pure products of varying sizes and shapes depending on the conditions of the synthesis. A combustion technique at low temperatures was successfully used by Bai et al. (2017) to synthesize ZnO nanoparticles employing a variety of fuels like glycine, citric acid, urea, and radish (*raphanussativus*) extract [56].

10.6.11 Template/surface-mediated synthesis

A variety of fabrication strategies are employed in template-mediated synthesis including electrochemical [57], sol-gel [58], polymerization [59], and vapor deposition [60]. Since the reactions between the metal and the oxidizing agent may occur in different media, this synthesis method can be classified into both of the previously listed classes of synthesis. In this method, using a nanoporous template, the desired nanomaterial can be fabricated within the pores or channels.

- Biological synthesis

Several disadvantages are associated with well-established classical chemical methods for nanomaterial synthesis, such as hazardous chemicals, high temperatures, chemical waste production, and so on. Biological methods can be used for metal oxide particle synthesis to overcome these challenges. This method is environment-friendly, non-toxic, and a simple method of synthesis of nanomaterials where plants and microorganisms are used as capping agents. As compared to organisms and plants, plants are readily available, easy to handle, and do not require special conditions, so they are excellent candidates for the biosynthesis of transition metal oxide nanocomposites [61, 62]. For this kind of synthesis, various parts of plants, including the roots, bark, leaves, and fruits, can be utilized [63, 64].

10.6.12 Metal Sulfides

Most of the synthesis methods for metal oxide particles can also apply to the synthesis of metal sulfide nanostructures. These include solution-based synthesis [65] such as the hydrothermal method [66]; vapor phase synthesis like chemical vapor deposition (CVD) [67], and physical vapor deposition (PVD) [68]; high-pressure autoclave processes [69, 70], and so on. Among these methods, synthesis by deposition is the most popular. The thermal evaporation methods (PVD and CVD), arc discharge, and laser-assisted growth are all members of this group.

10.6.13 Metal and non-metal nitrides, oxynitrides

Metal nitrides can be created synthetically using either one of two methods: (i) the physical approach or (ii) the chemical method [71]. The physical method entails physical vapor deposition [72] and laser and plasma methods [73, 74], whereas the chemical method requires metals reacting with nitrogen (N_2) sources (NH_3 or N_2 gas) at higher temperatures (800–2000 °C). With a wide variety of nitrides, chemical procedures are considered to be more flexible.

10.6.14 Metal phosphides and black phosphorous

For phosphide synthesis, there are numerous common methods including (i) vapor–solid phosphorization reaction, (ii) solution-phase phosphorization reaction, (iii) solid-state reduction (iv) calcination from phosphate and phosphonate. Metal phosphides are usually synthesized from the elements at high temperatures, and the source of phosphorus is the most important for this synthesis. For vapor–solid phosphorization, red phosphorus [75, 76], NaH_2PO_2 [77], and $(NH_4)_2HPO_4$ [78] were used as the phosphorus source. Organophosphorus such as phytic acid [79] and trioctylphosphine (TOP) were used in the case of solution-phase synthesis. For solid-state reduction, phosphorus sources included $NH_4H_2PO_4$ [80] and phosphoric acid (H_3O_4P). Red phosphorus (RP) was typically used as the starting material for the majority of conventional synthesis techniques that produced black phosphorous (BP) under intense heat and pressure [81]. Thermal annealing was initially used to evaporate RP powder and then transformed into BP by further heat treatment at 650 °C.

10.6.15 Metal oxyhalides and halides

Since these types of ceramic materials are bismuth-based, bismuth nitrate pentahydrate $[Bi(NO_3)_3{\cdot}5H_2O]$ is most commonly used as the bismuth source for the synthesis of this type of particles. Commonly, hydrothermal and solvothermal conditions are used for this type of particle production [82, 83]. Solid-state reaction, wet chemical route, ion exchange method, microemulsion, and microwave method are also used for the synthesis of such particles [84].

10.6.16 Magnetically recoverable ferrites

There are various synthesis methods in the literature for the preparation of magnetically recoverable ferrites, and most of them have been previously described in the metal oxide synthesis section. Some of them are the co-precipitation method [85], thermal decomposition [86, 87], microemulsion [88], sol-gel, sonochemical [89, 90], hydrothermal [91, 92], electrochemical [93], and so on.

10.6.17 Carbon-based

Carbon dots can be synthesized using a variety of techniques, including hydrothermal/solvothermal, microwave, electrochemical, microplasma, and chemical oxidation [94]. Carbon nanotubes (CNTs) with specific sizes could be produced utilizing a variety of techniques that are often classified into five types – carbon arc-discharge, laser ablation [95], sonochemical and hydrothermal [96, 97], electrolysis [98], and chemical vapor deposition (CVD) [99]. Graphenes can be prepared by various methods like cleavage and exfoliation, CVD by heating or enhanced by plasma (PECVD), chemical methods, in situ thermal decomposition on the substrates, and longitudinally tearing carbon nanotubes (CNTs) [100]. Graphene oxide (GO) can be produced by the oxidation of graphite, and then this graphite oxide can be exfoliated to create GO. GO was first synthesized by Brodie [101], Staudenmaier [102], and Hummers and Offeman

[103], each of whom derived graphite oxide via the oxidation of graphite using various techniques. The initial two procedures were modified by Hummers and Offerman in a variety of ways to make them safer, including the use of sodium nitrate and the addition of $KMnO_4$ as an oxidizer. The Hummers' approach is typically utilized (or, in most cases, somewhat modified) to create GO due to its safety and scalability.

10.6.18 Perovskite-Based

The majority of the synthesis techniques applied for metal oxide particles can also be applied to this kind of particle synthesis [104]. The procedures of electrospinning and colloidal-crystal-template synthesis are also available for synthesizing these kinds of materials. Using a hydrothermal process, perovskite oxides mostly based on titanates, such as $SrTiO_3$ [105], $BaTiO_3$ [106], and $PbTiO_3$ [107], are being produced. The $LaMn_{1-x}FexO_3$ ($x = 0$, 0.1, and 0.2) was synthesized by Geetha et al. (2018) [108] and is an example of a perovskite oxide material synthesized using a co-precipitation process. Another example of perovskite oxide nanoparticles is $LaMO_3$ (M = Fe, Co, Mn), which was synthesized using a simple homemade combustion apparatus using glycine as the fuel and metal nitrates as the oxidizers [109].

10.7 CHARACTERIZATION TECHNIQUES

In the study of these materials, characterization techniques are crucial. The different types of materials discussed here were produced using various synthesis techniques, as well as various reaction environments and precursors, thus each one exhibits a unique crystal form and structure, particle size and distribution, morphology, elemental configuration, and practical applications. Therefore, it is crucial to characterize synthesized material in terms of its various properties like size, shape, morphology, and chemical composition for its possible uses. Some important techniques used for the characterization and their details are as follows.

10.7.1 X-Ray Diffraction Technique (XRD)

Studies using the X-ray diffractometer (XRD) were carried out to check the structural characteristics, primarily the crystal structure, phase, crystallinity, crystallite size, symmetry of the unit cell, lattice constants of particles and physical properties of the synthesized materials. ZnO nanoparticles synthesized by Mahamuni et al. (2019) [110] using polyol synthesis methods have been characterized using XRD techniques. The analysis results revealed details on the purity of the phases and revealed that the nanoparticles are crystallite in nature with a wurtzite structure and hexagonal symmetry. Iron-doped and undoped zinc oxide nanoparticles were synthesized by Roy et al. (2016) [30] and were characterized by XRD. The XRD results (Figure 10.2) revealed the successful synthesis of nanoparticles which were hexagonal wurtzite in structure, and it also demonstrates that even after doping, ZnO's fundamental structure remains unaltered.

Halide perovskite materials such as $CsPbIBr_2$ thin films that were synthesized by Faridi et al. (2022) [111] were also characterized by XRD. The significant peaks seen from XRD results showed that the $CsPbIBr_2$ thin films deposited on glass substrates are polycrystalline. Without annealing, the initial sample's crystalline size was determined to be 27.35 nm. The tendency of the crystal size to grow with temperature was seen, but at the annealing temperature of 200 °C, the crystal size was reduced by raising the annealing temperature, possibly due to recrystallization.

10.7.2 Electron Microscopy (SEM and TEM)

Material morphology and particle size, mainly for the nano-sized particles, have a significant impact on their properties and applications. To examine the morphology and size of the produced materials, an electron microscope is used. The two types of electron microscopy that are typically applied by researchers for this purpose are transmission electron microscopy (TEM) and scanning electron microscopy (SEM). SEM and TEM share a lot of commonalities. However, the primary difference between SEM and TEM is that SEM produces images by detecting reflected or knocked-off electrons, whereas TEM generates images by using transmitted electrons that are traveling through the material.

The SEM micrograph of TiO_2 nanoparticles, from the study by Sharfudeen et al. (2017) [112], exhibited cluster appearances with crystalline natures. Nevertheless, spherical structures smaller than 20 nm in size with uneven surface morphologies suggested a larger grain size since the increasing temperature caused both crystalline and grain growth. In a few cases, stony appearances have been observed, which may have been caused by the accumulation of TiO_2 nanoparticles (NPs).

ZnO NPs produced by a green synthesis method employing *E. prostrata* leaf extract as a capping agent are shown in a transmission electron microscopy (HRTEM) image; the NPs have an average size of 29 nm and a size range of 16–85 nm, as well as a triangular, radial, hexagonal, rod, or rectangular form [113].

In their research, Mirikaram et al. (2021) [114] synthesized ZnO-GO composites by mixing different quantities of GO with ZnO using ultrasonication. SEM analysis depicted the surface morphology of various forms of the synthesized materials. Results indicated that, in contrast to pure ZnO, ZnO-GO composites, composed primarily of spherical and rod-shaped particles, had more aggregated structures.

Using the hydrothermal method and *E. globulus* leaf extract, Balaji et al. (2020) [115] developed biogenic ceria (CeO_2) nanoparticles (NPs) in different sizes, with diameters of 50 nm, 20 nm, 10 nm, and 5 nm. An HRTEM image confirms the size of the 20 nm-diameter CeO_2 NPs.

10.7.3 Energy Dispersive X-ray (EDX)

EDX can be used to determine the elemental composition of any produced materials. EDX, combined with other characterization techniques like SEM, can give comprehensive structural information on any material. In their study, Roy et al. (2016) [30] used EDX analysis to confirm the degree of iron doping in the ZnO crystal and, consequently, the atomic weight percentage of the nanoparticle.

The appearance of Zn and O in the EDX results of all forms of ZnO nanoparticles produced by Mahamuni et al. (2019) [110] indicates the production of pure ZnO nanocrystals. The pure nanoparticle exhibited fairly well antimicrobial and antibiofilm activity.

10.7.4 Fourier transform infrared spectrometer (FTIR)

FTIR spectroscopy can be used to study functional groups. Different peaks related to various functional groups can provide data about the material, and it also has a huge impact on inclusive structural research.

Alamdari et al. (2020) [116] compared the FTIR spectrums of the zinc oxide nanoparticles prepared using *S. ebulus* leaf extract and that of the otherwise biosynthesized ZnO NPs. Peaks associated with alkaloids, flavonoids, and phenolic compounds, at 3429 (O-H), 1527–1670 (N-H), 1076 (C-O), and 880–1380 (RCOO) cm^{-1}, respectively, were observed whereas the broad stretching band at 3429 cm^{-1} denoted the presence of hydrogen-bonded groups. The results indicated the existence of flavonoid derivatives in the *S. ebulus* extract.

By employing a surfactant-free hydrothermal method, Botsa et al. (2020) [117] synthesized a hybrid nanocomposite of nickel tungstate and reduced graphene oxide ($NiWO_4$-RGO). These synthesized materials were characterized using FTIR along with additional characterization methods. For pure $NiWO_4$, the observed FTIR spectra showed absorption bands between 3390 cm^{-1} and 3429 cm^{-1} and between 1560 cm^{-1} and 1630 cm^{-1}, respectively, as well as weak absorption bands between 835 cm^{-1} and 850 cm^{-1}. The first two strong bands indicated the presence of -OH stretching and bending vibrations due to the absorption of water molecules, while the weak bands indicated the stretching of W-O bonds in the WO_4 tetrahedral crystal. However, the FTIR spectra of $NiWO_4$-RGO composites showed peaks almost at 1160 cm^{-1} and 1550 cm^{-1}, which correspond to the stretching frequencies of C-O-H and C-C, respectively. This indicates that the GO has been significantly reduced during the process.

10.8 OTHER CHARACTERIZATION TECHNIQUES

Few other characterization techniques are also done for the characterization of advanced ceramic materials than those described earlier, such as UV-visible spectroscopy, atomic force microscopy (AFM), Brunauer-Emmett-Teller (BET), thermogravimetric analysis (TGA), or differential thermal analysis (DTA).

In their study, Roy et al. (2016) [30] used the UV-vis absorption spectroscopy technique in the 200–800 nm wavelength range to determine the optical characteristics of synthesized doped and undoped ZnO. The obtained spectra clearly showed strong absorbance present at wavelengths below 380 nm and very low absorption in the visible spectrum. All varieties of ZnO spectra exhibited strong bands at 369 nm, indicating ZnO nanostructures' free exciton absorption. The most significant results indicated that, as the extent of doping increased, the absorption edge shifted toward the red, resulting in a bandgap reduction that may be brought on by defects and crystallite size.

Mahamuni et al. (2019) synthesized ZnO in DEG and TEG medium and examined the thermal decomposition pattern and the existence of absorbed polyols by

TGA analysis [110]. For this, all samples were heated from 50 °C to 900 °C, at the rate of 10 °C/min. The result indicated two successive decompositions of all the samples. Evaporation of water adsorbed on the surface caused the initial loss of weight, which occurred in the range of 145–270 °C, and the next stage of weight loss was detected in the range of 245–490 °C because of the loss of adsorbed DEG/TEG molecules; this was subsequently validated by FTIR.

TG and DTA methods were used to investigate the loss in weight, change in enthalpy, and activation energy of CuO and ZnO nanoparticles at temperatures between 30 °C and 1000 °C [118]. This demonstrates that CuO nanoparticles possess only a limited degree of thermal instability at high temperatures, but ZnO nanoparticles synthesized via the green route display excellent thermal stability over the 30–1000 °C temperature range and exhibit significantly less weight loss than CuO nanoparticles. CuO nanoparticles exhibit endothermic peaks on their DTA curves, whereas ZnO nanoparticles exhibit exothermic peaks.

The AFM analysis technique allows imaging and analysis in three dimensions, which is different from other microscopy methods. The objective of the experiment was to find out the size, shape, and/or size distribution of nanoparticles. Rao et al. (2007) [119] in their study, employed TiO_2, ZrO_2, and Al_2O_3 nanoparticles that were applied on various substrates in different concentrations and were imaged by AFM in dynamic mode. From the AFM results, it has been found that due to the hydrophilic nature of mica, when ZrO_2 nanoparticles are deposited on mica, they appear to form aggregates, and contrarily they seem evenly distributed on the Si substrate. Also, it was discovered that chemically treating silicon makes its surface more hydrophobic. Another observation was that, whereas TiO_2 and ZrO_2 nanoparticles do not aggregate due to a chemical treatment on their surface, Al_2O_3 nanoparticles stay bare and, as a result, tend to agglomerate and attach better to each other than to the substrate.

Furthermore, the Brunauer-Emmett-Teller (BET) methodology can be used to analyze the specific surface area and porosity of different types of materials.

According to the work reported by Esmaeili et al. (2014) [120], GO has a specific surface area of about 53 m^2/g. These surface characteristics may be significantly vital for the adsorption of the desired adsorbate as well as the redox reactions carried out on the graphene materials by photogenerated active species (such as charge carriers and reactive oxygen species). This may enhance the photocatalytic activity.

ZnO nanoparticles with a size of less than 10 nm (7.4 nm) were successfully synthesized in the work by Shamhari et al. (2018) [121]. The specific surface area determined with a BET analyzer was 101.32 m^2/g. Similar findings were also reported in previous studies by Bian et al. (2011) [122], who got 105 m^2/g surface area for 4 nm ZnO NPs, and it was also confirmed by TEM.

10.9 FUTURE DIRECTIONS OF RESEARCH

Regarding photocatalytic ceramic materials, the basic challenge to the researchers is to get a catalyst material that would be effective in the visible light region, not only on the laboratory scale but also on the commercial scale; not only that, the material should be non-toxic, recyclable, and cost-effective. To date, most of the experiments have been conducted using simulated sunlight in the laboratories – very few numbers of experiments have been done using actual sunlight on a large scale. This research should be done in future for the actual application of solar photocatalysis for environmental remediation.

For many years, metal-oxide-based photocatalysts are ruling the scenario of photocatalysts. As discussed, they have certain limitations – so one of the trends in the research will be non-metal/non-metal-oxide-based photocatalysts [11].

The most eligible candidates in such areas are graphene-based and perovskite-based photocatalysts. They are not yet fully explored, and new findings are obtained in these areas every day. Minimal dispersibility but difficulty in recovery from treated wastewater is a great challenge of using graphene-based photocatalysts in wastewater treatment. 2D/3D composite structures of graphene-based materials and synergistic effects of metal oxides and graphene in a composite are being explored to develop an efficient, economical, and eco-friendly photocatalyst for energy and environmental use [123].

Photocatalytic glass ceramic (PGC) is another interesting area of future research. There are many advantages of PGC, including transparency and easy moldability. However, a challenge for PGC research is to balance transparency and crystallinity; another challenge is its use in real-life scenarios [124].

10.10 CONCLUSION

This chapter reviews various aspects of photocatalytic advanced ceramic materials for energy and environmental applications. The chapter describes the types of advanced ceramic materials and various techniques of their synthesis and characterization with examples. Advanced ceramic materials include many prosperous candidates for photocatalysts. They play significant roles in harvesting solar energy as well as in the remediation of environmental pollution. Future scope of research in the field of advanced ceramic materials includes hybrid materials, perovskites, graphene-based photocatalysts, as well as photocatalytic glass ceramics.

REFERENCES

[1] https://www.wundermold.com/what-difference-between-traditional-advanced ceramics/. accessed on February 2, 2023

[2] https://ourworldindata.org/worlds-energy-problem; The world's energy problem; Max Roser. accessed on February 2, 2023

[3] https://www.iea.org/reports/world-energy-outlook-2022/outlook-for-energy-demand. accessed on February 2, 2023

[4] https://www.who.int/news-room/fact-sheets/detail/household-air-pollution-and-health. accessed on February 3, 2023

[5] https://www.who.int/news-room/fact-sheets/detail/drinking-water. accessed on February 3, 2023

[6] https://www.who.int/news-room/fact-sheets/detail/soil-transmitted-helminth-infections. accessed on February 3, 2023

[7] Hoffmann MR, Martin ST, Choi W, Bahnemann DW (1995) Environmental applications of semiconductor photocatalysis. *Chem Rev* 95:69–96

[8] https://www.preciseceramic.com/blog/an-overview-of-the-advanced-ceramic-materials/. accessed on February 5, 2023

[9] Buscema M, Groenendijk DJ, Blanter SI, Steele GA, van der Zant HSJ, Castellanos-Gomez A (2014) Fast and broadband photoresponse of few-layer black phosphorus field-effect transistors. *Nano Lett* 14:3347–3352. https://doi.org/10.1021/nl500 8085
[10] Kefeni KK, Mamba BB, Msagati TAM (2017) Application of spinel ferrite nanoparticles in water and wastewater treatment: A review. *Sep Purif Technol* 188:399–422. https://doi.org/10.1016/j.seppu r.2017.07.015
[11] Kumar A, Thakur PR, Sharma G, Naushad M, Rana A, Mola GT, Stadler FJ (2019) Carbon nitride, metal nitrides, phosphides, chalcogenides, perovskites and carbides nanophotocatalysts for environmental applications. *Environ Chem Lett* 17:655–682. https://doi.org/10.1007/s10311-018-0814-8
[12] Guillon O (2021) Ceramic materials for energy conversion and storage: A perspective. *Int J Ceramic Eng Sci* 3:100–104
[13] https://ceramics.org/about/what-are-engineered-ceramics-and-glass/ceramics-and-glass-in-energy February 10, 2023
[14] Sajna MS, Simon SM, Unnikrishnan NV, Sadasivuni KK (2022) An overview of Graphene-based 2D/3D nanostructures for photocatalytic applications. *Top Catal.* https://doi.org/10.1007/s11244-021-01539-5
[15] Nahar S, Zain MFM, Kadhum AAH, Hasan HA, Hasan MR (2017) Advances in photocatalytic CO2 reduction with water: A review. *Materials* 10(6):629. https://doi.org/10.3390/ma10060629
[16] Kudo A, Miseki Y (2009) Heterogeneous photocatalyst materials for water splitting. *Chem Soc Rev* 38:253–278
[17] Iizuka K, Wato T, Miseki Y, Saito K, Kudo, A (2011) Photocatalytic reduction of carbon dioxide over agco-catalyst-loaded ALa4Ti4O15 (A = Ca, Sr, and Ba) using water as a reducing reagent. *J Am Chem Soc* 133:20863–20868
[18] Pozzo RL, Baltanás M, Cassano A (1997) Supported titanium oxide as photocatalyst in water decontamination: State of the art. *Catal Today* 39:219–231
[19] Shan AY, Ghazi TIM, Rashid SA (2010) Immobilisation of titanium dioxide onto supporting materials in heterogeneous photocatalysis: A review. *Appl Catal A Gen* 389:1–8
[20] Tototzintle MJ (2015) *Desarrollo de nuevas estrategias basadas en fotocatálisis solar para la regeneración de aguas de unaIndustria agro-alimentari.* Ph.D. Thesis, Universidad de Almería, Almería, Spain
[21] Danfá S, Martins RC, Quina MJ, Gomes J (2021) Supported TiO2 in ceramic materials for the photocatalytic degradation of contaminants of emerging concern in liquid effluents: A review. *Molecules* 26:5363. https://doi.org/10.3390/molecules26175363
[22] Morjène L, TasbihiM, Schwarze M, Schomäcker R, Aloulou F, Seffen M (2020) A composite of clay, cement, and wood as natural support material for the immobilization of commercial titania (P25, P90, PC500, C-TiO2) towards photocatalytic phenol degradation. *Water Sci Technol* 81:1882–1893
[23] Mustafa G, Wyns K, Janssens S, Buekenhoudt A, Meynen V (2018) Evaluation of the fouling resistance of methyl grafted ceramic membranes for inorganic foulants and co-effects of organic foulants. *Sep Purif Technol* 193:29–37
[24] Huang CY, Ko CC, Chen LH, Huang CT, Tung KL, Liao, YC (2018) A simple coating method to prepare superhydrophobic layers on ceramic alumina for vacuum membrane distillation. *Sep Purif Technol* 198:79–86
[25] Mishra A, Mehta A, Basu S (2018) Clay supported TiO2 nanoparticles for photocatalytic degradation of environmental pollutants: A review. *J Environ Chem Eng* 6:6088–6107
[26] Szczepanik B (2017) Photocatalytic degradation of organic contaminants over clay-TiO2 nanocomposites: A review. *Appl Clay Sci* 141:227–239
[27] Radeka M, Markov S, Lončar E, Rudić O, Vučetić S, Ranogajec J (2014) Photocatalytic effects of TiO2 mesoporous coating immobilized on clay roofing tiles. *J Eur Ceram Soc* 34:127–136

[28] Carneiro J, Teixeira V, Azevedo S, Fernandes F, Neves J (2012) Development of photocatalytic ceramic materials through the deposition of TiO2 nanoparticles layers. *J Nano Res* 18–19:165–176
[29] Higarashi MM, Jardim WF (2000) Photocatalytic treatment of pesticide-contaminated soil using solar light and titanium dioxide. *Am Lab* 32(11):26–26
[30] Roy A, Maitra S, Ghosh S, Chakrabarti S (2016) Sonochemically synthesized iron-doped zinc oxide nanoparticles: Influence of precursor composition on characteristics. *Mater Res Bull* 74:414–420
[31] Das P, Roy A, Chakrabarti S (2017) Photocatalytic degradation of the nanocomposite film comprising polyvinyl chloride (PVC) and sonochemically synthesized iron-doped zinc oxide: A comparative study of performances between sunlight and UV radiation. *J Polym Environ* 25:1231–1241
[32] Roy A, Chakrabarti S, Maitra S (2022) Simultaneous solar photo-degradation of PVC-Fe-doped ZnO-nanocomposite flakes and methylene blue dye in water. *J Environ Eng Landsc Manage* 30(2):268–275
[33] Sudarmonoharan S, Rao ML (2004) Sonochemical synthesis of nanomaterials, in: H.S. Nalwa (Eds.), *Encyclopedia of Nanoscience and Nanotechnology*, American Scientific Publishers 10:67–82
[34] Meshram SP, Adhyapak PV, Pardeshi SK, Mulla IS, Amalnerkar DP (2017) Sonochemically generated cerium doped ZnO nanorods for highly efficient photocatalytic dye degradation. *Powder Technol* 318:120–127
[35] LaGrow AP, Besenhard MO, Hodzic A, Sergides A, Bogart LK, Gavriilidis A, Nguyen Thanh NTK (2019) Unravelling the growth mechanism of the co-precipitation of iron oxide nanoparticles with the aid of synchrotron X-Ray diffraction in solution. *Nanoscale* 11:6620–6628
[36] Gan YX, Jayatissa AH, Yu Z, Chen X, Li M (2020) Hydrothermal synthesis of nanomaterials. *J Nanomater* 8917013. https://doi.org/10.1155/2020/8917013
[37] Komarnenia S, Noha YD, Kima JY, Kima SH, Katsukib H, Naturforsch Z (2010) Solvothermal/hydrothermal synthesis of metal oxides and metal powders with and without microwaves. *Zeitschrift Naturforschung B* 65b:1033–1037
[38] Corr SA, Gun'ko YK, Douvalis AP, Venkatesan M, Gunning RD, Nellist PD (2008) From nanocrystals to nanorods: New iron oxide-silica nanocomposites from metallorganic precursors. *J Phys Chem C* 112:1008–1018
[39] Bokov D, Jalil AT, Chupradit S, Suksatan W, Ansari MJ, Shewael IH, Valiev GH, Kianfar E (2021) Nanomaterial by sol-gel method: Synthesis and application. *Adv Mater Sci Eng* 5102014. https://doi.org/10.1155/2021/5102014
[40] Bayal N, Jeevanandam P (2014) Synthesis of TiO2-MgO mixed metal oxide nanoparticles via a sol-gel method and studies on their optical properties. *Ceram Int* 40:15463–15477
[41] Verma A, Dwivedi R, Prasad R, Bartwal KS (2013) Microwave-assisted synthesis of mixed metal-oxide nanoparticles. *J Nanopart* 737831. https://doi.org/10.1155/2013/737831
[42] Bilecka I, Djerdj I, Niederberger M (2008) One-minute synthesis of crystalline binary and ternary metal oxide nanoparticles. *Chem Commun* 7:886–888
[43] Roy A, Bhattacharya J (2011) Microwave-assisted synthesis and characterization of CaO nanoparticles. *Int J Nanosci* 10:413–418
[44] Malik MA, Wani MY, Hashim MA (2012) Microemulsion method: A novel route to synthesize organic and inorganic nanomaterials: 1st nano update. *Arab J Chem* 5:397–417
[45] Wu W, He Q, Jiang C (2008) Magnetic iron oxide nanoparticles: Synthesis and surface functionalization strategies. *Nanoscale Res Lett* 3:397–415.
[46] Solans C, Izquierdo P, Nolla J, Azemar N (2005) Nano-emulsions. *Curr Opin Colloid Interface Sci* 10:102–110
[47] Sanchez-Dominguez M, Boutonnet M, Solans C (2009) A novel approach to metal and metal oxide nanoparticle synthesis: The oil-in-water microemulsion reaction method. *J Nanopar Res* 11:1823–1829

[48] Zhang RB, Gao L (2002) Preparation of nanosized titania by hydrolysis of alkoxide titanium in micelles. *Mater Res Bull* 37:1659–1666
[49] Bumajdad A, Madkour M (2015) In situ growth of ZnO nanoparticles in precursor-insensitive water-in-oil microemulsion as soft nanoreactors. *Nanoscale Res Lett* 10:10
[50] Kumar A, Saxena A, De A, Shankar R, Mozumdar S (2013) Facile synthesis of size-tunable copper and copper oxide nanoparticles using reverse microemulsions. *RSC Adv* 3:5015–5021
[51] Kim M, Osone S, Kim T, Higashi H, Seto T (2017) Synthesis of nanoparticles by laser ablation: A review. *KONA Powder Part J* 34:80–90. https://doi.org/10.14356/kona.2017009
[52] Rivera-Chaverra MJ, Restrepo-Parra E, Acosta-Medina CD, Mello A, Ospina R, Synthesis of oxide iron nanoparticles using laser ablation for possible hyperthermia applications. *Nanomaterials* 10:10
[53] Terasako T, Yagi M, Ishizaki M, Senda Y, Matsuura H, Shirakata S (2007) Growth of zinc oxide films and nanowires by atmospheric-pressure chemical vapor deposition using zinc powder and water as source materials. *Surf Coatings Techn* 201:8924–8930
[54] El Kasmi A, Tian Z-Y, Vieker H, Beyer A, Chafik T (2016) Innovative CVD synthesis of Cu2O catalysts for CO oxidation. *Appl Catal B Environ* 186:10–18
[55] Kumar V, Kumar S (2011) Synthesis and characterization of ZnO nanoparticles using combustion method. *Inter Conf Adv Cond Nano Mat* 1393:331–332
[56] Bai DS, Kumar VR, Suvarna RP (2017) Synthesis and characterization of zinc oxide nanoparticles by solution combustion method: DC conductivity studies. *Indian J Adv Chem Sci* 5(3):137–141
[57] Whitney TM, Searson PC, Jiang JS, Chien CL (1993) Fabrication and magnetic properties of arrays of metallic nanowires. *Science* 261:1316–1319
[58] Hulteen J (1997) A general template-based method for the preparation of nanomaterials. *J Mater Chem* 7:1075–1087
[59] Martin CR (1996) Membrane-based synthesis of nanomaterials. *Chem Mater* 8:1739–1746
[60] Cepak VM, Hulteen JC, Che G, Jirage KB, Lakshmi BB, Fisher ER, Martin CR, Yoneyama H (1997) Chemical strategies for template syntheses of composite micro- and nanostructures. *Chem Mater* 9:1065–1067
[61] Yulizar Y, Bakri R, Apriandanu DO, Hidayat T (2018) ZnO/CuO nanocomposite prepared in one-pot green synthesis using seed bark extract of *Theobroma cacao*. *Nano-Struct Nano-Obj* 16:300–305
[62] Matinise N, Kaviyarasu K, Mongwaketsi N, Khamlich S, Kotsedi L, Mayedwa N, Maaza M (2018) Green synthesis of novel zinc iron oxide (ZnFe2O4) nanocomposite via *Moringa Oleifera* natural extract for electrochemical applications. *Appl Surf Sci* 446:66–73
[63] Liu Q, Ma P, Liu P, Li H, Han X, Liu L (2020) Green synthesis of stable Fe, Cu oxide nanocomposites from loquat leaf extracts for removal of Norfloxacin and Ciprofloxacin. *Water Sci Technol* 81:694–708
[64] Elemike EE, Onwudiwe DC, Singh M (2020) Eco-friendly synthesis of copper oxide, zinc oxide and copper oxide–zinc oxide nanocomposites and their anticancer applications. *J Inorg Organomet Polym Mater* 30:400–409
[65] Wu Y, Wadia C, Ma W, Sadtler B, Alivisatos AP (2008) Synthesis and photovoltaic application of copper (I) sulfide nanocrystals. *Nano Lett* 8(8):2551–2555
[66] Liu J, Xue DF (2011) Rapid and scalable route to CuS biosensors: A microwave-assisted Cu-complex transformation into CuS nanotubes for ultrasensitive nonenzymatic glucose sensor. *J Mater Chem* 21:223–228
[67] Jiang Y, Zhang WJ, Jie JS, Meng XM, Zapien JA, Lee ST (2006) Homoepitaxial growth and lasing properties of ZnS nanowire and nano ribbonarrays. *Adv Mater* 18:1527–1532

[68] Fang XS, Bando Y, Shen GZ, Ye CH, Gautam UK, Costa PMFJ, Zhi CY, Tang CC, Golberg D (2007) Ultrafine ZnS nanobelts as field emitters. *Adv Mater* 19: 2593–2596
[69] Cao F, Liu RX, Zhou L, Song SY, Lei YQ, Shi WD, Zhao FG, Zhang HJ (2010) One-pot synthesis of flowerlike Ni7S6 and its application in selective hydrogenation of chloronitrobenzene. *J Mater Chem* 20:1078–1085
[70] Zhang L, Yu JC, Mo M, Wu L, Li Q, Kwong KW (2004) A general solution-phase approach to oriented nanostructured films of metal chalcogenides on metal foils: The case of nickel sulfide. *J Am Chem Soc* 126(26):8116–8117
[71] Giordano C, Antonietti M (2011) Synthesis of crystalline metal nitride and metal carbide nanostructures by sol–gel chemistry. *Nano Today* 6:366–380
[72] Andrievski R (1997) Films of interstitial phases: Synthesis and properties. *J Mater Sci* 32:4463–4484
[73] Vissokov G, Grancharov I, Tsvetanov T (2003) On the plasma-chemical synthesis of nanopowders. *Plasm Sci Technol* 5:2039–2050
[74] Schaaf P, Kahle M, Carpene E (2005) Reactive laser synthesis of carbides and nitrides. *Appl Surf Sci* 247:607–615
[75] Yu F, Zhou H, Huang Y, Sun J, Qin F, Bao J, Goddard WA (2018) High-performance bifunctional porous non-noble metal phosphide catalyst for overall water splitting. *Nat Commun* 9:2551. https://doi.org/10.1038/s41467-018-04746-z
[76] Meng T, Qin J, Xu D, Cao M. (2019) Atomic heterointerface-induced local charge distribution and enhanced water adsorption behavior in a cobalt phosphide electrocatalyst for self-powered highly efficient overall water splitting. *ACS Appl Mater Interfaces* 11(9):9023–9032
[77] Du C, Yang L, Yang F, Cheng G, Luo W (2017) Nest-like NiCoP for highly efficient overall water splitting. *ACS Catal* 7(6):4131–4137
[78] Zhou P, Zhang Y, Ye B, Qin S, Zhang R, Chen T, Xu H, Zheng L, Yang Q (2019) MoP/Co2P hybrid nanostructure anchored on carbon fiber paper as an effective electrocatalyst for hydrogen evolution. *Chem Cat Chem* 11(24):6086–6091
[79] Zhao J, Pu Z, Jin H, Zhang Z, Liu J, Mu S (2020) Phosphorous-doped carbon coordinated iridium diphosphide bifunctional catalyst with ultralow iridium amount for efficient all-pH-value hydrogen evolution and oxygen reduction reactions. *J Catal* 383:244–253
[80] Yu J, Wu X, Zhong Y, Yang G, Ni M, Zhou W, Shao Z (2018) Multifold nanostructuring and atomic-scale modulation of Cobalt Phosphide to significantly boost Hydrogen production. *Chem Eur J* 24(52):13800–6
[81] Zhu XD, Xie Y, Liu YT. (2018) Exploring the synergy of 2D MXene-supported black phosphorus quantum dots in hydrogen and oxygen evolution reactions. *J Mater Chem A* 6(43):21255–260
[82] Jiang J, Zhao K, Xiao X, Zhang L (2012) Synthesis and facet-dependent photoreactivity of BiOCl single-crystalline nanosheets. *J Am Chem Soc* 134(10):4473–4476
[83] Li T, Lin L, Wei H, Liang G, Kuang X, Liu T (2016) Solvothermal synthesis of three-dimensional microspherical bismuth oxychloride self-assembled by microspheres. *Physica E Low Dimens Syst Nanostruct* 76:198–202
[84] Han Q (2021) Advances in preparation methods of bismuth-based photocatalysts. *Chem Eng J* 414:414
[85] Novakova AA, Lanchinskaya VY, Volkov AV, Gendler TS, Kiseleva TY, Moskvina MA, Zezin SB (2003) Magnetic properties of polymer nanocomposites containing iron oxide nanoparticles. *J Magn Magn Mater* 258:354–357
[86] Hyeon T, Lee SS, Park J, Chung Y, Na HB (2001) Synthesis of highly crystalline and monodisperse maghemite nanocrystallites without a size-selection process. *J Am Chem Soc* 123:12798–12801
[87] Woo K, Hong J, Choi S, Lee HW, Ahn JP, Kim CS, Lee SW (2004) Easy synthesis and magnetic properties of iron oxide nanoparticles. *Chem Mater* 16(14):2814–2818

[88] Chin AB, Yaacob II (2007) Synthesis and characterization of magnetic iron oxide nanoparticles via w/o microemulsion and Massart's procedure. *J Mater Processtechnol* 191(1–3):235–237

[89] Bang JH, Suslick KS (2007) Sonochemical synthesis of nanosized hollow hematite. *J Am Chem Soc* 129(8):2242–2243

[90] Pinkas J, Reichlova V, Zboril R, Moravec Z, Bezdicka P, Matejkova J (2008) Sonochemical synthesis of amorphous nanoscopic iron(III) oxide from Fe (acac)3. *Ultrason Sonochem* 15(3):257–264

[91] Giri S, Samanta S, Maji S, Ganguli S, Bhaumik A (2005) Magnetic properties of α-Fe2O3 nanoparticle synthesized by a new hydrothermal method. *J Magn Magn Mater* 285: 296–302

[92] Hu X, Yu JC, Gong J (2007) Fast production of self-assembled hierarchical α-Fe2O3 nanoarchitectures. *J Phys Chem C* 111(30):11180–11185

[93] Cabrera L, Gutierrez S, Menendez N, Morales MP, Herrasti P (2008) Magnetite nanoparticles: Electrochemical synthesis and characterization. *Electrochim Acta* 53(8):3436–3441

[94] Liu Z, Ling Q, Cai Y, Xu L, Su J, Yu K, Wu X, Xu J, Hu B, Wang X (2022) Synthesis of carbon-based nanomaterials and their application in pollution management. *Nanoscale Adv* 4:1246–1262

[95] Prasek J, Drbohlavova J, Chomoucka J, Hubalek J, Jasek O, Adam V, Kizek R (2011) Methods for carbon nanotubes synthesis. *J Mater Chem* 21:15872–15884

[96] Manafi S, Nadali H, Irani HR (2008) Low temperature synthesis of multi-walled carbon nanotubes via a sonochemical/hydrothermal method. *Mater Lett* 62:4175–4176

[97] Jeong SH, Ko JH, Park JB, Park W (2004) A sonochemical route to single-walled carbon nanotubes under ambient conditions. *J Am Chem Soc* 126:15982–15983

[98] Novoselova IA, Oliinyk NF, Volkov SV, Konchits AA, Yanchuk IB, Yefanov VS, Kolesnik SP, Karpets MV (2008) Electrolytic synthesis of carbon nanotubes from carbon dioxide in molten salts and their characterization. *Phys E* 40:2231–2237

[99] Prasek J, Drbohlavova J, Chomoucka J, Hubalek J, Jasek O, Adam V, Kizek R (2011) Methods for carbon nanotubes synthesis. *J Mater Chem* 21:15872–15884

[100] Sabzehmeidani MM, Mahnaee S, Ghaedi M, Heidari H, Roy VA (2021) Carbon based materials: A review of adsorbents for inorganic and organic compounds. *Mater Adv* 2:598–627

[101] Brodie BC. XIII (1859) On the atomic weight of graphite. *Philos Trans Royal Soc London* 149:249–259

[102] Staudenmaier L (1898) Verfahren zur Darstellung der Graphitsäure. *Ber Dtsch Chem Ges* 31(2):1481–1487

[103] Hummers WS, Offeman RE (1958) Preparation of graphitic oxide. *J Am Chem Soc* 80(6): 1339–1339

[104] Kumar D, Yadav S, Singh CB, Yadav RS, Rai SB, Singh AK (2023) Impact of Sr2+ doping on the structural, dielectric, ferroelectric and optical properties of YFeO3 perovskite phosphor. *J Alloys Compd* 945:945

[105] Ma TY, Li H, Ren TZ, Yuan ZY (2012) Mesoporous SrTiO3 nanowires from a template-free hydrothermal process. *RSC Adv* 2(7):2790–2796

[106] Wang Y, Xu H, Wang X, Zhang X, Jia H, Zhang L, Qiu J (2006) A general approach to porous crystalline TiO2, SrTiO3, and BaTiO3 spheres. *J Phys Chem B* 110:13835–13840

[107] Xu G, Bai H, Huang X, He W, Li L, Shen G, Han G (2015) Self-assembled 3D flower-like perovskite PbTiO3 nanostructures and their application in the catalytic oxidation of CO. *J Mater Chem A* 3:547–554

[108] Geetha N, Kumar S and Prakash D (2018) Synthesis and characterization of Lamn1-XFexO3 (X=0, 0.1, 0.2) by coprecipitation route. *J Phys Chem Biophys* 8(3):1–6

[109] Zhu C, Nobuta A, Nakatsugawa I, Akiyama T (2013) Solution combustion synthesis of LaMO3 (M= Fe, Co, Mn) perovskite nanoparticles and the measurement of their electrocatalytic properties for air cathode. *Int J Hydrog Energy* 38(30):13238–48
[110] Mahamuni PP, Patil PM, Dhanavade MJ, Badiger MV, Shadija PG, Lokhande AC, Bohara RA (2019) Synthesis and characterization of zinc oxide nanoparticles by using polyol chemistry for their antimicrobial and antibiofilm activity. *BB Rep* 17:71–80
[111] Faridi AW, Imran M, Tariq GH, Ullah S, Noor SF, Ansar S, Sher F (2023) Synthesis and characterization of high-efficiency halide perovskite nanomaterials for light-absorbing applications. *Ind Eng Chem Res* 62 (11):4494–4502
[112] Sharfudeen BF, Latheef AF, Ambrose RV (2017) Synthesis and characterization of TiO2 nanoparticles and investigation of antimicrobial activities against human pathogens. *J Pharm Sci Res* 9(9):1604
[113] Chung IM, Abdul Rahuman A, Marimuthu S, Vishnu Kirthi A, Anbarasan K, Rajakumar G (2015) An investigation of the cytotoxicity and caspase-mediated apoptotic effect of green synthesized zinc oxide nanoparticles using *Ecliptaprostrata* on human liver carcinoma cells. *Nanomaterials* 5(3):1317–1330
[114] Mirikaram N, Pérez-Molina Á, Morales-Torres S, Salemi A, Maldonado-Hódar FJ, Pastrana-Martínez LM (2021) Photocatalytic perfomance of ZnO-graphene oxide composites towards the degradation of vanillic acid under solar radiation and visible-LED. *Nanomaterials* 11(6):1576
[115] Balaji S, Mandal BK, Vinod Kumar Reddy L, Sen D (2020) Biogenic ceria nanoparticles (CeO2 NPs) for effective photocatalytic and cytotoxic activity. *Bioengineering* 7(1):26
[116] Alamdari S, Sasani Ghamsari M, Lee C, Han W, Park HH, Tafreshi MJ, Afarideh H, Ara MH (2020) Preparation and characterization of zinc oxide nanoparticles using leaf extract of Sambucus ebulus. *Appl Sci* 10(10):3620
[117] Botsa SM, Babu MJ, Suresh P, Kalyani P, Venkateswararao B, Muralikrishna R (2020) Spherical NiWO4-reduced graphene oxide nanocomposite for effective visible light driven photocatalytic activity for the decolourisation of organic pollutants. *Arab J Chem* 13(11):8489–8497
[118] Shashanka R (2021) Investigation of optical and thermal properties of CuO and ZnO nanoparticles prepared by *Crocus sativus* (Saffron) flower extract. *J Iran Chem Soc* 18(2):415–427
[119] Rao A, Schoenenberger M, Gnecco E, Glatzel T, Meyer E, Brändlin D, Scandella L (2007) Characterization of nanoparticles using atomic force microscopy. *J Phys Conf Ser* 61(1):971
[120] Esmaeili A, Entezari M (2014) Facile and fast synthesis of graphene oxide nanosheets via bath ultrasonic irradiation. *J Colloid Interface Sci* 432:19–25
[121] Shamhari NM, Wee BS, Chin SF, Kok KY (2018) Synthesis and characterization of zinc oxide nanoparticles with small particle size distribution. *Acta Chim Slovenica* 65(3):578–585
[122] Bian SW, Mudunkotuwa IA, Rupasinghe T, Grassian VH (2011) Aggregation and dissolution of 4 nm ZnO nanoparticles in aqueous environments: Influence of pH, ionic strength, size, and adsorption of humic acid. *Langmuir* 27(10):6059–6068
[123] Sajna MS, Simon SM, Unnikrishnan NV, Sadasivuni KK (2022) An overview of graphene-based 2D/3D nanostructures for photocatalytic applications. *Top Catal* 14:1–25
[124] Wang J, Wang M, Tian Y, Deng W (2022) A review on photocatalytic glass ceramics: Fundamentals, preparation, performance enhancement and future development. *Catalysts* 12(10):1235

11 Supplementary Cementing Materials

From Waste to High-Performance Cement Composites

Arthur Behenck Aramburu and Rafael de Avila Delucis

11.1 OVERVIEW OF SUPPLEMENTARY CEMENTING MATERIALS

The production of Ordinary Portland Cement (OPC) involves a series of processes that require a substantial amount of energy and emit large volumes of quantities of greenhouse gases, particulate matter, and other harmful pollutants. The release of carbon dioxide (CO_2) in the atmosphere during the calcination process is the primary contributor to pollution in the production of Portland cement (Emetere and Dania 2019). The generation of CO_2 and waste is even more worrying given that the world consumption of cement has increased exponentially in recent decades (Figure 11.1) (Panesar and Zhang 2020a; Duchesne 2021). Various strategies have been developed to mitigate the environmental impact of OPC production, including the decrease in dependence on fossil fuels during the manufacturing process, exploring alternative materials, and adopting more sustainable manufacturing methods. In this sense, the utilization of supplementary cementing materials is considered one of the most prevalent and well-established methods for reducing the environmental impact of Portland cement production (Panesar 2019).

Supplementary cementing materials (SCMs) are materials that are added to cement composites to enhance their properties and performance (Juenger and Siddique 2015). SCMs can be pozzolanic or non-pozzolanic and are often derived from wastes of various industrial and agricultural processes (Kadri et al. 2010). According to ASTM C618, pozzolans are inorganic materials that possess pozzolanic or hydraulic activity, or both, and contribute to the properties of a cementitious mixture. In fact, the combination of Portland cement and pozzolanic materials produces more durable and sustainable composites, since this incorporation reduces the amount of Portland cement required and, consequently, results in lower carbon emissions, reduced energy consumption, and a decrease in the amount of waste generated (Borosnyói 2016; Franco de Carvalho et al. 2019).

 DOI: 10.1201/9781003731672-11

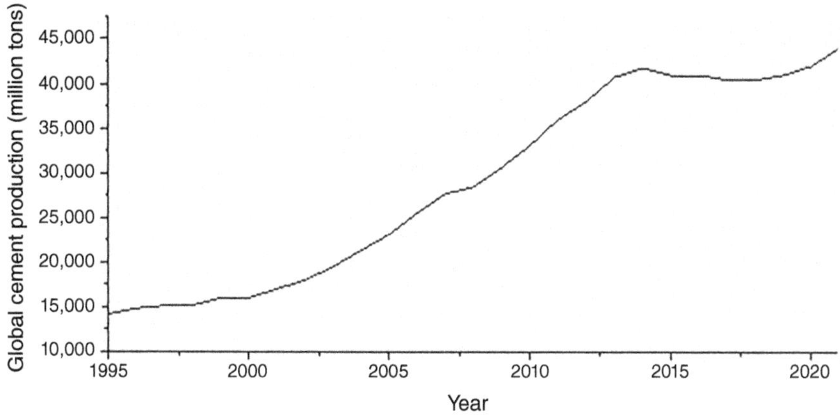

FIGURE 11.1 Global cement production in recent decades according to the US Geological Survey.

Before understanding the importance of cementitious SCM in cementitious composites, it is necessary to understand how these materials react with the hydration products of cement and contribute to the microstructure of the cement composite. In this sense, OPC has a hydraulic activity that results in a chemical reaction with water to set and harden (Xu et al. 2017). In this sense, OPC is formed by two main calcium silicates – tricalcium silicate (C_3S) and dicalcium silicate (C_2S) –, and after hydration, the main products of this reaction are calcium silicate hydrate (C-S-H) and calcium hydroxide (CH). The hydration reaction of the calcium silicates of OPCs and its products are presented in Equations 1 and 2, wherein the letters in brackets represent the abbreviations widely used in cement chemistry (Marchon and Flatt 2016; Scrivener et al. 2019):

$$3CaO.SiO_2\left[C_3S\right]+6H_2O\left[6H\right]=\left(CaO\right)_x.\left(SiO_2\right)_x.(H_2O)_x\left[C-S-H\right]+ 3Ca\left(OH\right)_2\left[3CH\right] \quad (11.1)$$

$$2CaO.SiO_2\left[C_2S\right]+4H_2O\left[4H\right]=\left(CaO\right)_x.\left(SiO_2\right)_x.(H_2O)_x\left[C-S-H\right]+ Ca\left(OH\right)_2\left[CH\right] \quad (11.2)$$

C-S-H is majorly responsible for most of the properties of these composites and is the most desired among all the hydration products. On the other hand, calcium hydroxide (CH) is inherently formed during the same reaction, but it is considered undesirable due to its limited contribution to strength and its involvement in various materials-related distress mechanisms. In this sense, the most remarkable property of pozzolanic SCMs is its reaction mechanism with water and the undesirable

CH to generate additional C-S-H products (Equation 3), increasing the durability and mitigating the materials-related distress mechanisms (Lemonis et al. 2015; El-Diadamony et al. 2018; Celik et al. 2019).

$$Pozzolan + Ca(OH)_2[CH] + H_2O[H] = (CaO)_x.(SiO_2)_x.(H_2O)_x[C-S-H] \quad (11.3)$$

However, these reactions are simplifications, as the kinetics of SCM reactions are affected by multiple factors, including the chemical composition, fineness, quantity of reactive phases like glass or zeolites, and the composition of the solution (Taylor 1993). The detailed impact of these parameters remains largely unknown due to the challenge of quantitatively measuring SCM reaction (Richardson and Groves 1993). Indeed, SCMs typically have lower calcium content compared to OPC, thus, the hydrates formed during the hydration process exhibit notable disparities. For instance, the C-S-H produced from silica-rich SCM systems differs significantly from the C-S-H generated from Portland cement, for example, in silica-rich SCM systems with Ca/Si ratios exceeding 1.5, the resulting C-S-H structures display disordered characteristics and contain regions reminiscent of calcium hydroxide (Lothenbach et al. 2011).

Furthermore, SCMs can also improve their durability through reducing permeability, leading to a reduction of corrosion in reinforced structures and increasing resistance to chemical attack, mainly in aggressive environments such as marine or chloride-exposed structures. Finally, SCMs can enhance the overall sustainability of concrete by reducing its carbon footprint, conserving natural resources, and reducing the generated waste (Park et al. 2021; Gilfillan and Marland 2021).

11.2 INDUSTRIAL WASTE ASHES

Industrial solid waste is a substantial contributor to the total volume of global waste, and it is characterized by its diverse forms and complex nature. Improper handling of industrial heavy metals-rich wastes can result in leachate that penetrates soil and groundwater, leading to irreversible environmental damage and posing a threat to human health (Li et al. 2022). In fact, the use of industrial waste ashes as SCMs is highly widespread in the cement industry, primarily due to its potential to mitigate the environmental impact associated with industrial production.

Industrial waste ashes are by-products of various industrial processes, such as coal-fired power plants, steel mills, and silicon metal production, that have pozzolanic properties. Several types of industrial waste ashes are commonly used as SCMs, including fly ash (Li et al. 2022), silica fume (Zhang et al. 2016a), and slag (Özbay et al. 2016); however, the use of industrial waste ashes also requires careful consideration of their physical and chemical characteristics, since the size of the particle and chemical composition of the SCM can alter the properties of the concrete. Additionally, the potential for leaching of heavy metals and other contaminants from the SCM must be evaluated to ensure the safety of the concrete and the environment (Zhao et al. 2022).

11.2.1 Fly Ash

Fly ash (FA) is a by-product of combustion from coal-fired power plants and is among the most commonly utilized SCMs (Khan et al. 2019). During and after burning, mineral particles present in the coal undergo liquefaction, evaporation, consolidation, or agglomeration, wherein these processes result in the formation of amorphous FA grains, which take on a spherical shape due to surface tension forces when they rapidly cool in the post-combustion stage (Li et al. 2022). The entrapped volatile matter within these grains expands at high temperatures, leading to the formation of a hollow cenosphere that, based on the composition of the residues and the heating/cooling conditions, may even crystallize or turn into glass (Kutchko and Kim 2006). In general, the FA used as SCM is fine, powdery, consists mainly of silica and significant quantities of alumina, and has limited calcium oxide amounts, which is highly dependent on the origin of the industrial material. Fly ashes can be classified as high-calcium fly ashes (Class C) and low-calcium ashes (Class F) (Sua-iam and Makul 2015; Provis et al. 2015).

The use of FA in concrete, both in low (<30%) and high (>30%) volumes, has significantly influenced the approach of the global concrete industry (Herath et al. 2020). FA Class F is the most commonly used type and typically contains 15–35% alumina. When blended with OPC, it leads to the formation of high levels of Al-rich phases. Consequently, the overall amount of portlandite in the hydrated mixture is decreased. The reactivity of fly ash is somewhat restricted, and moderate quantities of fly ash destabilize portlandite while facilitating the formation of additional C-S-H phases (Wardhono 2018).

Incorporating FA into cement composites can enhance the workability of the materials by providing plasticizing properties. In fact, Lee et al. (2003) reported the plasticizing effect of FA can be attributed to several factors, which diminish the flocculation of OPC particles. Furthermore, the slower reaction rate of FA decelerates the development of hydration products in the early stages of cement hydration. Moreover, the spherical shape of fly ash grains facilitates the mobility of close particles, thereby enhancing the workability of the material.

Regarding mechanical properties, multiple research studies have indicated that incorporating FA in composites can initially hinder the development of their strength (Li et al. 2022). Indeed, as concrete ages, FA exhibits a beneficial effect on the development of mechanical properties, which may be related to its pozzolanic property, wherein it reacts with CH produced during cement hydration. This reaction results in the creation of additional hydrates, including additional C-S-H gel, which contributes to the progressive enhancement of concrete strength over time (Supit and Shaikh 2015). Lower concentrations (up to 15–20%) of FA in composites generally lead to an improvement in mechanical strength; however, Hemalatha and Ramaswamy (2017) affirm that fly ash is suitable for high replacement (>50%) in concrete and results in better strength, durability, permeability, and shrinkage properties.

Overall, the use of FA as an SCM in concrete production has numerous benefits, including reduced carbon emissions, improved workability, increased strength, and enhanced durability. Extensive research has been conducted on FA, demonstrating its efficacy in increasing the characteristics of concrete. Its utilization in the construction industry represents a sustainable solution that can simultaneously mitigate the environmental impact of cement production and improve the performance of cement composites

11.2.2 Granulated blast-furnace slag

Granulated blast furnace slag (GBFS) is a commonly used SCM in concrete production and is obtained during the production of pig iron in a blast furnace, wherein the rapid cooling of the molten slag with water results in the formation of granulated slag, which is a fine, granular, mostly noncrystalline, glassy material with hydraulic properties (Barnett et al. 2006). The conversion of molten slag can be achieved by either cooling it in a pond or by employing vigorous water jets. Ground granulated blast furnace slag (GBFS) primarily consists of calcium silicates and alumino-silicates. Compared to FA, GBFS has a higher concentration of calcium oxide, but it contains significantly lower amounts of aluminum oxide (Ahmad et al. 2022). Moreover, the magnesium content of this residue is usually between 7% and 15%, leading to the formation of phase solids that vary from the behavior of FA systems (Özbay et al. 2016). Like the others studied SCMs, when mixed with cement, GBFS reacts with the CH formed during hydration to form additional C-S-H and results in enhanced strength and durability of the concrete, providing these desirable properties.

The utilization of GBFS in cement composite production, along with other SCMs, offers substantial benefits. It helps reduce the unit cost of concrete production while simultaneously enhancing rheological and workability properties (Aghaeipour and Madhkhan 2017). GBFS has also shown positive effects on segregation resistance and has a positive impact on long-term compressive strength, particularly beyond 40 days. Also, it contributes to increased flexural strength at both early and later ages of concrete (Barnett et al. 2006). The benefits of using GBFS also include improved abrasion resistance (Bilir et al. 2017), reduced water (Wang et al. 2023), chloride ion (Wang et al. 2019), gas permeability, and better resistance to sulfate and alkali-silica reactions (Zhang et al. 2019). Finally, GBFS has undergone thorough research and has proven to be highly effective in enhancing the properties of concrete. Its utilization in the construction industry serves as a sustainable solution that can effectively reduce the impact associated with OPC manufacturing while simultaneously improving the overall performance of concrete.

11.2.3 Silica fume

Silica fume (SF), or micro silica, is a residue of silicon and ferrosilicon alloy manufacturing processes that involves the reduction of high-purity quartz at high temperatures (≈ 2000 °C) in an electric arc furnace (Zhang et al. 2016a). SF is produced through the condensation of silicon monoxide vapor in the upper sections of a furnace, resulting in the formation of microspheres of amorphous silica. SF is an ultrafine powder consisting predominantly of SiO_2 and is commonly utilized as an SCM in applications requiring high-performance and high-strength concretes since its small particle sizes provide high reactivity and a large surface area (Tavares et al. 2020).

The utilization of SF as SCM has several benefits, including reduced heat liberation during hydration, enhanced mechanical strength, a more refined distribution of pore sizes, and the ability to consume lime. These properties contribute to the improved performance and durability of the concrete (Adil et al. 2020). SF particles have the capability to fill in the pores within the hardened cement matrix, leading to an increase in the density and strength of the cement paste. By occupying these spaces, SF particles contribute to a more compact and solid structure, enhancing the overall mechanical properties of the concrete (Xu et al. 2020). Previous studies have demonstrated that the utilization of silica fume in concrete can lead to several benefits, such as reduced bleeding, lower porosity, and decreased permeability. To attain the desired workability, superplasticizers are commonly employed since the ultrafine nature of silica fume necessitates additional water. Incorporating silica fume can help achieve the desired hardened properties of self-compacting lightweight concrete and enhance crack resistance in cementitious materials (Farahani et al. 2015; Lei et al. 2016; Xu et al. 2020). Despite the various benefits of silica fume, its ultrafine particles pose health risks, making handling the dry powder form difficult (Faraj et al. 2021).

In conclusion, the utilization of silica fume as a supplementary cementitious material (SCM) in concrete production offers numerous advantages, with improved strength being a significant benefit. Its application in the construction industry represents a sustainable solution that not only helps reduce the environmental impact of cement production but also enhances the overall performance of concrete.

11.3 AGRICULTURAL WASTE ASHES

The utilization of agricultural waste ashes as SCM has received significant interest in recent years. These materials, which are readily available and cost-effective, have shown promise in mitigating the environmental impact related to OPC production. Their use as SCMs offers a sustainable alternative that can contribute to more environmentally friendly cement production practices (Thomas et al. 2021b). Agricultural waste ashes are by-products of agricultural processes, such as rice milling, sugarcane processing, and wheat straw burning. These waste materials contain high amounts of silica and other pozzolanic compounds, which can be activated by alkaline solutions and used as SCMs in concrete manufacturing (Aprianti et al. 2015).

The application of agricultural waste ashes as SCMs aligns with the principles of the circular economy, where waste materials are reused to create new value and reduce environmental impact (He et al. 2020). By using agricultural waste ashes as SCMs, the amount of waste generated by agricultural processes can be reduced, as well as the reduce the demand for virgin materials. This approach can also create economic opportunities for farmers and waste management companies, which can sell these wastes to the construction industry (Paul et al. 2019).

Overall, the application of agricultural waste ashes as SCMs in OPC composite production has the potential to transform the building materials industry and reduce its environmental impact. This approach can create a more sustainable and circular economy, where waste materials are reused to create new value and reduce environmental harm. Future research is necessary to determine the full potential of agricultural waste ashes as SCMs and optimize their use in concrete production (Martirena and Monzó 2018).

11.3.1 Rice husk ash

Rice husks are a natural protective covering that develops during the growth of rice grains. Rice-producing countries have a substantial amount of agricultural solid waste material, which lacks any economic worth when isolated during the refining procedure (Krishna et al. 2016). However, this material can be transformed into rice husk ashes (RHA) through the combustion process (600–800 °C) and milling, resulting in a very fine material with a high amount of amorphous silica (85–90%) and other pozzolanic compounds, making it a promising SCM for cement composites production. In fact, many researchers have demonstrated that the addition of RHA to OPC composites can improve its properties and performance (Alex et al. 2016).

The implementation of RHA in OPC composites improves various hydraulic properties, strength, permeability, durability, and volumetric stability of concrete over a long period. Recent studies have demonstrated remarkable outcomes, indicating the potential to achieve concretes reaching 55 MPa and 72 MPa after 7 and 28 days of curing, respectively, through the application of SCMs such as RHA (Aprianti et al. 2015; Wang et al. 2021). Indeed, numerous studies have investigated the techniques and methodologies for manufacturing high-strength concrete using RHA. These studies have explored various concrete mix designs incorporating RHA, and the findings consistently indicate that RHA concrete exhibits superior compressive strengths at different ages, including an extended period of up to 90 days. In relation to normal concrete lacking SCMs, RHA concrete consistently demonstrates higher compressive strengths (Jamil et al. 2016; Sandhu and Siddique 2017). The desirable properties of concretes, such as mechanical strength and durability, are improved by the effects of RHA in the OPC matrix. RHA has the capacity to reduce the presence of pores in the matrix and decrease the continuous pores structure. This transformation leads to a more homogeneous and denser microstructure of the OPC systems, resulting in enhanced overall performance (Hu et al. 2020).

Moreover, concrete with RHA as an additional cementitious material exhibits excellent water and chloride ion resistance, crucial for durability and corrosion prevention (Gursel et al. 2016; Zareei et al. 2017). Introducing RHA into concrete addresses the existing challenges related to RHA disposal. In conclusion, the utilization of RHA as SCM in OPC composite manufacturing provides numerous advantages, particularly in terms of significantly enhanced mechanical behavior and overall performance. Nonetheless, additional research is necessary to optimize RHA utilization in concrete production and fully explore its potential as an SCM (Fapohunda et al. 2017).

11.3.2 Sugarcane bagasse ash

Sugarcane bagasse ash (SBA) is obtained as a by-product of the sugar industry and is produced when bagasse, the fibrous waste material left behind after sugarcane is extracted, is burned at a temperature of around 600–800 °C (Shafiq et al. 2018; Thomas et al. 2021a). Along with other lignocellulosic wastes, the ashes of this material are rich in amorphous silica content and are promising SCM due to their pozzolanic properties. In fact, SBA has been found to be a useful pozzolanic material due to its abundance of alumina and amorphous silica; however, the presence of these oxides can vary depending on several factors such as the burning parameters (temperature and method), the type of soil of cultivation of the sugarcane, and the characteristics of the raw material (Aprianti et al. 2015).

According to Ahmad et al. (2021), by incorporating SBA into cement-based composites, the need for a reduced amount of water to reach the desired workability is achieved due to the surface texture of SBA, similar to glass. Furthermore, the author suggests that the addition of SBA, up to an optimal level, within a replacement range of 5–15%, can marginally decrease the weight of the OPC composite due to the lower density of SBA while simultaneously enhancing its mechanical strength. However, excessive SBA addition may have a harmful effect on the mechanical properties of the composite. Additionally, the incorporation of SBA as an SCM in cement-based composites also improves their resistance to chloride penetration (Jahanzaib Khalil et al. 2021; Wu et al. 2022). Therefore, utilizing the optimal content of SBA in cement composite can enhance their strength properties and improve their resistance to environmental factors, contributing to sustainable construction at a lower cost (Bahurudeen et al. 2015). Likewise, like other agricultural waste-based SCMs, the utilization of SBA in cement composites not only aids in reducing global CO_2 emissions but also enhances the overall value of this feedstock cultivation. Additionally, it contributes to the advancement of sustainable and circular solutions within the construction industry (Yogitha et al. 2020).

11.3.3 Wood waste ash

Wood ash (WA) is both inorganic and organic matter residue that remains after the combustion of wood and wood products, such as sawdust, bark, chips, and unbleached wood fiber (Sigvardsen et al. 2021). The disposal of this waste can lead to environmental pollution, but recent research has shown that it can also be used as an SCM in the manufacturing of OPC composites. The characteristics of wood ash can vary when it is used as a partial replacement for OPC, wherein the properties can also differ. Some studies have observed high pozzolanic activity in wood ash, which can be attributed to its high silicon, aluminum, and iron oxide content or its amorphous silica content (Batt and Garg 2017; Martínez-García et al. 2022).

Sigvardsen et al. (2021) affirm that WWA exhibits hydraulic properties when mixed with water, wherein the residue presents setting, hardening, and developing strength. Regarding its addition to cement composites, Siddique (2012) reported that WWA can be used to replace a portion of OPC in composites. However, WWA may adversely impact concrete slump and increase water absorption. While the strength properties may decrease with higher WWA content, they tend to improve as time passes due to pozzolanic reactions (Chowdhury et al. 2015b, 2015a; Hamid and Rafiq 2021). Despite these challenges, WWA can still be used to make structural-grade concrete and precast products. In this sense, Sigvardsen et al. (2019) affirm that

WWA derived from the combustion of wood chips is the most appropriate type of WWA with hydraulic activity for use as SCM. Conversely, wood ash obtained from high-temperature combustion is considered the optimal choice as a filler due to its exceptional characteristics. In conclusion, the use of wood waste ash as a supplementary cementing material is still limited if compared to RHA and SBA; thus, further research efforts are needed to optimize the use of WWA in concrete and to investigate its long-term behavior.

11.3.4 Other lignocellulosic ashes

While most of the research is focused on the previously discussed ashes, other types of feedstocks, have received attention. Like the previously discussed lignocellulosic ashes, these materials also possess characteristics that make them suitable as SCMs, and their use can offer both economic and environmental benefits. In this context, exploring the properties and potential applications of other lignocellulosic ashes is an important area of research that can contribute to the development of sustainable and efficient processes for biomass conversion.

In this sense, Silva et al. (2021) affirm that bamboo leaf ash (BLA) possesses exceptional pozzolanic properties, satisfying the requirements stipulated in standard specifications, such as high surface area, granulometry, and pozzolanic activity, demonstrating the immense potential of the material as a SCM in the production of OPC composites with even 30% of BLA replacement. Moreover, Ahmad et al. (2023) related that corn cob ash (CCA) also may be used as SCM in the construction industry and can be beneficial for waste reduction and sustainable development. The authors affirm that cement can be replaced by CCA for up to 10% without impacting concrete qualities. Finally, Ahmad et al. (2023) reviewed the applications and effects of palm oil fuel ash (POFA) on concrete characteristics, which is the result of burning waste materials like palm oil fiber, palm oil husk, and palm kernel shell. The authors concluded that cement composites containing POFA generally exhibit better durability, mechanical strength, and other characteristics compared to concrete with conventional matrices.

11.4 NATURAL POZZOLANS AND MINERAL POWDERS

Natural pozzolans are silica-rich materials found in nature that do not possess significant cementitious properties. However, when finely ground and exposed to moisture, they undergo a chemical reaction with CH at normal temperatures, resulting in the formation of compounds that exhibit cementitious properties (Dedeloudis et al. 2018; Panesar 2019). Natural pozzolans have been used as building materials in construction for thousands of years, particularly in regions where volcanic activity and deposits are prevalent. In fact, the ancient Romans were known to use these natural pozzolans to produce concrete mixtures with improved strength and durability (Weber et al. 2007).

Natural pozzolans are typically found in the form of volcanic ash, pumice, and diatomaceous earth, as well as certain types of clay and shale (Duchesne 2021). The chemical structure of natural pozzolans can vary widely, depending on the geological origin, and this can affect their reactivity when combined with cement. Some natural pozzolans contain high amounts of reactive silica and alumina, which enhances their pozzolanic activity, while others contain more inert materials that may require additional processing or activation to achieve the desired cementitious properties (Cobîrzan et al. 2015; Omrane et al. 2017; Shukla et al. 2020; Omrane and Rabehi 2020).

11.4.1 Limestone filler

Limestone filler (LF) is a carbonate sedimentary rock of calcium carbonates, mostly calcite and aragonite, that originates from chemical and organic precipitation (Panesar and Zhang 2020b). In the manufacturing of OPC, natural limestone is crushed and finely ground to produce the clinker (Vizcaíno Andrés et al. 2015; Rodriguez and Tobon 2020). In fact, limestone is one of the most common components of cement clinker, however, the use of limestone as filler, after clinker assembling, has also been extensively studied. The physical effects of limestone on cement include modification of the particle size distribution, resulting in changes in the pores sizes, dilution, and nucleation of reactions, wherein these reactions can improve the properties of concrete (Pillai et al. 2019; Panesar and Zhang 2020b).

In this sense, Lothenbach et al. (2008) affirm that the blend of OPC and limestone has been shown to influence the initial hydration reaction, as well as the composition of the hydrate assemblage, in cement pastes. According to the authors, the inclusion of limestone in the cement matrices promotes the formation of monocarbonate instead of monosulfate at room temperature. The primary effect of limestone on the cement system is the inhibition of tricalcium aluminate hydration during the initial stages of OPC hydration. This suppression occurs due to the chemical reaction between limestone, monosulfate, and CH (Pillai et al. 2019; Sharma et al. 2021).

Moreover, the incorporation of LF as a replacement for OPC has been shown to have positive and negative effects on the characteristics of these composites. In fact, Panesar and Zhang (2020a) have stated that the use of LF as a replacement in cement composites can have beneficial effects at levels of up to 10%, wherein LF has been found to enhance compressive strength and decrease chloride penetration. However, the authors affirm that higher levels of LF incorporation generally result in reduced strength, increased porosity, and greater permeability in cement composites. While LF has similar carbonation resistance to SCMs, it is inferior in improving resistance to sulfate attack and freeze/thaw action and alleviating alkali-silica reaction at the same replacement percentage (Panesar and Zhang 2020a; Nguyen et al. 2020a, 2020b). Moreover, incorporating LF as a substitute for cement exhibits a similar impact as SCMs in decreasing the environmentally harmful potential of OPC materials (Sharma et al. 2021).

11.4.2 Calcinated clays

Clay is a flexible natural substance comprising minerals, organic matter, and metal oxides. Its formation primarily occurs through chemical and physical modifications or through processes involving decomposition and recrystallization. These processes can result in a range of clay types with different properties, making it a valuable material for various applications. In recent years, calcined clays have garnered considerable attention as a type of SCM due to their potential to enhance the properties of OPC systems while reducing their environmental footprint. Calcined clays are obtained by subjecting natural clay minerals to a process of calcination at temperatures ranging between 650 °C and 900 °C (Sabir et al. 2001; Jaskulski et al. 2020).

One of the most prominent clays used for SCM is Metakaolin, which will be discussed further in the next section; however, numerous other clays have been studied in recent years (Jaskulski et al. 2020; Khandelwal and Rhee 2022; Shi et al. 2022). In this sense, Tran et al. (2019) studied the use of natural zeolite in cement composite production, wherein the authors stated that natural zeolite has potential as an additive in the building materials industry due to its advantageous characteristics, such as high SiO_2/Al_2O_3 content, ion exchange, dehydration/rehydration, and adsorption, as well as its porous framework structure. Conversely, the authors also noted that incorporating zeolite in concrete may lead to a decrease in mechanical strength and workability. However, the addition of natural zeolites to concrete can enhance various other properties, such as the reduction of shrinkage, improving resistance to chloride penetration, enhancing water permeability, improving sulfate resistance, and increasing carbonation resistance.

Furthermore, Bullerjahn et al. (2020) stated that the pozzolanic reactivity of aluminosilicates can be enhanced by subjecting them to calcination in the presence of dolomite clay. According to the authors, this improved reactivity is attributed to the dolomite decomposition process, which occurs in two stages within a wide temperature range that coincides with the dehydroxylation of aluminosilicates and the formation of inert phases. Similarly, Shi et al. (2022) conducted a study on the utilization of calcined attapulgite clay as an SCM and found that attapulgite calcined at 500 °C displays the highest level of hydration activity. Furthermore, the authors suggested further research to investigate the potential of this SCM in enhancing the durability of cement-based materials, as well as increasing the resistance against chloride ions penetration.

The use of calcined clays as an SCM in concrete offers several benefits, including improved mechanical and durability properties. Since calcined clays are a natural and abundant resource, their use provides a sustainable solution to mitigate the impact of OPC production; however, further research is needed to optimize the use of calcined clays in concrete and to investigate their long-term performance.

11.4.3 Metakaolin

Kaolin is a fine-grained natural clay that can be converted to Metakaolin (MK), a highly reactive natural pozzolan, through calcining and dehydroxylating the material at temperatures ranging from 500 °C to 900 °C (Khatib et al. 2018). Thus, MK, differently of most SCM materials, is not a product that comes from an industrial or agricultural waste activity, nor is it entirely natural, since it is produced for this end. The burn of kaolin clays in this range of temperatures results in the loss of water from the clay and the collapse of its structure, producing the amorphous alumino-silicate material (i.e., MK) (Siddique and Klaus 2009). This process, known as dehydroxylation, plays a vital role in the production of this highly reactive pozzolanic material utilized in diverse applications. Although silicon dioxide and alumina oxide are the major constituents of MK, their composition can vary depending on the source of kaolin (Zhang et al. 2016b; Dill 2016).

The pozzolanic reaction between metakaolin and CH, besides the well-described formation of C-S-H, also produces alumina-containing phases, improving the mechanical behavior and durability of OPC composites. In fact, Siddique and Klaus (2009) state that the inclusion of this SCM not only improves the mechanical behavior in the early age and long-term strength characteristics of concrete but also decreases water infiltration through capillary action. Additionally, the incorporation of this SCM leads to substantial alterations by altering the chemical composition of the pore solution phase in the hydrated substance, consequently modifying the pore structure of OPC systems. This modification results in reduced permeability and enhanced resistance to the diffusion of detrimental ions that could potentially degrade the matrix (Rakhimova and Rakhimov 2019; Zhan et al. 2020; Pillay et al. 2022). Finally, the incorporation of high-reactivity metakaolin in concrete can be beneficial in mitigating harmful expansion caused by silica-alkali reactions. The effectiveness of this approach may vary depending on the characteristics of the aggregate used (Wianglor et al. 2017; Rakhimova and Rakhimov 2019).

11.5 TRENDS IN SUPPLEMENTARY CEMENTING MATERIALS

Besides the already discussed materials, the need for sustainable and environmentally friendly cement composites focused on circular economy has resulted in the study of new materials being explored as potential SCMs. In this sense, most of these materials originated from the previously presented origins, that is, industrial, agricultural, and natural origins. Nevertheless, there has been an emergence of new materials, such as glass powder and dehydrated cementitious powder, that have garnered considerable attention. These materials have demonstrated great potential in improving the mechanical and durability characteristics of concrete while concurrently mitigating the environmental repercussions linked to cement production.

11.5.1 Glass powder

Glass is widely recognized for its remarkable properties, including high inherent strength, transparency, low permeability, chemical resistance, and inertness (Islam et al. 2017). However, despite its theoretical recyclability, there are limitations regarding the necessary quality standards for glass recycling. Consequently, the non-recyclable portion is typically discarded into landfills, representing environmental challenges (Jiang et al. 2019). In this sense, glass powder (GP) has emerged as a novel SCM and is produced by grinding recycled glass into a fine powder, which can then be utilized as a partial replacement for OPC in cement composites production (Afshinnia and Rangaraju 2016).

In this sense, GP has a small particle size range and is suitable for replacing cement in high volumes (<20%) (Aliabdo et al. 2016). In fact, GP can be used as a filler in high-performance concretes, replacing SF and quartz powder for better performance and cost optimization (Elaqra and Rustom 2018). Moreover, GP can be used to implement in alkali-activated materials systems to act as a mineral additive in eco-friendly cement. Aliabdo et al. (2016) research focused on the utilization of GP in cement composites and related that this waste has been found to meet the ASTM C618 standards for pozzolanic materials in terms of physical, mechanical, and chemical properties. Additionally, the use of GP in cement-based products can improve the mechanical strength of OPC composites since it causes the decrease of pore size in cement pastes (Afshinnia and Rangaraju 2016; Güllü et al. 2017; Tayeh et al. 2020). However, the increase in GP replacement level in cement composites can decrease the 28-day mechanical strength of concrete when used by more than 15% (Aliabdo et al. 2016).

In conclusion, the implementation of GP as SMC in concrete production offers several benefits, including improved mechanical and durability properties, reduced environmental impact, and a sustainable solution to waste disposal. Further research is needed to optimize the use of GP and to investigate their long-term performance.

11.5.2 Dehydrated cementitious powder

Dehydrated cementitious powder (DCP) is a type of SCM that has recently been studied as a potential replacement for OPC in composites. It has been observed that when exposed to high temperatures, hardened OPC pastes undergo a series of physical and chemical processes (Qian et al. 2020). In the main process, the hydration products undergo the release of both bonded water and free water. Additionally, certain hydration products like ettringite, C-S-H, and CH gradually dehydrate when submitted to higher temperatures (Letelier et al. 2017; Abadel et al. 2023). The process of dehydration in cement pastes is typically completed at temperatures of 800 °C or higher. It is important to note that the extent of the dehydration process is influenced by the rate at which the heating occurs, as not all reactions take place instantaneously. Consequently, DCP is a complex material composed of three main components, the initial unhydrated cement, partially dehydrated products, and residual hydration products. When DCP comes into contact with water again, it has the potential to undergo rehydration (Song et al. 2018).

In this sense, Shui et al. (2009) affirm that the introduction of DCP in cement composites implies an increase in water requirement, wherein the water volume is in proportion with the dehydration temperatures used in the process. Moreover, a decrease in the setting time of DCP is observed as the dehydration temperature rises. The authors also affirm that, in the later stages of the hydration process, the mechanical strength of rehydrated DCP can be influenced by the pre-heating dehydration temperature. Similarly, Qian et al. (2020) studied the incorporation of DCP into concrete and affirmed that a replacement of a part of cement content with DCP leads to a more compact cement matrix. In this sense, when more than 25% of the cement is substituted with DCP, deleterious effects may be observed in the mechanical strength, workability, and durability of the composite material. Finally, the authors related that the added DCP can react with the free water and cause a reduction in the total water requirement for hydration, which can hinder the promotion of hydration kinetics and the generation of a dense structure. Nonetheless, the particles present in DCP can exhibit nucleation effects, potentially enhancing cement hydration and contributing to the development of a denser microstructure.

11.6 BLENDS OF SUPPLEMENTARY CEMENTITIOUS MATERIALS

The blending of SCMs (Figure 11.2) in cement composite production has received improved interest in recent years for their potential to have a synergistic effect, where the combined effect of the individual SCMs is greater than their individual effects. Blending SCMs is widely studied since the presence of SCMs can have an impact on the type and number of hydrate products formed in OPC systems, which can ultimately affect the properties of these composites. At typical substitution levels, the composition of the C-S-H phase is usually altered, resulting in a lower Ca/Si ratio, despite the presence of portlandite. In SCMs containing Al_2O_3, the C-S-H phase may also contain a significant amount of this oxide. Additionally, the presence of SCMs may increase the amount of hydrates containing aluminate, and the inclusion of magnesium in the SCM may result in the appearance of a phase similar to hydrotalcite (Lothenbach et al. 2011). In conclusion, the incorporation of SCMs into Portland cement creates a more complex system where the hydraulic reaction of the SCM and hydration of Portland cement occur concurrently and may affect each other's reactivity. Given that most SCMs have a slower reaction rate than clinker phases, it can be challenging to monitor their reaction progress.

The relation of SiO_2, CaO, and Al_2O_3 content in the SCM can vary in a large range, wherein there are differences in pore solutions between blended systems and pure Portland materials, with a tendency toward lower pH levels that can prevent alkali-silica reaction. This behavior makes the blending of SCMs an attractive approach to mitigate the adverse results of its incorporation in Portland cement systems.

In this sense, Fernández et al. (2018) affirm that the optimization of ternary blends is heavily influenced by the composition of OPC, thus highlighting the significant role of OPC in the effect of SCMs in ternary blends. The authors related that a combination of BFS and FA achieved a synergistic effect mainly due to the influence of the OPC composition, wherein the initial formation of hydrates is more pronounced in the ternary blends. Similarly, Andrade et al. (2019) reported that, after long curing periods (>90 days), pastes produced with a combination of nanosilica and metakaolin or silica fume exhibited a higher C–S–H content than those prepared with only one type of SCM.

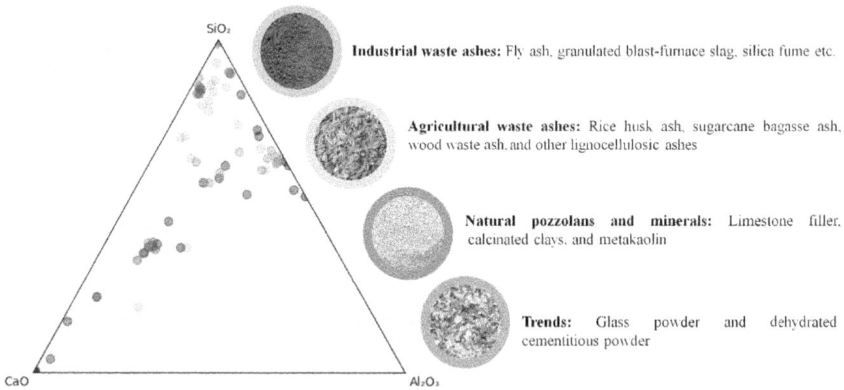

FIGURE 11.2 CaO–Al_2O_3–SiO_2 ternary diagram of different sources of SCMs.

11.7 CONCLUDING REMARKS

The reduction of the amount of OPC required in concrete due to the use of SCM can significantly contribute to the reduction of carbon emissions related to cement production. This reduction in carbon emissions is particularly significant when considering the global scale of concrete production, wherein SCMs also reduce the amount of waste generated by implementing industrial by-products and agricultural residues as supplementary materials. Overall, while there are still challenges associated with understanding the reaction mechanisms and optimizing the use of these materials, the use of SCMs is widespread worldwide.

In this context, the utilization of industrial waste ashes has become widely accepted as a means to mitigate the environmental consequences of cement manufacturing. Nevertheless, the incorporation of these industrial waste ashes as SCMs in concrete necessitates a thorough assessment of their chemical and physical characteristics, particle size distribution, and potential for leaching heavy metals and other contaminants.

Meanwhile, the use of agricultural waste ashes, such as rice husk ash, sugarcane bagasse ash, and wood waste ash, as SCMs has been gaining attention in recent years. These by-products of agricultural and forestry processes contain high amounts of silica and other pozzolanic compounds, which can be activated by alkaline solutions and used in concrete production. The use of agricultural waste ashes as SCMs in the construction industry offers a significant opportunity to minimize waste generated from agricultural and forestry activities, as well as decrease the demand for raw materials. This approach is directed related to circular economy principles, which strive to reduce the environmental impact by repurposing waste materials to generate new value.

Although less currently employed, natural pozzolans are silica-rich materials that possess the ability to chemically react with CH, resulting in the formation of compounds with cementitious behavior. These natural pozzolans can be found in various forms, including volcanic ash, pumice, diatomaceous earth, clay, and shale. Among these, limestone is widely employed both in clinker production and as a filler material.

Finally, despite challenges in understanding reaction mechanisms and optimizing SCM usage, their widespread adoption is observed worldwide, aligning with circular economy principles by repurposing waste materials to create new value. In fact, SCMs hold significant potential in addressing environmental challenges, optimizing resource utilization, and reducing waste in cement production.

REFERENCES

Abadel AA, Alghamdi H, Alharbi YR, et al (2023) Investigation of alkali-activated slag-based composite incorporating dehydrated cement powder and red mud. *Materials (Basel)* 16:1551. https://doi.org/10.3390/ma16041551

Adil G, Kevern JT, Mann D (2020) Influence of silica fume on mechanical and durability of pervious concrete. *Constr Build Mater* 247:118453. https://doi.org/10.1016/j.conbuildmat.2020.118453

Afshinnia K, Rangaraju PR (2016) Impact of combined use of ground glass powder and crushed glass aggregate on selected properties of Portland cement concrete. *Constr Build Mater* 117:263–272. https://doi.org/10.1016/j.conbuildmat.2016.04.072

Aghaeipour A, Madhkhan M (2017) Effect of ground granulated blast furnace slag (GGBFS) on RCCP durability. *Constr Build Mater* 141:533–541. https://doi.org/10.1016/j.conbuildmat.2017.03.019

Ahmad J, Arbili MM, Alabduljabbar H, Deifalla AF (2023) Concrete made with partial substitution corn cob ash: A review. *Case Stud Constr Mater* 18:e02100. https://doi.org/10.1016/j.cscm.2023.e02100

Ahmad J, Kontoleon KJ, Majdi A, et al (2022) A comprehensive review on the ground granulated blast furnace slag (GGBS) in concrete production. *Sustainability* 14:8783. https://doi.org/10.3390/su14148783

Ahmad W, Ahmad A, Ostrowski KA, et al (2021) Sustainable approach of using sugarcane bagasse ash in cement-based composites: A systematic review. *Case Stud Constr Mater* 15:e00698. https://doi.org/10.1016/j.cscm.2021.e00698

Alex J, Dhanalakshmi J, Ambedkar B (2016) Experimental investigation on rice husk ash as cement replacement on concrete production. *Constr Build Mater* 127:353–362. https://doi.org/10.1016/j.conbuildmat.2016.09.150

Aliabdo AA, Abd Elmoaty AEM, Aboshama AY (2016) Utilization of waste glass powder in the production of cement and concrete. *Constr Build Mater* 124:866–877. https://doi.org/10.1016/j.conbuildmat.2016.08.016

Aprianti E, Shafigh P, Bahri S, Farahani JN (2015) Supplementary cementitious materials origin from agricultural wastes – A review. *Constr Build Mater* 74:176–187. https://doi.org/10.1016/j.conbuildmat.2014.10.010

Bahurudeen A, Kanraj D, Gokul Dev V, Santhanam M (2015) Performance evaluation of sugarcane bagasse ash blended cement in concrete. *Cem Concr Compos* 59:77–88. https://doi.org/10.1016/j.cemconcomp.2015.03.004

Barnett SJ, Soutsos MN, Millard SG, Bungey JH (2006) Strength development of mortars containing ground granulated blast-furnace slag: Effect of curing temperature and determination of apparent activation energies. *Cem Concr Res* 36:434–440. https://doi.org/10.1016/j.cemconres.2005.11.002

Batt AS, Garg A (2017) Partial replacement of wood ash with ordinary Portland cement and foundry sand as fine aggregate. *J Civ Environ Eng* 07. https://doi.org/10.4172/2165-784X.1000272

Bilir T, Yüksel I, Topcu IB, Gencel O (2017) Effects of bottom ash and granulated blast furnace slag as fine aggregate on abrasion resistance of concrete. *Sci Eng Compos Mater* 24:261–269. https://doi.org/10.1515/secm-2015-0101

Borosnyói A (2016) Long term durability performance and mechanical properties of high performance concretes with combined use of supplementary cementing materials. *Constr Build Mater* 112:307–324. https://doi.org/10.1016/j.conbuildmat.2016.02.224

Bullerjahn F, Zajac M, Pekarkova J, Nied D (2020) Novel SCM produced by the co-calcination of aluminosilicates with dolomite. *Cem Concr Res* 134:106083. https://doi.org/10.1016/j.cemconres.2020.106083

Celik K, Hay R, Hargis CW, Moon J (2019) Effect of volcanic ash pozzolan or limestone replacement on hydration of Portland cement. *Constr Build Mater* 197:803–812. https://doi.org/10.1016/j.conbuildmat.2018.11.193

Chowdhury S, Maniar A, Suganya OM (2015a) Strength development in concrete with wood ash blended cement and use of soft computing models to predict strength parameters. *J Adv Res* 6:907–913. https://doi.org/10.1016/j.jare.2014.08.006

Chowdhury S, Mishra M, Suganya O (2015b) The incorporation of wood waste ash as a partial cement replacement material for making structural grade concrete: An overview. *Ain Shams Eng J* 6:429–437. https://doi.org/10.1016/j.asej.2014.11.005

Cobîrzan N, Balog A-A, Moşonyi E (2015) Investigation of the natural pozzolans for usage in cement industry. *Procedia Technol* 19:506–511. https://doi.org/10.1016/j.protcy.2015.02.072

da Silva Andrade D, da Silva Rêgo JH, Morais PC, et al (2019) Investigation of C-S-H in ternary cement pastes containing nanosilica and highly-reactive supplementary cementitious materials (SCMs): Microstructure and strength. *Constr Build Mater* 198:445–455. https://doi.org/10.1016/j.conbuildmat.2018.10.235

Dedeloudis C, Zervaki M, Sideris K, et al (2018) Natural Pozzolans. *Proper Fresh Hard Concr Contain Suppl Cement Mater*: 181–231. https://doi.org/10.1007/978-3-319-70606-1_6

Dill HG (2016) Kaolin: Soil, rock and ore. *Earth Sci Rev* 161:16–129. https://doi.org/10.1016/j.earscirev.2016.07.003

Duchesne J (2021) Alternative supplementary cementitious materials for sustainable concrete structures: A review on characterization and properties. *Waste Biomass Valorization* 12:1219–1236. https://doi.org/10.1007/s12649-020-01068-4

El-Diadamony H, Amer AA, Sokkary TM, El-Hoseny S (2018) Hydration and characteristics of metakaolin pozzolanic cement pastes. *HBRC J* 14:150–158. https://doi.org/10.1016/j.hbrcj.2015.05.005

Elaqra H, Rustom R (2018) Effect of using glass powder as cement replacement on rheological and mechanical properties of cement paste. *Constr Build Mater* 179:326–335. https://doi.org/10.1016/j.conbuildmat.2018.05.263

Emetere ME, Dania E (2019) Short review on air pollution from cement factories. *J Phys Conf Ser* 1299:012033. https://doi.org/10.1088/1742-6596/1299/1/012033

Fapohunda C, Akinbile B, Shittu A (2017) Structure and properties of mortar and concrete with rice husk ash as partial replacement of ordinary Portland cement – A review. *Int J Sustain Built Environ* 6:675–692. https://doi.org/10.1016/j.ijsbe.2017.07.004

Farahani A, Taghaddos H, Shekarchi M (2015) Prediction of long-term chloride diffusion in silica fume concrete in a marine environment. *Cem Concr Compos* 59:10–17. https://doi.org/10.1016/j.cemconcomp.2015.03.006

Faraj RH, Sherwani AFH, Jafer LH, Ibrahim DF (2021) Rheological behavior and fresh properties of self-compacting high strength concrete containing recycled PP particles with fly ash and silica fume blended. *J Build Eng* 34:101667. https://doi.org/10.1016/j.jobe.2020.101667

Fernández, García Calvo JL, Alonso MC (2018) Ordinary Portland Cement composition for the optimization of the synergies of supplementary cementitious materials of ternary binders in hydration processes. *Cem Concr Compos* 89:238–250. https://doi.org/10.1016/j.cemconcomp.2017.12.016

Franco de Carvalho JM, Melo TV de, Fontes WC, et al (2019) More eco-efficient concrete: An approach on optimization in the production and use of waste-based supplementary cementing materials. *Constr Build Mater* 206:397–409. https://doi.org/10.1016/j.conbuildmat.2019.02.054

Gilfillan D, Marland G (2021) CDIAC-FF: Global and national CO2 emissions from fossil fuel combustion and cement manufacture: 1751–2017. *Earth Syst Sci Data* 13:1667–1680. https://doi.org/10.5194/essd-13-1667-2021

Güllü H, Canakci H, Al Zangana IF (2017) Use of cement based grout with glass powder for deep mixing. *Constr Build Mater* 137:12–20. https://doi.org/10.1016/j.conbuildmat.2017.01.070

Gursel AP, Maryman H, Ostertag C (2016) A life-cycle approach to environmental, mechanical, and durability properties of "green" concrete mixes with rice husk ash. *J Clean Prod* 112:823–836. https://doi.org/10.1016/j.jclepro.2015.06.029

Hamid Z, Rafiq S (2021) An experimental study on behavior of wood ash in concrete as partial replacement of cement. *Mater Today Proc* 46:3426–3429. https://doi.org/10.1016/j.matpr.2020.11.776

He J, Kawasaki S, Achal V (2020) The utilization of agricultural waste as agro-cement in concrete: A review. *Sustainability* 12:6971. https://doi.org/10.3390/su12176971

Hemalatha T, Ramaswamy A (2017) A review on fly ash characteristics – Towards promoting high volume utilization in developing sustainable concrete. *J Clean Prod* 147:546–559. https://doi.org/10.1016/j.jclepro.2017.01.114

Herath C, Gunasekara C, Law DW, Setunge S (2020) Performance of high volume fly ash concrete incorporating additives: A systematic literature review. *Constr Build Mater* 258:120606. https://doi.org/10.1016/j.conbuildmat.2020.120606

Hu L, He Z, Zhang S (2020) Sustainable use of rice husk ash in cement-based materials: Environmental evaluation and performance improvement. *J Clean Prod* 264:121744. https://doi.org/10.1016/j.jclepro.2020.121744

Islam GMS, Rahman MH, Kazi N (2017) Waste glass powder as partial replacement of cement for sustainable concrete practice. *Int J Sustain Built Environ* 6:37–44. https://doi.org/10.1016/j.ijsbe.2016.10.005

Jahanzaib Khalil M, Aslam M, Ahmad S (2021) Utilization of sugarcane bagasse ash as cement replacement for the production of sustainable concrete – A review. *Constr Build Mater* 270:121371. https://doi.org/10.1016/j.conbuildmat.2020.121371

Jamil M, Khan MNN, Karim MR, et al (2016) Physical and chemical contributions of rice husk ash on the properties of mortar. *Constr Build Mater* 128:185–198. https://doi.org/10.1016/j.conbuildmat.2016.10.029

Jaskulski R, Jóźwiak-Niedźwiedzka D, Yakymechko Y (2020) Calcined clay as supplementary cementitious material. *Materials (Basel)* 13:1–36. https://doi.org/10.3390/ma13214734

Jiang Y, Ling TC, Mo KH, Shi C (2019) A critical review of waste glass powder – Multiple roles of utilization in cement-based materials and construction products. *J Environ Manage* 242:440–449. https://doi.org/10.1016/j.jenvman.2019.04.098

Juenger MCG, Siddique R (2015) Recent advances in understanding the role of supplementary cementitious materials in concrete. *Cem Concr Res* 78:71–80. https://doi.org/10.1016/j.cemconres.2015.03.018

Kadri EH, Aggoun S, De Schutter G, Ezziane K (2010) Combined effect of chemical nature and fineness of mineral powders on Portland cement hydration. *Mater Struct Constr* 43:665–673. https://doi.org/10.1617/s11527-009-9519-6

Khan MI, Usman M, Rizwan SA, Hanif A (2019) Self-consolidating lightweight concrete incorporating limestone powder and fly ash as supplementary cementing material. *Materials (Basel)* 12. https://doi.org/10.3390/ma12183050

Khandelwal S, Rhee KY (2022) Evaluation of pozzolanic activity, heterogeneous nucleation, and microstructure of cement composites with modified bentonite clays. *Constr Build Mater* 323:126617. https://doi.org/10.1016/j.conbuildmat.2022.126617

Khatib JM, Baalbaki O, ElKordi AA (2018) Metakaolin. *Waste Suppl Cem Mater Concr Characterisation Prop Appl* 493–511. https://doi.org/10.1016/B978-0-08-102156-9.00015-8

Krishna NK, Sandeep S, Mini KM (2016) Study on concrete with partial replacement of cement by rice husk ash. *IOP Conf Ser Mater Sci Eng* 149:012109. https://doi.org/10.1088/1757-899X/149/1/012109

Kutchko B, Kim A (2006) Fly ash characterization by SEM–EDS. *Fuel* 85:2537–2544. https://doi.org/10.1016/j.fuel.2006.05.016

Lee CY, Lee HK, Lee KM (2003) Strength and microstructural characteristics of chemically activated fly ash–cement systems. *Cem Concr Res* 33:425–431. https://doi.org/10.1016/S0008-8846(02)00973–0

Lei D-Y, Guo L-P, Sun W, et al (2016) A new dispersing method on silica fume and its influence on the performance of cement-based materials. *Constr Build Mater* 115:716–726. https://doi.org/10.1016/j.conbuildmat.2016.04.023

Lemonis N, Tsakiridis PE, Katsiotis NS, et al (2015) Hydration study of ternary blended cements containing ferronickel slag and natural pozzolan. *Constr Build Mater* 81:130–139. https://doi.org/10.1016/j.conbuildmat.2015.02.046

Letelier V, Tarela E, Muñoz P, Moriconi G (2017) Combined effects of recycled hydrated cement and recycled aggregates on the mechanical properties of concrete. *Constr Build Mater* 132:365–375. https://doi.org/10.1016/j.conbuildmat.2016.12.010

Li G, Zhou C, Ahmad W, et al (2022) Fly ash application as supplementary cementitious material: A review. *Materials (Basel)* 15:1–23. https://doi.org/10.3390/ma15072664

Lothenbach B, Le Saout G, Gallucci E, Scrivener K (2008) Influence of limestone on the hydration of Portland cements. *Cem Concr Res* 38:848–860. https://doi.org/10.1016/j.cemconres.2008.01.002

Lothenbach B, Scrivener K, Hooton RD (2011) Supplementary cementitious materials. *Cem Concr Res* 41:1244–1256. https://doi.org/10.1016/j.cemconres.2010.12.001

Marchon D, Flatt RJ (2016) Mechanisms of cement hydration. *In: Science and Technology of Concrete Admixtures*. Elsevier, pp 129–145.

Martínez-García R, Jagadesh P, Zaid O, et al (2022) The present state of the use of waste wood ash as an eco-efficient construction material: A review. *Materials (Basel)* 15:5349. https://doi.org/10.3390/ma15155349

Martirena F, Monzó J (2018) Vegetable ashes as supplementary cementitious materials. *Cem Concr Res* 114:57–64. https://doi.org/10.1016/j.cemconres.2017.08.015

Nguyen QD, Afroz S, Castel A (2020a) Influence of calcined clay reactivity on the mechanical properties and chloride diffusion resistance of limestone calcined clay cement (LC3) concrete. *J Mar Sci Eng* 8:301. https://doi.org/10.3390/jmse8050301

Nguyen QD, Kim T, Castel A (2020b) Mitigation of alkali-silica reaction by limestone calcined clay cement (LC3). *Cem Concr Res* 137:106176. https://doi.org/10.1016/j.cemconres.2020.106176

Omrane M, Kenai S, Kadri E-H, Aït-Mokhtar A (2017) Performance and durability of self compacting concrete using recycled concrete aggregates and natural pozzolan. *J Clean Prod* 165:415–430. https://doi.org/10.1016/j.jclepro.2017.07.139

Omrane M, Rabehi M (2020) Effect of natural pozzolan and recycled concrete aggregates on thermal and physico-mechanical characteristics of self-compacting concrete. *Constr Build Mater* 247:118576. https://doi.org/10.1016/j.conbuildmat.2020.118576

Özbay E, Erdemir M, Durmuş HI (2016) Utilization and efficiency of ground granulated blast furnace slag on concrete properties – A review. *Constr Build Mater* 105:423–434. https://doi.org/10.1016/j.conbuildmat.2015.12.153

Panesar DK (2019) *Supplementary Cementing Materials*. Elsevier Ltd.

Panesar DK, Zhang R (2020a) Performance comparison of cement replacing materials in concrete: Limestone fillers and supplementary cementing materials – A review. *Constr Build Mater* 251:118866. https://doi.org/10.1016/j.conbuildmat.2020.118866

Panesar DK, Zhang R (2020b) Performance comparison of cement replacing materials in concrete: Limestone fillers and supplementary cementing materials – A review. *Constr Build Mater* 251:118866. https://doi.org/10.1016/j.conbuildmat.2020.118866

Park S, Wu S, Liu Z, Pyo S (2021) The role of supplementary cementitious materials (Scms) in ultra high performance concrete (uhpc): A review. *Materials (Basel)* 14:1–24. https://doi.org/10.3390/ma14061472

Paul SC, Mbewe PBK, Kong SY, Šavija B (2019) Agricultural solid waste as source of supplementary cementitious materials in developing countries. *Materials (Basel)* 12. https://doi.org/10.3390/ma12071112

Pillai RG, Gettu R, Santhanam M, et al (2019) Service life and life cycle assessment of reinforced concrete systems with limestone calcined clay cement (LC3). *Cem Concr Res* 118:111–119. https://doi.org/10.1016/j.cemconres.2018.11.019

Pillay DL, Olalusi OB, Kiliswa MW, et al (2022) Engineering performance of metakaolin based concrete. *Clean Eng Technol* 6:100383. https://doi.org/10.1016/j.clet.2021.100383

Provis JL, Palomo A, Shi C (2015) Advances in understanding alkali-activated materials. *Cem Concr Res* 78:110–125. https://doi.org/10.1016/j.cemconres.2015.04.013

Qian D, Yu R, Shui Z, et al (2020) A novel development of green ultra-high performance concrete (UHPC) based on appropriate application of recycled cementitious material. *J Clean Prod* 261:121231. https://doi.org/10.1016/j.jclepro.2020.121231

Rakhimova NR, Rakhimov RZ (2019) Reaction products, structure and properties of alkali-activated metakaolin cements incorporated with supplementary materials – A review. *J Mater Res Technol* 8:1522–1531. https://doi.org/10.1016/j.jmrt.2018.07.006

Richardson IG, Groves GW (1993) The incorporation of minor and trace elements into calcium silicate hydrate (C·S·H) gel in hardened cement pastes. *Cem Concr Res* 23:131–138. https://doi.org/10.1016/0008-8846(93)90143-W

Rodriguez C, Tobon JI (2020) Influence of calcined clay/limestone, sulfate and clinker proportions on cement performance. *Constr Build Mater* 251:119050. https://doi.org/10.1016/j.conbuildmat.2020.119050

Sabir B, Wild S, Bai J (2001) Metakaolin and calcined clays as pozzolans for concrete: A review. *Cem Concr Compos* 23:441–454. https://doi.org/10.1016/S0958-9465(00)00092–5

Sandhu RK, Siddique R (2017) Influence of rice husk ash (RHA) on the properties of self-compacting concrete: A review. *Constr Build Mater* 153:751–764. https://doi.org/10.1016/j.conbuildmat.2017.07.165

Scrivener K, Ouzia A, Juilland P, Kunhi Mohamed A (2019) Advances in understanding cement hydration mechanisms. *Cem Concr Res* 124:105823. https://doi.org/10.1016/j.cemconres.2019.105823

Shafiq N, Hussein AAE, Nuruddin MF, Al Mattarneh H (2018) Effects of sugarcane bagasse ash on the properties of concrete. Proc Inst Civ Eng – Eng Sustain 171:123–132. https://doi.org/10.1680/jensu.15.00014

Sharma M, Bishnoi S, Martirena F, Scrivener K (2021) Limestone calcined clay cement and concrete: A state-of-the-art review. *Cem Concr Res* 149:106564. https://doi.org/10.1016/j.cemconres.2021.106564

Shi T, Liu Y, Zhang Y, et al (2022) Calcined attapulgite clay as supplementary cementing material: Thermal treatment, hydration activity and mechanical properties. *Int J Concr Struct Mater* 16. https://doi.org/10.1186/s40069-022-00499-8

Shui Z, Xuan D, Chen W, et al (2009) Cementitious characteristics of hydrated cement paste subjected to various dehydration temperatures. *Constr Build Mater* 23:531–537. https://doi.org/10.1016/j.conbuildmat.2007.10.016

Shukla A, Gupta N, Gupta A, et al (2020) Natural pozzolans a comparative study: A review. *IOP Conf Ser Mater Sci Eng* 804:012040. https://doi.org/10.1088/1757-899X/804/1/012040

Siddique R (2012) Utilization of wood ash in concrete manufacturing. *Resour Conserv Recycl* 67:27–33. https://doi.org/10.1016/j.resconrec.2012.07.004

Siddique R, Klaus J (2009) Influence of metakaolin on the properties of mortar and concrete: A review. *Appl Clay Sci* 43:392–400. https://doi.org/10.1016/j.clay.2008.11.007

Sigvardsen NM, Geiker MR, Ottosen LM (2021) Reaction mechanisms of wood ash for use as a partial cement replacement. *Constr Build Mater* 286:122889. https://doi.org/10.1016/j.conbuildmat.2021.122889

Sigvardsen NM, Kirkelund GM, Jensen PE, et al (2019) Impact of production parameters on physiochemical characteristics of wood ash for possible utilisation in cement-based materials. *Resour Conserv Recycl* 145:230–240. https://doi.org/10.1016/j.resconrec.2019.02.034

Silva LHP, Tamashiro JR, Guedes de Paiva FF, et al (2021) Bamboo leaf ash for use as mineral addition with Portland cement. *J Build Eng* 42:1–9. https://doi.org/10.1016/j.jobe.2021.102769

Song H, Jeong Y, Bae S, et al (2018) A study of thermal decomposition of phases in cementitious systems using HT-XRD and TG. *Constr Build Mater* 169:648–661. https://doi.org/10.1016/j.conbuildmat.2018.03.001

Sua-iam G, Makul N (2015) Rheological and mechanical properties of cement–fly ash self-consolidating concrete incorporating high volumes of alumina-based material as fine aggregate. *Constr Build Mater* 95:736–747. https://doi.org/10.1016/j.conbuildmat.2015.07.180

Supit SWM, Shaikh FUA (2015) Durability properties of high volume fly ash concrete containing nano-silica. *Mater Struct Constr* 48:2431–2445. https://doi.org/10.1617/s11527-014-0329-0

Tavares LRC, Junior JFT, Costa LM, et al (2020) Influence of quartz powder and silica fume on the performance of Portland cement. *Sci Rep* 10:21461. https://doi.org/10.1038/s41598-020-78567-w

Tayeh BA, Al Saffar DM, Aadi AS, Almeshal I (2020) Sulphate resistance of cement mortar contains glass powder. *J King Saud Univ Eng Sci* 32:495–500. https://doi.org/10.1016/j.jksues.2019.07.002

Taylor HFW (1993) Nanostructure of C·S·H: Current status. *Adv Cem Based Mater* 1:38–46. https://doi.org/10.1016/1065-7355(93)90006-A

Thomas BS, Yang J, Bahurudeen A, et al (2021a) Sugarcane bagasse ash as supplementary cementitious material in concrete – A review. *Mater Today Sustain* 15:100086. https://doi.org/10.1016/j.mtsust.2021.100086

Thomas BS, Yang J, Mo KH, et al (2021b) Biomass ashes from agricultural wastes as supplementary cementitious materials or aggregate replacement in cement/geopolymer concrete: A comprehensive review. *J Build Eng* 40:102332. https://doi.org/10.1016/j.jobe.2021.102332

Tran YT, Lee J, Kumar P, et al (2019) Natural zeolite and its application in concrete composite production. *Compos Part B Eng* 165:354–364. https://doi.org/10.1016/j.compositesb.2018.12.084

Vizcaíno Andrés LM, Antoni MG, Alujas Diaz A, et al (2015) Effect of fineness in clinker-calcined clays-limestone cements. *Adv Cem Res* 27:546–556. https://doi.org/10.1680/jadcr.14.00095

Wang J, Che Z, Zhang K, et al (2023) Performance of recycled aggregate concrete with supplementary cementitious materials (fly ash, GBFS, silica fume, and metakaolin): Mechanical properties, pore structure, and water absorption. *Constr Build Mater* 368:130455. https://doi.org/10.1016/j.conbuildmat.2023.130455

Wang J, Xiao J, Zhang Z, et al (2021) Action mechanism of rice husk ash and the effect on main performances of cement-based materials: A review. *Constr Build Mater* 288:123068. https://doi.org/10.1016/j.conbuildmat.2021.123068

Wang Y, Zhang S, Niu D, et al (2019) Effects of silica fume and blast furnace slag on the mechanical properties and chloride ion distribution of coral aggregate concrete. *Constr Build Mater* 214:648–658. https://doi.org/10.1016/j.conbuildmat.2019.04.149

Wardhono A (2018) Comparison study of class F and class C fly ashes as cement replacement material on strength development of non-cement mortar. *IOP Conf Ser Mater Sci Eng* 288:012019. https://doi.org/10.1088/1757-899X/288/1/012019

Weber J, Gadermayr N, Kozłowski R, et al (2007) Microstructure and mineral composition of Roman cements produced at defined calcination conditions. *Mater Charact* 58:1217–1228. https://doi.org/10.1016/j.matchar.2007.04.025

Wianglor K, Sinthupinyo S, Piyaworapaiboon M, Chaipanich A (2017) Effect of alkali-activated metakaolin cement on compressive strength of mortars. *Appl Clay Sci* 141:272–279. https://doi.org/10.1016/j.clay.2017.01.025

Wu N, Ji T, Huang P, et al (2022) Use of sugar cane bagasse ash in ultra-high performance concrete (UHPC) as cement replacement. *Constr Build Mater* 317:125881. https://doi.org/10.1016/j.conbuildmat.2021.125881

Xu X, Lin X, Pan X, et al (2020) Influence of silica fume on the setting time and mechanical properties of a new magnesium phosphate cement. *Constr Build Mater* 235:117544. https://doi.org/10.1016/j.conbuildmat.2019.117544

Xu Z, Li W, Sun J, et al (2017) Research on cement hydration and hardening with different alkanolamines. *Constr Build Mater* 141:296–306. https://doi.org/10.1016/j.conbuildmat.2017.03.010

Yogitha B, Karthikeyan M, Muni Reddy MG (2020) Progress of sugarcane bagasse ash applications in production of Eco-Friendly concrete – Review. *Mater Today Proc* 33:695–699. https://doi.org/10.1016/j.matpr.2020.05.814

Zareei SA, Ameri F, Dorostkar F, Ahmadi M (2017) Rice husk ash as a partial replacement of cement in high strength concrete containing micro silica: Evaluating durability and mechanical properties. *Case Stud Constr Mater* 7:73–81. https://doi.org/10.1016/j.cscm.2017.05.001

Zhan P, He Z, Ma Z, et al (2020) Utilization of nano-metakaolin in concrete: A review. *J Build Eng* 30:101259. https://doi.org/10.1016/j.jobe.2020.101259

Zhang T, Tian W, Guo Y, et al (2019) The volumetric stability, chloride binding capacity and stability of the Portland cement-GBFS pastes: An approach from the viewpoint of hydration products. *Constr Build Mater* 205:357–367. https://doi.org/10.1016/j.conbuildmat.2019.02.026

Zhang Z, Zhang B, Yan P (2016a) Comparative study of effect of raw and densified silica fume in the paste, mortar and concrete. *Constr Build Mater* 105:82–93. https://doi.org/10.1016/j.conbuildmat.2015.12.045

Zhang ZH, Zhu HJ, Zhou CH, Wang H (2016b) Geopolymer from kaolin in China: An overview. *Appl Clay Sci* 119:31–41. https://doi.org/10.1016/j.clay.2015.04.023

Zhao Q, Pang L, Wang D (2022) Adverse effects of using metallurgical slags as supplementary cementitious materials and aggregate: A review. *Materials (Basel)* 15. https://doi.org/10.3390/ma15113803

12 Synergistic Advancements in Biopolymer-Ceramic Composites and Their Applications

Akhila Mathew, Siju C. R., and Jijo V. J.

12.1 INTRODUCTION

Ceramics gained attention over the last decades owing to their versatile properties and found applications in medical, environmental, energy storage, automotive, and industrial fields. They are fabricated mainly from silica, alumina, zirconia, zeolite, and titanium dioxide. Single-component materials may not be able to meet all the requirements for the aforementioned applications. Hence, polymers are combined with them to enhance their properties and widen the applicability of ceramics. Generally, ceramics are brittle, and thus combining with polymers provides a way to overcome this drawback. The polymer component improves the mechanical properties and provides toughness and flexibility, and the ceramic part imparts heat resistance, strength, and stiffness to materials. Compared to other composites such as ceramic-ceramic composites, metal-matrix composites, and others, polymer-ceramic composites are known for their fracture toughness, wear resistance, fatigue resistance, higher hardness, and so on (Kumar et al., 2022). For instance, polymer-ceramic composites such as polyvinylidene fluoride (PVDF)-ceramic composites have found application in energy storage. Polymer-ceramics are now considered as good dielectric materials due to the easy processability, flexibility, and high dielectric strength of polymers and the high dielectric constant of ceramics.

The combination of organic polymers and ceramics leads to materials that offer a unique combination of properties (Arbatti et al., 2007). Owing to the good pyroelectric and dielectric performance of polyvinylidene fluoride (PVDF) and poly (vinylidene fluoridetrifluoroethylene) [P(VDF-TrFE)], they are used along with ceramics to produce dielectric materials (Chan et al., 1999). Polytetrafluoroethylene (PTFE) is another example of a polymer that is used along with ceramics to obtain materials with excellent dielectric and mechanical properties. PTFE has $\varepsilon_r = 2.1$ and $\tan \delta = 10^{-4}$ at 800 MHz and is known for its chemical resistance. However, they have low dielectric constant and coefficient of thermal expansion. PTFE/$CaTiO_3$,

 DOI: 10.1201/9781003731672-12

PTFE/TeO_2, PTFE/$Sr_2Ce_2Ti_5O_{16}$ are some examples of PTFE composites incorporated with various ceramics (Hu et al., 2011; Subod et al., 2007). Investigation of dielectric properties of epoxy resin/$BaTiO_3$, cyanate resin/$BaTiO_3$, and bismaleimide resin/$BaTiO_3$ composites showed enhanced dielectric properties compared with pure $BaTiO_3$ (Chao et al., 2008). Owing to biocompatibility, bioactivity, controlled degradation, and similarity with bones and tissues, polymer-ceramics also find application in the biomedical field (Miyazaki et al., 2016). Compared to biomaterials produced from ceramics or polymers alone, their composites are more suitable due to the combined effect of ceramics and polymers. Polymers can be easily molded, and they closely resemble collagen, a component of bones and tissue. For instance, (S. Nath, 2009) developed high-density polyethylene-Al_2O_3-hydroxyapatite (HAp) composites for biomedical applications. HDPE/20 vol%, HAp/20 vol%, and Al_2O_3 showed an improved hardness of 226.5 MPa and elastic modulus of 6.2 GPa. Polyurethane/calcium phosphate composites, which are used as weight-bearing implants, are another example of polymer-ceramic composite for biomedical applications (Yoshii et al., 2012). However, many of the polymers used for the aforementioned applications are petroleum-derived, non-biodegradable, and non-resorbable and cause severe pollution in the environment. Hence, an alternative to these fossil-derived polymers is highly necessary.

Biopolymers are large macromolecules derived from living organisms such as plants, animals, and microbes. Similar to petroleum-derived polymers, biopolymers are formed by connecting their monomers through covalent bonds (Singh et al., 2021). Based on the preparation of biopolymers, they can be classified as biopolymers directly extracted from natural renewable materials, chemically modified bio-based monomeric or polymeric materials, and those produced from bacterial fermentation (Khalil et al., 2019). Biopolymers are directly extracted from renewable sources such as starch, cellulose, hemicellulose, chitin, lignin, proteins, lipids, nucleic acids, alginate, collagen, gelatin, and so on. Many of the directly extracted biopolymers are from agricultural feedstocks such as corn, soya, wheat, and potatoes. Chemically modified biopolymers comprise cellulose acetate, carboxymethyl cellulose, hydroxypropyl methylcellulose, chitosan, polylactic acid, polyvinyl alcohol, bio-polyethylene, polybutylene succinate, and so on. Biopolymers produced through bacterial fermentation are polyhydroxy alkanoates, bacterial cellulose, xanthan, dextran, pullulan, and so on (Babu et al., 2013; Rao et al., 2014; Balart et al., 2021; Khalil et al., 2019).

Considering the large-scale environmental pollution caused by polymers from fossil fuels, the use of biopolymers has accelerated in recent years because of their environmental compatibility, biodegradability, non-toxicity, and abundance in nature. Moreover, biopolymers pave the way for reducing the consumption of non-renewable resources and replacing the conventional petroleum polymers (Balart et al., 2021). They offer a platform for different applications such as in medical devices, food packaging, environmental, agricultural, and construction fields. In recent years, for the processing and synthesis of ceramics, biopolymers are used as binders, pore formers, foaming agents, adhesives, consolidators, and gelling agents (George et al., 2020). Moreover, in some cases, biopolymers put forward room for enhancing the properties of ceramics by acting as reinforcements in blends and composites. This chapter gives a brief idea about the different uses of biopolymers in the production of ceramics and the application of ceramics incorporated with biopolymers in biomedical, environmental, and energy storage applications.

12.2 BIOPOLYMERS IN CERAMICS

12.2.1 Starch

Starch is a biopolymer found in most plants, consisting of two D-glucopyranose entities – amylose and amylopectin, linked together with α (1→4) and α (1→6) glycosidic bonds. They are abundantly found in wheat, corn, rice, potatoes, tapioca, etc. (Jane, 1995). Starch is often used with ceramics as binders, gelling agents, additives, and pore-forming agents. Since starch can be easily burned during firing due to the presence of C, H, and O, they are used as a pore-forming agent in ceramics. Starch consolidation is one of the methods to prepare porous ceramics. Various ceramics such as alumina and zirconia are prepared using this technique. Lyckfeldt and Ferreira (1998) developed porous alumina ceramics using starch as both consolidator/binder and pore former. Simple and complex porous alumina with ultimate porosities between 23 % and 70 % were fabricated. The average size of the small pores connecting the large pores varied between 0.5 μm and 9.5 μm, which were modulated by total solids loading and starch content in the originally prepared slips. During water processing, ceramics with chemically modified starch showed stable properties, and they have superior dimensional control and regularity than native starch. Pabst et al. (2001) studied slip casting of porous ceramics using starch and gelatin as additives. Through starch consolidation, porous ceramics with total porosities up to 35 % and with pore size comparable to that of dry starch were produced. The starch-consolidated samples were 7 μm in size at interconnections between pores. Starch consolidation casting of oxide ceramics was feasible only for porosity control between 25 % and 50 %. Depending on the type of starch, the pore size of ceramics varied. Ceramics based on potato starch had the largest pore size of 50 μm and higher, while those of corn starch had the smallest pore size (14 μm), and wheat starch had an intermediate size (20 μm) (Gregorová & Pabst, 2007). The size and shape of tapioca starch and corn starch are similar with narrow size distribution. Most of the starch types such as corn, wheat, rice, and tapioca are isometric in nature and have an average aspect ratio of 1, while potato starch is anisometric (Gregorová et al., 2006). Porous alumina ceramics of different porosity made from these starch types were found to fit second-order polynomials and 1/T-type relations. The relation $k_r = \exp(-1.5\phi/(1-\phi))$ describes the thermal conductivity of these porous ceramics, where ϕ is the porosity (Živcová et al., 2009). Gel casting along with a pore-forming agent technique yielded porous alumina ceramics of ultra-high porosity. The addition of starch as a pore-forming agent fostered the particle packing efficiency as well as increased the nominal solid loading. With the increase in starch content from 0 % to 30 %, the porosity of the ceramics increased, and the thermal conductivity decreased. However, the performance of ceramics declined with the addition of starch above 40 % (Li et al., 2013a).

Ceramics are prone to cracking while sintering due to shrinking, binders are used along with it to reduce the cracks. Starch was used by many researchers as binders during the processing of ceramics due to its gelling ability, adhesive properties, and easy-to-burn-out nature. Almeida et al. (2009) studied the effect of starch content and sintering temperature on alumina casting using cassava starch as a consolidator, binder, and pore former. The samples have open porosity in the range of 13–55% based on the amount of starch present in the precursor suspension. Starch was used as a binder and pore former for the production of porous silicon nitride ceramics through extrusion. In the case of ceramics extruded without starch, the extrusion force was very high and increased with time, which then resulted in difficulty to extrude a large amount out of the die. This is because the whole applied stress was taken up by the liquid in the paste, and the separation of liquid from paste occurred which led to the movement of liquid through the particles. However, when starch is used as a binder, the movement of water through Si_3N_4 particles is reduced due to the high chemical affinity of starch for water. Owing to the retention of water in the matrix, the ceramic particles remain lubricated and lead to stable extrusion force (Jiang et al., 2008). Using corn starch as a binder for alumina and calcium carbonate ceramics increased the porosity of the ceramics with an increase in starch content and closed pores in which a few of them were interconnected with each other. Without additives the pore size ranges from 10 μm – 25 μm; for 5 wt %, it is 15–28 μm; for 10 wt %, it ranges from 17–28 μm; and for 15 wt %, it is 20–55 μm (Alag & Zamel, 2018).

12.2.2 Proteins

Proteins are defined as macromolecules with long chains of amino acid residues. Based on the conformation, they are categorized as fibrous and globular proteins (Whitford, 2013). They are widely used to fabricate ceramic foams through protein foaming. Proteins being amphiphilic in nature, they are prone to foaming. The addition of proteins to ceramic slurry results in the absorption of proteins on the air–water interface and stabilizes air bubbles leading to the formation of slurry foam. On heating, proteins in the slurry will gel by denaturing and form a 3D network to stabilize the pore structure (Yin et al., 2009). Using egg yolk both as consolidating and foaming agent for developing porous alumina having a pore size in the range of 100–500 μm was done by Fadli and his co-worker (Fadli & Sopyan, 2009). In this case, the porosity changed from 43.5 % to 50.4 %, and the volume shrinkage of sintered alumina increased from 7.83 % to 29.3 % as the drying temperature increased from 1100 °C to 1800 °C. In addition, the density changed from 2.27 g cm^{-3} at 1100 °C to 1.83 g cm^{-3} at 1800 °C, showing that a lower drying temperature brings about lower foaming capacity and hence reduced pore formation. Fadli et al. in 2011 reported a new protein foaming consolidation method for the preparation of porous alumina ceramics. They again used egg yolk as both a consolidating and foaming agent and studied the change in physical properties of ceramics with initial egg yolk content in the slurry. Ceramics with big pore sizes were obtained resulting from low viscosity and high foaming capacity when the alumina-to-yolk ratio was low. The pore size also depended on the slurry drying temperature proving that high temperature gives big pore size and high foaming capacity. The mean pore size varied from 200–1000 μm with shape changing from spherical to polygonal, and the density between

0.7 g cm^{-3} and 1.8 g cm^{-3} was influenced by the dispersant concentration in the alumina slurry. With the increase in porosity from 40% to 71% the compressive strength was also increased from 1.1–5.7 MPa, respectively (Fadli & Sopyan, 2011). Porous zirconia ceramics with homogeneous pore distribution having thinner pore walls were developed by Zhang et al. (2008) using albumen through microwave heating, while samples obtained through conventional heating had visible cracks and delamination. He et al. (2009) exploited the direct protein foaming method to produce 3D interconnected porous alumina ceramics. They used whey protein isolate (WPI) as a foaming agent and compared it with those prepared by using egg white protein (EWP) as a foaming agent. Even though both ceramic foams have similar density and porosity, they differ in pore size. While EWP samples had small pores with even distribution, WPI samples had large pore sizes. Moreover, EWP foamed alumina ceramics had more compressive strength than WPI foamed ceramics owing to their different foaming stabilities. A comparison of the protein direct foaming method and the sponge replica method was studied by He et al. in 2012 using egg white protein as a foaming agent in both methods. Results indicated that the samples developed through protein foaming are stronger than those obtained by the sponge replica method (He et al., 2012). Yin et al. (2013) showed that egg white protein and fish collagen were used to produce Si_3N_4 ceramic foams. The combinational addition of 2 % fish collagen along with 6 % egg protein resulted in a 23 % increase in pore size than those with 8 % egg white alone. With the increase in fish collagen content, the rate of increase in average pore size increased, while open porosity decreased. In addition, the pores become regular in shape after the addition of fish collagen.

Lyckfeldt et al. (2000) used proteins as binders/gelling agents for aqueous ceramic slurries. They used the globular proteins bovine serum albumin (BSA), albumen (egg white powder), and whey protein concentrate (WPC) with Si_3N_4, Al_2O_3, and ZrO_2 powders for producing ceramics. The resulting slurries demonstrated that WPC showed a faster gelling process, less foam stabilizing ability, least viscosity increase, and fewer cracks during drying. While considering the sintering process, bovine serum albumin showed the best result. On the other hand, a higher amount of albumin in ceramic slurries results in shear thickening behavior; hence, the amount is limited to 5–8 wt%. Egg white as an environment-friendly low-cost binder for ceramics was carried out by Dhara and Bhargava (2001). Unlike other binders such as methacrylamide, ovalbumin did not need any chemical additives for gelation. Based on several parameters such as egg white protein–water ratio, ceramic solid loading, foaming time, and sintering temperature, the porosity and foam structure of the ceramics varied. Highly loaded ceramic slurries with shear thinning behavior were prepared using egg white. The polar nature of egg white made it adsorbed on alumina powder surfaces. Garrn et al. (2004) used bovine serum albumin as binders for producing ceramic foams. Stable protein-ceramic foam structures of 50–300μm cell diameter and 8–20 % relative densities were produced owing to the combination of foaming and increased stiffness.

12.2.3 Polyvinyl Alcohol

Polyvinyl alcohol (PVA) is a synthetic biopolymer that has applications as an additive, binder, dispersing agent, and so on. It belongs to the family of polyvinyl esters and is formed by the hydrolysis of polyvinyl acetate (Baker et al., 2012). Ecsedi et al. (2009) utilized PVA as a binder and pore former for alumina ceramics and found that the porosity of the ceramics varied with PVA content. The addition of PVA led to the production of unsupported membranes of thickness 20–40 μm as crack-free layers. Porosity and pore size of samples increased in the range of 30.9–84.6 % with the addition of PVA having a weight ratio in the range of 1.25–3.75 g PVA/2.5 g Al_2O_3. With the addition of PVA to hydroxyapatite in porous ceramics production through freeze casting, the ceramics resulted have porous walls and were made up of small lamellar pores, while the samples produced without PVA had non-interconnected macroscopic lamellar pores. The morphology and pore size differ with variation in PVA concentration and were of open porosity. Small three-dimensional reticulate pores were formed for 12 wt % PVA. The size of lamellar pores decreased and the porosity increased with an increase in PVA content (Zuo et al., 2010a). Peko et al. (2010) studied the effect of PVA on the microstructure and permeability of freeze-cast porous alumina and found that both Darcian and non-Darcian permeabilities decreased with the addition of PVA and were qualitatively consistent with a decrease in pore size. For samples without PVA, fine secondary dendrites were present on directionally grown ice crystals. In the presence of PVA, constitutional supercooling occurs, and PVA gets gelatinized before the ice crystals start to nucleate and grow (Zuo et al., 2010b).

12.3 ENVIRONMENTAL APPLICATIONS

12.3.1 Removal of Toxic Chemicals, Metals, and Dyes

The presence of toxic chemicals, organic dyes, and heavy metals such as chromium, cadmium, mercury, and arsenic in water bodies is detrimental to the environment and human health and is a serious problem globally. Hence, researchers worldwide have been focused on finding possible solutions to remove these hazardous contaminants. Currently, there are several approaches to get rid of this issue; among these, the use of ceramic membranes and adsorbents has been recognized as an effective method to eliminate these contaminants from aqueous environments. Even though these ceramic membranes and adsorbents are capable of removing metal contaminants and dyes, composite ceramic membranes and adsorbents are more efficient due to the enhanced properties of the composite from the synergistic effect of the constituents. Of these, biopolymer composite ceramics are prominent due to their non-toxic, biodegradable, and biocompatible nature. Combining biopolymers with ceramics provides physical support as well as easy accessibility to metal/dye binding sites. Moreover, these biocomposites can be regenerated through the adsorption/desorption cycle (Boddu et al., 2003). Chitosan, alginate, cellulose etc. are the common biopolymers used along with ceramics for wastewater treatment.

12.3.1.1 Chitosan

Chitosan/Ceramic Composite-Based Adsorbents

Chitosan, a natural amino polysaccharide obtained by the alkaline deacetylation of chitin extracted from crustaceans, fungi, insects, annelids, coelenterates, etc. (Shepherd et al., 1997), is an effective adsorbent. Reactive amino and hydroxyl groups present in chitosan make it suitable for binding metals and dyes (Crini, 2006). However, they may agglomerate or form gels in aqueous solutions when they are used in their natural form. Also, their active binding sites for metals may not be readily available and have low porosity. Hence, coating chitosan over ceramics is a possible solution to overcome these limitations. By coating chitosan on alumina, more binding sites are exposed, and porosity is increased, thereby increasing the adsorption capacity. Boddu et al. (2008b) showed that chitosan-coated alumina has more adsorption capacity than activated alumina as well as chitosan separately. The adsorption capacity of the composite is 56.5 mg/g and 93.5 mg/g for As (III) and As (V), respectively, while for activated alumina it is 3.5 mg/g and 15.9 mg/g (Lin & Wu, 2001) and for chitosan, it is 0.0 mg/g and 58.0 mg/g (Mcafee et al., 2001). Removal of Cr (VI) was also studied by Boddu et al. (2003), and the composite biosorbent exhibits an increase in surface area, which leads to greater adsorption of hexavalent chromium. The mechanism of metal ion uptake involves electrostatic attraction between the protonated amino groups $\left(NH_3^+\right)$ of chitosan at lower pH and the anionic metal species, followed by adsorption and chelation. Cu (II) and Ni (II) ions were also removed with the composite chitosan biosorbent (Boddu et al., 2008a). To enhance the stability of the resulting composite, oxalic acid is used as the cross-linking agent between chitosan and alumina. Through ester linkage between alumina and one carboxyl group, oxalic acid forms a surface chelate, while with the other carboxyl group, it forms an ionic bond with the amine group of chitosan.

Activated alumina-chitosan composite cross-linked with oxalic acid was also used for defluoridation from an aqueous solution. The results showed an increased defluoridation than those with alumina and chitosan separately and are due to the combined effect of chitosan and alumina. Adsorption sites of F – are Al-O of alumina and -NH of chitosan (Li et al., 2013). Calderon et al. removed anionic azo dye Reactive Red (RR195) from wastewater using the oxalic acid-chitosan/alumina ceramic biocomposite. The electrostatic interaction between the amino group of chitosan and ionic sulfonated groups of dye led to the adsorption of dye on the composite.

Chitosan/Ceramic Composite-Based Membranes

Ceramic membranes are widely used for the removal of heavy metals and oils from aqueous solutions. They are known for their excellent corrosion resistance, high-temperature applicability, stability in the high-pressure application, low fouling, long lifespan, and high flux (He et al., 2019). Compared with ceramic membranes of zirconia, alumina, silica etc., biopolymer-incorporated membranes of these ceramics are more promising for eliminating contaminants from wastewater due to their improved removal efficiency. Jana et al. (2011) synthesized chitosan-based ceramic ultrafiltration membranes using clay and kaolin for the removal of mercury and arsenic from wastewater. For the separation of these metal ions, they used a polymer-enhanced ultrafiltration technique using polyvinyl alcohol as the chelating agent. On coating cross-linked chitosan over the ceramic support, the pore size of the membranes decreased and was suitable for ultrafiltration. By virtue of the decreased pore size, the viscous flux of the membranes decreased, and the Knudsen flux increased. These membranes showed chemical stability and are less prone to corrosion (Jana et al., 2012). Removal of Cr (VI) ions with chitosan-coated ceramic membranes was greater compared with the membranes comprising ceramics alone. They were known for their good mechanical stability and ultrafine structure. On analyzing the surface area, pore volume and pore size of the kaolin/clay ceramic membranes and chitosan-coated kaolin/clay ceramic membranes, the surface area and pore volume of the chitosan-coated membrane increased compared to the ceramic membrane, and the pore size decreased (Bat-Amgalan et al., 2022). After being coated with chitosan, volcanic ash ceramic membranes exhibited good mechanical strength. In addition, the composite membranes were stable in water, indicating they were not destroyed when the filtration process went on. The molecular weight of chitosan plays a significant role in permeate flux and swelling degree. The higher the molecular weight, the lower the permeate flux and swelling degree (Apriyanti et al., 2018).

Chitosan, being a suitable matrix for separation membranes, Mukherjee et al. (2018) incorporated chitosan as an intermediate matrix in graphene oxide-alumina ceramic membranes, employed in the membrane bioreactor process. They developed this membrane for atrazine removal, and the resulting membranes showed less accumulation of foulants and increased hydrophilicity. Cross-linking chitosan with graphene oxide enhanced the interfacial adhesion and imparted stability to the membranes. The film-forming capacity was also improved due to the presence of hydroxyl and amine groups of chitosan.

12.3.1.2 Alginate

Alginate is a biopolymer typically extracted from brown seaweeds. They are unbranched 1–4 linked glycuronans comprising copolymers of β-D-mannuronate and its C5 epimer α-L-guluronate (Gacesa, 1988). Taking advantage of the existence of carboxyl and hydroxyl groups in alginate, they are suitable for the removal of heavy metals. Integrating alginate with ceramics intensifies the adsorption ability of the ceramics in wastewater treatment. Athanasekou et al. (2009) modified tubular nanofiltration membranes made of multilayer silica, zirconia, and alumina with alginate for separating Cd^{2+} ions. Among these, silica membranes showed higher retention efficiency, and alginate-modified membranes showed higher retention than the unmodified ones. They modified the membranes with both the low molecular weight (MW) and high molecular weight (MW) alginates. Since the high MW alginates were deposited only in the larger pores and the low MW alginates could penetrate into all pores, Cd^{2+} removal with low MW-modified membranes is more efficient. Hybrid sorbents of silica and alumina ceramics chemically modified with alginate exhibited enhanced removal of Cd^{2+} about 4.5 times higher than the alginate beads. Silane-alginate modification on the porous alumina produced sorbents with improved stability (Athanasekou et al., 2010). For the removal of radioactive Co (II) ions, sodium alginate-hydroxyapatite-CNT nanocomposite beads were developed by Karkeh-Abadi et al. (2016). The adsorption capacities varied with time, pH, amount of adsorbent, temperature, and initial concentration of adsorbate. Adsorption occurred because of the electrostatic attraction between the hydroxyl groups, phosphate groups, donor of electrons in CNT, and the positively charged Co (II) ions, which acts as an electron acceptor. The adsorption process followed a pseudo second-order model and fits in Freundlich isotherm. Grafting alginates on ultrafiltration and nanofiltration ceramic membranes for wastewater treatment revealed a 25–60 % of enhancement in Cd^{2+} binding capacity and increased stability. Only 46 % of alginate was released during the Cd^{2+} removal by this membrane while the physically modified showed 72 % release. Moreover, the grafted membranes showed a 180% improvement in retention efficiency than the unmodified membrane (Athanasekou et al., 2010a). Brandes et al. (2014) fabricated free-shapable ceramic membranes for micro and ultrafiltration having adjustable pore size through gel casting and using calcium iodate as a cross-linking agent. Compared with alumina ceramics of similar pore size which are commercially available, the resulting membranes are chemically and mechanically stable which is attributed to the strong and irreversible binding of alumina with alginates and Ca^{2+} ions. Moreover, the hydroxyl groups of alumina and carboxyl groups of alginates formed covalent bonds. Bin Bandar et al. (2021) used sodium alginate as a binder in ceramic membranes for desalination. Vacuum membrane distillation showed an average flux of 13.10 kg/m^2 h with 98.96 % rejection for raw well water, and it showed a stable performance for 4h.

12.3.1.3 Cellulose

Cellulose is the most abundant natural biopolymer, and it has been extensively used along with ceramics for purification of water. Kannan et al. (2014) synthesized cellulose-strontium titanate ceramic biocomposite for the removal of Cr (VI). It was found that the composites could quantitatively (99.5%) remove Cr (VI). Cellulose ceramic membranes for removal of Cr (VI) and Pb (II) from contaminated water were synthesized by Choudhury et al. (2018). They prepared a novel clay alumina ceramic membrane comprising hydroxyethyl cellulose and CuO nanoparticles. The resulting composite membranes showed higher water permeability as well as metal ion rejection efficiency. The incorporation of CuO nanoparticles into the biopolymer enhanced the stability of the membranes and reduced the pore diameter toward the high-pressure ultrafiltration ceramic substrate.

Cellulose acetate-ceramic composite membranes from low-cost kaolin using the dip coating method were prepared by Nandi et al. (2009). The composite membranes with 2 wt % and 4 wt % of cellulose acetate were appropriate for microfiltration, while the composites with compositions 6 wt % and 8 wt % were capable of ultrafiltration applications. Kaur et al. (2018) studied the influence of pH and temperature on the dip coating of cellulose acetate ceramic composite membranes for ultrafiltration. The membranes developed at neutral pH, that is, pH ≈ 7, were suitable for the removal of oil from water, and these membranes showed optimum results of pore size 35 nm and porosity 47 %. At pH 2–6, the polymer layers are weakly bonded, and at pH above 6, they are strongly bonded to the ceramic. 99.6 % oil rejection at 2 bar transmembrane pressure was observed for 100 mg/L solution. The studies revealed that with an increase in pH, the pore size, porosity, and permeability decreased. For this polymer-ceramic composite membrane, Knudsen flux was dominant, while viscous flux was dominant for ceramic support. Cellulose acetate-coated ceramic membranes being hydrophilic show antifouling properties and hence can be used for oil-water separation (Mittal et al., 2011). The membranes had an effective pore size of 28 nm and a porosity of 0.56. Separation of oil-water emulsions at a lower concentration of oil was studied, and at an initial oil concentration of 200 mg/L at 138 kPa maximum rejection of 93 % was attained. Higher flux decline resulted from high pressure and high initial oil concentration.

12.3.2 Microbiological Purification

Diseases caused by microbes present in water and air are increasing day by day. Low-cost water filters for microbial purification become essential. Ceramic filters were a good choice, since they are very cheap and efficient (Farrow et al., 2018). However, to improve the antimicrobial properties, surface modification of ceramic filters was done. Chitosan is known for its antiviral, antifungal, and antimicrobial properties as well as its non-toxic nature. They were used with ceramics for microbiological purification of wastewater (Alshahrani et al., 2020). Xu et al. (2020) functionalized clay-based ceramic disk filters with chitosan/TiO_2 nanocomposites for removing bacteria from drinking water.

12.3.3 Carbon Dioxide Capture

The level of carbon dioxide in the atmosphere has increased in recent years and has become a serious environmental concern. A diverse range of methods are available for reducing the level of carbon dioxide which involves carbon dioxide capture and storage. Use of ceramic materials such as zeolites and microporous silica has proven to be a good choice for carbon dioxide capture (Drioli et al., 2011). However, it is found that combining basicity enhances the adsorption of carbon dioxide by ceramic materials due to the acidic nature of carbon dioxide (Chatti et al., 2009). Using biopolymers is an excellent method to impart basicity to ceramics, and hence Thote et al. (2012) prepared N-doped mesoporous alumina using chitosan as a biopolymer template for carbon dioxide capture. They reported that the corresponding mesoporous alumina showed higher adsorption of carbon dioxide than commercially available mesoporous alumina because of the presence of nitrogen from the chitosan template.

12.3.4 Humidity Sensors

Recently, the demand for humidity sensors has significantly increased in various fields of application – instrumentation, industrial processing, agriculture, environmental control, domestic applications, etc. (Chen & Lu, 2005). Humidity sensors based on ceramic materials and polymers are extensively used in the aforementioned applications. Ceramic humidity sensors are environmentally adaptable and chemically and thermally stable, while polymer humidity sensors have high sensitivity and small humidity hysteresis (Wang et al., 2002). Hence, several works utilized the advantages of both ceramics and polymers in humidity sensing. Herein, we discuss sensors comprising biopolymers and ceramic materials.

Hashim and Hadi (2018a) integrated polyvinyl alcohol/carboxymethyl cellulose blend with niobium carbide and titanium carbide for humidity sensing. PVA is known for its high storage capacity, dielectric strength, and electrical properties. PVA on contact with humidity attracts -OH groups, which in turn decrease the resistance and make it suitable for application in humidity sensors (Ahmed et al., 2019). With the increase in the concentration of niobium carbide and titanium carbide, the electrical conductivity, absorbance, absorption coefficient, extinction coefficient, dielectric constants, and optical conductivity of the composite were increased. On the other hand, transmittance and energy bandgap were decreased. Moreover, the composites showed a high sensitivity for relative humidity, enabling the composite as a potential material for humidity sensors (Hashim & Hadi, 2018b). Similarly, Ahmed et al. synthesized PVA/TiO_2 nanocomposites and doped them with SiC to obtain humidity sensors with high sensitivity, corrosion resistance, and flexibility. The sensing mechanism is based on adsorption and desorption of humidity. When the humidity level is low, the uncurling of polymer chains becomes slow and hence the large distance of hoping sites. While at higher humidity, the change in resistance decreases suddenly due to the absorption of vapor by the composites. On absorption of water molecules, hydrogen ions dissociate from the surface and form bonds with surface lattice oxygen atoms to form hydroxyl groups, which in turn release free electrons that lead to electrical conduction (Agool et al., 2017).

12.4 ENERGY STORAGE APPLICATIONS

Energy storage materials are indispensable in this current era due to the sudden exhaustion of fossil fuels. The high dielectric constant and high dielectric permittivity of ceramics made it suitable for application in energy storage devices. However, they are brittle in nature, require high-temperature processing, and have lower dielectric breakdown strength, poor flexibility, and lower voltage coefficient (Zhang et al., 2010; Xie et al., 2013). Incorporating polymers with ceramics was found to be a solution to overcome these disadvantages. Polymers have good processing ability, adaptability for large mechanical forces, mechanical flexibility, and high breakdown strength, making them desirable for energy storage but have a low dielectric constant (Zhu & Wang, 2012). Hence, combining ceramics and polymers results in the production of good energy storage devices. On the other hand, polymers of synthetic origin are non-biodegradable and toxic to the environment; hence, polymer-ceramic composites made of biopolymers gained much attention. Kannan et al. (2014) prepared cellulose-strontium titanate ceramic biocomposite for embedded capacitor application. Depending on the concentration of $SrTiO_3$ filler in the cellulose matrix, the dielectric properties of the resulting biocomposite could be tuned. Wang et al. (2020) developed a composite polymer electrolyte with 3D cellulose/ceramic network having high lithium conductivity and mechanical strength for lithium batteries. Through self-assembly of hybrid cellulose fibers, ceramic nanofillers formed interconnected networks, and the cellulose skeleton supported the whole structure providing mechanical strength and spatial guidance to ceramic fillers. The resulting composite provides 3D continuous Li pathways, and the hydroxyl rich surface provides more Li^+ conduction through surface charge-governed ion transport than cellulose-free composite polymer electrolytes.

Joyce et al. (2014) incorporated hybrid films made of deoxyribonucleic acid (DNA) and cetyltrimethylammonium (CTMA) in ceramics for application in high-voltage capacitors as insulating dielectrics. Utilizing DNA in dielectrics is remarkable for their low cost, lightweight, and malleable to different low-temperature film fabrication processes (Norwood et al., 2009). The resulting hybrid film was stable in low temperature and showed frequency-dependent dielectric properties, which are attributed to the stabilization of the sol-gel network through the interaction of the cation of DNA-CTMA and anionic oxide groups of ceramics. The resulting hybrid films have a dielectric constant greater than that shown by synthetic polymer-based ceramic dielectrics and have high dielectric strength values in the range of 300–350 V/μm. DNA-ceramic hybrid films made of TiO_2 and $BaTiO_3$ showed an enhanced dielectric constant of 14.3 at 1 kHz for 40 % TiO_2 in comparison with DNA. Through the phosphate groups of DNA, binding of TiO_2 nanoparticles to DNA occurs. Analysis of temperature-dependent dielectric constants of both DNA-TiO_2 and DNA-$BaTiO_3$ ceramics showed a similar behavior. With an increase in the content of nanoparticles, an increasingly tighter band is obtained. The DNA-ceramic hybrid films showed DC resistivity higher than that of DNA at a temperature of 20 °C (Venkat et al., 2011).

Polyvinyl alcohol/barium zirconium titanate (BZT) polymer-ceramic composites were prepared, and the studies revealed that dielectric permittivity and dielectric loss increased with an increase in the volume fraction of BZT. Moreover, polyvinyl alcohol is a good insulating material having low electrical conductivity. The excellent dielectric properties of the composite made it a good choice for embedded capacitors. The combined effect of the high dielectric permittivity of ceramic and the dielectric strength of polymer makes the composite an excellent candidate for energy storage devices (Badapanda et al., 2013).

Fan et al. (2015) developed biopolymer nanocomposites with polylactic acid as matrix and core shell structured $BaTiO_3$ (BT) nanoparticles as high-dielectric constant fillers for energy storage. Polylactic acid has high breakdown strength and low electrical conductivity. However, they experience low dielectric constant (Matsugasaki et al., 2013). Hence, making composites provides a platform to overcome the defects of both components. Polydopamine-encapsulated BT (BT@PDA) and polylactic acid (PLA)-encapsulated BT@PDA (BT@PDA@PLA) nanoparticles were produced and compared with BT nanocomposites. The core-shell nanoparticle-based composites have a high dielectric constant, improved breakdown strength, and suppressed dielectric loss tangent than BT nanocomposites. Among the core-shell nanoparticles, BT@PDA@PLA showed even more dielectric constant (Fan et al., 2015). PLA is known for its piezoelectric properties owing to the carbonyl groups present, and it is suitable for energy storage and harvesting. Hanani (2020) designed lead-free ceramic/biopolymer composites with $Ba_{0.85}Ca_{0.15}Zr_{0.1}Ti_{0.9}O_3$ and PLA for energy storage and energy harvesting. The work involved functionalization of BCZT with polydopamine via core-shell structurization and embedding in PLA matrix. Bio-flexible piezoelectric nanogenerators made from this were able to generate an open circuit voltage of 14.4 V and a short circuit current of 0.55 μA. Moreover, it could handle a 1 μF capacitor to store energy of 3.92 μJ within 115 s.

12.5 BIOMEDICAL APPLICATIONS

Advancements in the medical field led to the use of ceramics for the repair and reformation of damaged parts of the human body. Ceramics are resistant to corrosion, have high compression resistance, and are biocompatible. On the other hand, they are brittle, have low fracture strength, are not mechanically reliable, hard to fabricate, and have high density, which makes them unsuitable to be used alone for application in bone replacement. To deal with the drawbacks of ceramics they are used along with polymers (Ramakrishna et al., 2001). Polymer-ceramics made with biopolymers found application in the biomedical field because of their non-toxic nature, biocompatibility, bioactivity, and easy processability. They are used in orthopedic applications, tissue engineering, dental applications, drug delivery systems, in coating and surface modifications. Hydroxyapatite is the most common ceramics used for the aforementioned applications due to its similarity with bones in the human body. It is the major component of mammalian hard tissue, found in abundance, shows osteoconductivity, and is cheap. They are used along with the biopolymers chitosan, cellulose, gelatin, collagen, polylactic acid, and so on to produce biomaterials. Compared to pure hydroxyapatite, polymer-hydroxyapatite blends were found to be easy to mold, handle, and apply during surgery. Chitosan/hydroxyapatite composite is one of the biopolymer ceramic composites which is widely used in the biomedical field (Pighinelli & Kucharska, 2013). Chitosan has good bioactivity, stability to pH variations, is non-antigenic, and is bio-functional, which makes it suitable for bone regeneration and tissue engineering (Kim et al., 2008). Oktay Yildirim (2004) developed chitosan/hydroxyapatite. Chitosan/hydroxyapatite whisker, chitosan/β-tricalciumphosphate/hydroxyapatite composites could be used as replacements for hips, knees, ligaments, teeth, and tendons. They are suitable for only low load-bearing applications. With increase in ceramic content, the density of the composites increased and the porosity decreased. Silk fibroin/hydroxyapatite composites are another biomaterial used for bone regeneration. Silk fibroin is easily processable due to its β-sheet structure and can be utilized to produce hydrogels, membranes, fibers, and microspheres for application in blood vessels, cartilage, bone, and skin (Saleem et al., 2020). Through freeze drying, Liu et al. synthesized nano-hydroxyapatite/silk fibroin scaffolds for bone tissue engineering. MTT assay of the resulting composite showed that they are biocompatible. Cell seeding was easily done due to the interconnected pore structure of the scaffold (Liu et al., 2012). In summary, biopolymer–ceramic composites combine the strength and thermal stability of ceramics with the flexibility, biodegradability, and functional groups of biopolymers, resulting in multifunctional materials with enhanced performance. Biopolymers such as starch, proteins, PVA, chitosan, alginate, and cellulose act as binders, pore formers, and modifiers to improve ceramic processing and tailor porosity. Environmentally, these composites enable efficient adsorption and membrane-based removal of heavy metals, dyes, oils, and microbes. In energy storage, they offer high dielectric strength, tunable permittivity, and better ion transport. Biomedical applications include scaffolds, bone substitutes, and implant coatings, supported by excellent biocompatibility and bioactivity.

REFERENCES

Agool, I. R., Kadhim, K. J., & Hashim, A. (2017). Fabrication of new nanocomposites: (PVA-PEG-PVP) blend-zirconium oxide nanoparticles) for humidity sensors. *International Journal of Plastics Technology*, *21*, 397–403.

Ahmed, H., Abduljalil, H. M., & Hashim, A. (2019). Analysis of structural, optical and electronic properties of polymeric nanocomposites/silicon carbide for humidity sensors. *Transactions on Electrical and Electronic Materials*, *20*(3), 206–217.

Alag, H. K., & Zamel, R. S. (2018). Studying the properties of porous alumina using starch as a binder. *Al-Nahrain Journal of Science*, *21*(3), 112–118.

Almeida, F. A., Botelho, E. C., Melo, F. C. L., Campos, T. M. B., & Thim, G. P. (2009). Influence of cassava starch content and sintering temperature on the alumina consolidation technique. *Journal of the European Ceramic Society*, *29*(9), 1587–1594.

Alshahrani, A. A., Algamdi, M. S., Alsohaimi, I. H., Nghiem, L. D., Tu, K. L., Al-Rawajfeh, A. E., & in het Panhuis, M. (2020). The rejection of mono-and di-valent ions from aquatic environment by MWNT/chitosan buckypaper composite membranes: Influences of chitosan concentrations. *Separation and Purification Technology*, *234*, 116088.

Apriyanti, E., Hadiyanto, H., & Wijayanto, W. (2018). Preparation and characterization of volcanic ash-chitosan composite ceramic membrane for clean water production. *Journal of Environmental Science and Technology*, *3*(11), 112–118.

Arbatti, M., Shan, X., & Cheng, Z. Y. (2007). Ceramic-polymer composites with high dielectric constant. *Advanced Materials*, *19*(10), 1369–1372.

Athanasekou, C. P., Papageorgiou, S. K., Kaselouri, V., Katsaros, F. K., Kakizis, N. K., Sapalidis, A. A., & Kanellopoulos, N. K. (2009). Development of hybrid alginate/ceramic membranes for Cd2+ removal. *Microporous and Mesoporous Materials*, *120*(1–2), 154–164.

Athanasekou, C. P., Romanos, G. E., Kordatos, K., Kasselouri-Rigopoulou, V., Kakizis, N. K., & Sapalidis, A. A. (2010). Grafting of alginates on UF/NF ceramic membranes for wastewater treatment. *Journal of Hazardous Materials*, *182*(1–3), 611–623.

Athanasekou, C. P., Romanos, G. E., Sapalidis, A. A., & Kanellopoulos, N. K. (2010). Ceramic-supported alginate adsorbent for the removal of heavy metal ions. *Adsorption Science & Technology*, *28*(3), 253–266.

Babu, R. P., O'Connor, K., & Seeram, R. (2013). Current progress on bio-based polymers and their future trends. *Progress in Biomaterials*, *2*, 1–16.

Badapanda, T., Senthil, V., Anwar, S., Cavalcante, L. S., Batista, N. C., & Longo, E. (2013). Structural and dielectric properties of polyvinyl alcohol/barium zirconium titanate polymer-ceramic composite. *Current Applied Physics*, *13*(7), 1490–1495.

Baker, M. I., Walsh, S. P., Schwartz, Z., & Boyan, B. D. (2012). A review of polyvinyl alcohol and its uses in cartilage and orthopedic applications. *Journal of Biomedical Materials Research Part B: Applied Biomaterials*, *100*(5), 1451–1457.

Balart, R., Garcia-Garcia, D., Fombuena, V., Quiles-Carrillo, L., & Arrieta, M. P. (2021). Biopolymers from natural resources. *Polymers*, *13*(15), 2532.

Bandar, K. B., Alsubei, M. D., Aljlil, S. A., Darwish, N. B., & Hilal, N. (2021). Membrane distillation process application using a novel ceramic membrane for Brackish water desalination. *Desalination*, *500*, 114906.

Bat-Amgalan, M., Miyamoto, N., Kano, N., Yunden, G., & Kim, H. J. (2022). Preparation and characterization of low-cost ceramic membrane coated with chitosan: Application to the ultrafine filtration of Cr (VI). *Membranes*, *12*(9), 835.

Boddu, V. M., Abburi, K., Randolph, A. J., & Smith, E. D. (2008a). Removal of copper (II) and nickel (II) ions from aqueous solutions by a composite chitosan biosorbent. *Separation Science and Technology*, *43*(6), 1365–1381.

Boddu, V. M., Abburi, K., Talbott, J. L., & Smith, E. D. (2003). Removal of hexavalent chromium from wastewater using a new composite chitosan biosorbent. *Environmental Science & Technology*, *37*(19), 4449–4456.

Boddu, V. M., Abburi, K., Talbott, J. L., Smith, E. D., & Haasch, R. (2008b). Removal of arsenic (III) and arsenic (V) from aqueous medium using chitosan-coated biosorbent. *Water Research*, *42*(3), 633–642.

Brandes, C., Treccani, L., Kroll, S., & Rezwan, K. (2014). Gel casting of free-shapeable ceramic membranes with adjustable pore size for ultra-and microfiltration. *Journal of the American Ceramic Society*, *97*(5), 1393–1401.

Chan, H. L. W., Cheung, M. C., & Choy, C. L. (1999). Study on BaTiO3/P (VDF-TrFE) 0–3 composites. *Ferroelectrics*, *224*(1), 113–120.

Chao, F., Liang, G., Kong, W., Zhang, Z., & Wang, J. (2008). Dielectric properties of polymer/ceramic composites based on thermosetting polymers. *Polymer Bulletin*, *60*, 129–136.

Chatti, R. V., Bansiwal, A. K., Thote, J. A., Kumar, V., Jadhav, P., Lokhande, S. K., Biniwale, R. B., Labhsetwar, N. K., & Rayalu, S. S. (2009). Amine loaded zeolites for carbon dioxide capture: Amine loading and adsorption studies. *Microporous and Mesoporous Materials*, *121*(1–3), 84–89.

Chen, Z., & Lu, C. (2005). Humidity sensors: A review of materials and mechanisms. *Sensor Letters*, *3*(4), 274–295.

Choudhury, P. R., Majumdar, S., Sahoo, G. C., Saha, S., & Mondal, P. (2018). High pressure ultrafiltration CuO/hydroxyethyl cellulose composite ceramic membrane for separation of Cr (VI) and Pb (II) from contaminated water. *Chemical Engineering Journal*, *336*, 570–578.

Crini, G. (2006). Non-conventional low-cost adsorbents for dye removal: A review. *Bioresource Technology*, *97*, 1061–1085.

Dhara, S., & Bhargava, P. (2001). Egg white as an environmentally friendly low-cost binder for gelcasting of ceramics. *Journal of the American Ceramic Society*, *84*(12), 3048–3050.

Drioli, E., Brunetti, A., & Barbieri, G. (2011). Ceramic membranes in carbon dioxide capture: Applications and potentialities. *Advances in Science and Technology*, *72*, 105–118.

Ecsedi, Z., Lazău, I., & Păcurariu, C. (2009). Microstructural analysis of the effects of polyvinyl alcohol content on the porosity of sol-gel derived alumina ceramics. *Microporous and Mesoporous Materials*, *118*(1–3), 453–457.

Fadli, A., & Sopyan, I. (2009). Preparation of porous alumina for biomedical applications through protein foaming-consolidation method. *Materials Research Innovations*, *13*(3), 327–329.

Fadli, A., & Sopyan, I. (2011). Porous ceramics with controllable properties prepared by protein foaming-consolidation method. *Journal of Porous Materials*, *18*, 195–203.

Fan, Y., Huang, X., Wang, G., & Jiang, P. (2015). Core–shell structured biopolymer@ BaTiO3 nanoparticles for biopolymer nanocomposites with significantly enhanced dielectric properties and energy storage capability. *The Journal of Physical Chemistry C*, *119*(49), 27330–27339.

Farrow, C., McBean, E., Huang, G., Yang, A., Wu, Y., Liu, Z., Dai, Z., Fu, H., Cawte, T., & Li, Y. (2018). Ceramic water filters: A point-of-use water treatment technology to remove bacteria from drinking water in Longhai City, Fujian Province, China. *Journal of Environmental Informatics*, *32*(2), 63–68.

Gacesa, P. (1988). Alginates. *Carbohydrate Polymers*, *8*(3), 161–182.

Garrn, I., Reetz, C., Brandes, N., Kroh, L. W., & Schubert, H. (2004). Clot-forming: The use of proteins as binders for producing ceramic foams. *Journal of the European Ceramic Society*, *24*(3), 579–587.

George, A., Sanjay, M. R., Srisuk, R., Parameswaranpillai, J., & Siengchin, S. (2020). A comprehensive review on chemical properties and applications of biopolymers and their composites. *International Journal of Biological Macromolecules*, *154*, 329–338.

Gregorová, E., & Pabst, W. (2007). Porosity and pore size control in starch consolidation casting of oxide ceramics – achievements and problems. *Journal of the European Ceramic Society*, *27*(2–3), 669–672.

Gregorová, E., Pabst, W., & Bohačenko, I. (2006). Characterization of different starch types for their application in ceramic processing. *Journal of the European Ceramic Society*, *26*(8), 1301–1309.

Hanani, Z. (2020). *Design of flexible lead-free ceramic/biopolymer composite for energy storage and energy harvesting applications* (Doctoral dissertation, Université de Bordeaux, Université Cadi Ayyad, Marrakech, Maroc).

Hashim, A., & Hadi, Q. (2018a). Structural, electrical and optical properties of (biopolymer blend/titanium carbide) nanocomposites for low-cost humidity sensors. *Journal of Materials Science: Materials in Electronics*, *29*, 11598–11604.

Hashim, A., & Hadi, Q. (2018b). Synthesis of novel (polymer blend-ceramics) nanocomposites: structural, optical and electrical properties for humidity sensors. *Journal of Inorganic and Organometallic Polymers and Materials*, *28*, 1394–1401.

He, X., Su, B., Tang, Z., Zhao, B., Wang, X., Yang, G., Qiu, H., Zhang, H., & Yang, J. (2012). The comparison of macroporous ceramics fabricated through the protein direct foaming and sponge replica methods. *Journal of Porous Materials*, *19*, 761–766.

He, X., Zhou, X., & Su, B. (2009). 3D interconnective porous alumina ceramics via direct protein foaming. *Materials Letters*, *63*(11), 830–832.

He, Z., Lyu, Z., Gu, Q., Zhang, L., & Wang, J. (2019). Ceramic-based membranes for water and wastewater treatment. *Colloids and Surfaces A: Physicochemical and Engineering Aspects*, *578*, 123513.

Hu, Y., Zhang, Y., Liu, H., & Zhou, D. (2011). Microwave dielectric properties of PTFE/CaTiO3 polymer ceramic composites. *Ceramics International*, *37*(5), 1609–1613.

Jana, S., Purkait, M. K., & Mohanty, K. (2012). Polymer enhanced ultrafiltration of mercury using chitosan impregnated ceramic membrane. *Desalination and Water Treatment*, *37*(1–3), 321–330.

Jana, S., Saikia, A., Purkait, M. K., & Mohanty, K. (2011). Chitosan based ceramic ultrafiltration membrane: Preparation, characterization and application to remove Hg (II) and As (III) using polymer enhanced ultrafiltration. *Chemical Engineering Journal*, *170*(1), 209–219.

Jane, J. (1995). Starch properties, modifications, and applications. *Journal of Macromolecular Science, Part A: Pure and Applied Chemistry*, *32*(4), 751–757.

Jiang, G. P., Yang, J. F., Gao, J. Q., & Niihara, K. (2008). Porous silicon nitride ceramics prepared by extrusion using starch as binder. *Journal of the American Ceramic Society*, *91*(11), 3510–3516.

Joyce, D. M., Venkat, N., Ouchen, F., Singh, K. M., Smith, S. R., Grabowski, C. A., Terry Murray, P., & Grote, J. G. (2014). Deoxyribonucleic acid-based hybrid thin films for potential application as high energy density capacitors. *Journal of Applied Physics*, *115*(11).

Kannan, B. R., Kalidhasan, S., Krishna Kumar, A. S., Rajesh, N., & Venkataraman, B. H. (2014). An integrated use of biopolymer-ceramic composites towards capacitor and environmental application. *Polymer-Plastics Technology and Engineering*, *53*(6), 626–630.

Karkeh-Abadi, F., Saber-Samandari, S., & Saber-Samandari, S. (2016). The impact of functionalized CNT in the network of sodium alginate-based nanocomposite beads on the removal of Co (II) ions from aqueous solutions. *Journal of Hazardous Materials*, *312*, 224–233.

Kaur, H., Bulasara, V. K., & Gupta, R. K. (2018). Influence of pH and temperature of dip-coating solution on the properties of cellulose acetate-ceramic composite membrane for ultrafiltration. *Carbohydrate Polymers*, *195*, 613–621.

Khalil, M. T., Fatima, I., Abbas, F., & Tahir, M. U. (2019). Synthesis of bio based polymers and its applications. *International Journal of Global Science Research*, *2*, 165–184.

Kim, I. Y., Seo, S. J., Moon, H. S., Yoo, M. K., Park, I. Y., Kim, B. C., & Cho, C. S. (2008). Chitosan and its derivatives for tissue engineering applications. *Biotechnology Advances*, *26*(1), 1–21.

Kumar, R., Singh, R., & Hashmi, M. S. J. (2022). Polymer-ceramic composites: A state of art review and future applications. *Advances in Materials and Processing Technologies*, *8*(1), 895–908.

Li, S., Wang, C. A., & Zhou, J. (2013). Effect of starch addition on microstructure and properties of highly porous alumina ceramics. *Ceramics International*, *39*(8), 8833–8839.

Li, W. Y., Liu, J., Chen, H., Deng, Y., Zhang, B., Wang, Z., Zhang, X., & Hong, S. (2013). Application of oxalic acid cross-linking activated alumina/chitosan biocomposites in defluoridation from aqueous solution: Investigation of adsorption mechanism. *Chemical Engineering Journal*, *225*, 865–872.

Lin, T.-F., & Wu, J.-K. (2001). Adsorption of arsenite and arsenate within activated alumina grains: Equilibrium and kinetics. *Water Research*, *35*, 2049–2057.

Liu, L., Liu, J., Wang, M., Min, S., Cai, Y., Zhu, L., & Yao, J. (2012). Preparation and characterization of nano-hydroxyapatite/silk fibroin porous scaffolds. *Journal of Biomaterials Science, Polymer Edition*, *19*(3), 325–338. https://doi.org/10.1163/156856208783721010

Lyckfeldt, O., Brandt, J., & Lesca, S. (2000). Protein forming – a novel shaping technique for ceramics. *Journal of the European Ceramic Society*, *20*(14–15), 2551–2559.

Lyckfeldt, O., & Ferreira, J. M. F. (1998). Processing of porous ceramics by 'starch consolidation'. *Journal of the European Ceramic Society*, *18*(2), 131–140.

Matsugasaki, N., Shinyama, K., & Fujita, S. (2013). Dielectric breakdown and mechanical properties of polylactic acid of different spherulite sizes. *IEEJ Transactions on Electrical and Electronic Engineering*, *8*, S106–S107.

Mcafee, B. J., Gould, W. D., Nadeau, J. C., & da Costa, A. C. (2001). Biosorption of metal ions using chitosan, chitin, and biomass of Rhizopus oryzae. *Separation Science and Technology*, *36*(14), 3207–3222.

Mittal, P., Jana, S., & Mohanty, K. (2011). Synthesis of low-cost hydrophilic ceramic-polymeric composite membrane for treatment of oily wastewater. *Desalination*, *282*, 54–62.

Miyazaki, T., Kawashita, M., & Ohtsuki, C. (2016). Ceramic-polymer composites for biomedical applications. In *Handbook of bioceramics and biocomposites* (pp. 287–300). Springer International Publishing.

Mukherjee, D., Dewanjee, A., Ghosh, S., & Majumdar, S. (2018). Development of graphene oxide/chitosan composite membrane on ceramic support for atrazine remediation by MBR process. *Environmental Science and Pollution Research*, *25*, 33334–33352.

Nandi, B. K., Uppaluri, R., & Purkait, M. K. (2009). Effects of dip coating parameters on the morphology and transport properties of cellulose acetate-ceramic composite membranes. *Journal of Membrane Science*, *330*(1–2), 246–258.

Nath, S., Bodhak, S., & Basu, B. (2009). HDPE-Al_2O_3-HAp composites for biomedical applications: processing and characterizations. *Journal of Biomedical Materials Research Part B: Applied Biomaterials*, *88*(1), 1–11.

Norwood, R. A., DeRose, C. T., Himmelhuber, R., Peyghambarian, N., Wang, J., Li, L., Ouchen, F., & Grote, J. G. (2009, August). Dielectric and electrical properties of sol-gel/DNA blends. In *Nanobiosystems: Processing, characterization, and applications II* (Vol. 7403, pp. 57–67). SPIE.

Pabst, W., Gregorová, E., Havrda, J., & Týanová, E. (2001). Gelatin casting and starch consolidation of alumina ceramics. *Ceramic Materials and Components for Engines*, 587–592.

Peko, C., Groth, B., & Nettleship, I. (2010). The effect of polyvinyl alcohol on the microstructure and permeability of freeze-cast alumina. *Journal of the American Ceramic Society*, *93*(1), 115–120.

Pighinelli, L., & Kucharska, M. (2013). Chitosan-hydroxyapatite composites. *Carbohydrate Polymers*, *93*(1), 256–262.

Ramakrishna, S., Mayer, J., Wintermantel, E., & Leong, K. W. (2001). Biomedical applications of polymer-composite materials: A review. *Composites Science and Technology*, *61*(9), 1189–1224.

Rao, M. G., Bharathi, P., & Akila, R. M. (2014). A comprehensive review on biopolymers. *Scientific Reviews & Chemical Communications*, *4*(2), 61–68.

Saleem, M., Rasheed, S., & Yougen, C. (2020). Silk fibroin/hydroxyapatite scaffold: A highly compatible material for bone regeneration. *Science and Technology of Advanced Materials*, *21*(1), 242–266.

Shepherd, R., Reader, S., & Falshaw, A. (1997). Chitosan functional properties. *Glycoconjugate Journal*, *14*, 535–542.

Singh, R., Gautam, S., Sharma, B., Jain, P., & Chauhan, K. D. (2021). Biopolymers and their classifications. In *Biopolymers and their industrial applications* (pp. 21–44). Elsevier.

Subodh, G., Joseph, M., Mohanan, P., & Sebastian, M. T. (2007). Low dielectric loss polytetrafluoroethylene/TeO2 polymer ceramic composites. *Journal of the American Ceramic Society*, *90*(11), 3507–3511.

Subodh, G., Pavithran, C., Mohanan, P., & Sebastian, M. T. (2007). PTFE/Sr2Ce2Ti5O16 polymer ceramic composites for electronic packaging applications. *Journal of the European Ceramic Society*, *27*(8–9), 3039–3044.

Thote, J. A., Chatti, R. V., Iyer, K. S., Kumar, V., Valechha, A. N., Labhsetwar, N. K., Biniwale, R. B., Yenkie, M. K. N., & Rayalu, S. S. (2012). N-doped mesoporous alumina for adsorption of carbon dioxide. *Journal of Environmental Sciences*, *24*(11), 1979–1984.

Venkat, N., Ouchen, F., Singh, K. M., Smith, S. R., Joyce, D. M., Miller, T., Yaney, P. P., Grote, J. G., & Naik, R. R. (2011, September). Bio-dielectrics based on DNA-ceramic hybrid films for potential energy storage applications. In *Nanobiosystems: Processing, characterization, and applications IV* (Vol. 8103, pp. 98–107). SPIE.

Wang, C., Huang, D., Li, S., Yu, J., Zhu, M., Liu, N., & Lu, Z. (2020). Three-dimensional-percolated ceramic nanoparticles along natural-cellulose-derived hierarchical networks for high Li+ conductivity and mechanical strength. *Nano Letters*, *20*(10), 7397–7404.

Wang, J., Lin, Q., Zhou, R., & Xu, B. (2002). Humidity sensors based on composite material of nano-BaTiO3 and polymer RMX. *Sensors and Actuators B: Chemical*, *81*(2–3), 248–253.

Whitford, D. (2013). *Proteins: Structure and function*. John Wiley & Sons.

Xie, L. Y., Huang, X. Y., Huang, Y. H., Yang, K., & Jiang, P. K. (2013). Core@ double-shell structured BaTiO3-polymer nanocomposites with high dielectric constant and low dielectric loss for energy storage application. *The Journal of Physical Chemistry C*, *117*, 22525–22537.

Xu, Z., Huang, G., An, C., Huang, J., Chen, X., Xin, X., Song, P., Feng, R., & Li, Y. (2020). Low-cost microbiological purification using a new ceramic disk filter functionalized by chitosan/TiO2 nanocomposites. *Separation and Purification Technology*, *248*, 116984.

Yildirim, O. (2004). *Preparation and characterization of chitosan/calcium phosphate based composite biomaterials*. Turkey: Izmir Institute of Technology.

Yin, L., Peng, H. X., Dhara, S., Yang, L., & Su, B. (2009). Natural additives in protein coagulation casting process for improved microstructural controllability of cellular ceramics. *Composites Part B: Engineering*, *40*(7), 638–644.

Yin, L. Y., Zhou, X. G., Yu, J. S., Wang, H. L., Zhao, S., Luo, Z., & Yang, B. (2013). Preparation of Si3N4 ceramic foams by simultaneously using egg white protein and fish collagen. *Ceramics International*, *39*(1), 445–448.

Yoshii, T., Dumas, J. E., Okawa, A., Spengler, D. M., & Guelcher, S. A. (2012). Synthesis, characterization of calcium phosphates/polyurethane composites for weight-bearing implants. *Journal of Biomedical Materials Research Part B: Applied Biomaterials*, *100*(1), 32–40.

Zhang, L., Wang, X., Liu, H., & Yao, X. (2010). Structural and dielectric properties of BaTiO3–CaTiO3–SrTiO3 ternary system ceramics. *Journal of the American Ceramic Society*, *93*(4), 1049–1055.

Zhang, P. Y., Liu, H. Y., Liu, J. C., Fan, W., & Liu, C. (2008). Consolidation of porous ceramics using protein foam by microwave heating. *Key Engineering Materials*, *368*, 1595–1596.

Zhu, L., & Wang, Q. (2012). Novel ferroelectric polymers for high energy density and low loss dielectrics. *Macromolecules*, *45*, 2937–2954.

Živcová, Z., Gregorová, E., Pabst, W., Smith, D. S., Michot, A., & Poulier, C. (2009). Thermal conductivity of porous alumina ceramics prepared using starch as a pore-forming agent. *Journal of the European Ceramic Society*, *29*(3), 347–353.

Zuo, K. H., Zeng, Y. P., & Jiang, D. (2010a). Effect of polyvinyl alcohol additive on the pore structure and morphology of the freeze-cast hydroxyapatite ceramics. *Materials Science and Engineering: C*, *30*(2), 283–287.

Zuo, K. H., Zeng, Y. P., & Jiang, D. (2010b). Effect of cooling rate and polyvinyl alcohol on the morphology of porous hydroxyapatite ceramics. *Materials & Design*, *31*(6), 3090–3094.

Index

Note: Page numbers in *italics* indicate a figure and page numbers in **bold** indicate a table on the corresponding page.

B

C

D

E

F

G

M

O

P

Q

R

S

U

V

W

X

Y

Z

For Product Safety Concerns and Information please contact our EU representative GPSR@taylorandfrancis.com
Taylor & Francis Verlag GmbH, Kaufingerstraße 24, 80331 München, Germany

www.ingramcontent.com/pod-product-compliance
Ingram Content Group UK Ltd.
Pitfield, Milton Keynes, MK11 3LW, UK
UKHW022220190926
475702UK00003B/460
* 9 7 8 1 0 3 2 8 0 4 5 8 3 *